Progress in the Chemistry of Organic Natural Products

The volumes of this classic series, now referred to simply as "Zechmeister" after its founder, Laszlo Zechmeister, have appeared under the Springer Imprint ever since the series' inauguration in 1938. It is therefore not really surprising to find out that the list of contributing authors, who were awarded a Nobel Prize, is quite long: Kurt Alder, Derek H. R. Barton, George Wells Beadle, Dorothy Crowfoot-Hodgkin, Otto Diels, Hans von Euler-Chelpin, Paul Karrer, Luis Federico Leloir, Linus Pauling, Vladimir Prelog, with Walter Norman Haworth and Adolf F. J. Butenandt serving as members of the editorial board.

The volumes contain contributions on various topics related to the origin, distribution, chemistry, synthesis, biochemistry, function or use of various classes of naturally-occurring substances ranging from small molecules to biopolymers.

Each contribution is written by a recognized authority in the field and provides a comprehensive and up-to-date review of the topic in question. Addressed to biologists, technologists, and chemists alike, the series can be used by the expert as a source of information and literature citations and by the non-expert as a means of orientation in a rapidly developing discipline.

All contributions are listed in PubMed.

A. Douglas Kinghorn · Heinz Falk ·
Simon Gibbons · Yoshinori Asakawa · Ji-Kai Liu ·
Verena M. Dirsch
Editors

Biologically Active Compounds of Malaysian Medicinal and Aromatic Plants

Volume 127

By
Nurhuda Manshoor · Fatimah Salim · Che Puteh Osman ·
Norizan Ahmat · Aisyah Salihah Kamarozaman ·
Nurul 'Izzah Mohd Sarmin ·
Sharifah Aminah Syed Mohamad

Springer

Editors
A. Douglas Kinghorn
College of Pharmacy
Ohio State University
Columbus, OH, USA

Heinz Falk
Institute of Organic Chemistry
Johannes Kepler University Linz
Linz, Austria

Simon Gibbons
Natural and Medical Sciences Research Center
University of Nizwa
Nizwa, Oman

Yoshinori Asakawa
Faculty of Pharmaceutical Sciences
Tokushima Bunri University
Tokushima, Japan

Ji-Kai Liu
School of Pharmaceutical Sciences
South Central University for Nationalities
Wuhan, China

Verena M. Dirsch
Department of Pharmaceutical Sciences
University of Vienna
Vienna, Austria

ISSN 2191-7043 ISSN 2192-4309 (electronic)
Progress in the Chemistry of Organic Natural Products
ISBN 978-3-031-86377-6 ISBN 978-3-031-86378-3 (eBook)
https://doi.org/10.1007/978-3-031-86378-3

This Springer imprint is published by the registered company Springer Nature Switzerland AG
The registered company address is: Gewerbestrasse 11, 6330 Cham, Switzerland

Contents

Contributors

Norizan Ahmat Centre of Foundation Studies, Universiti Teknologi MARA, Cawangan Selangor, Selangor, Malaysia

Aisyah Salihah Kamarozaman Centre of Foundation Studies, Universiti Teknologi MARA, Cawangan Selangor, Selangor, Malaysia

Nurhuda Manshoor Atta-ur-Rahman Institute for Natural Product Discovery (AuRIns), Universiti Teknologi MARA Selangor Branch, Selangor, Malaysia

Nurul 'Izzah Mohd Sarmin Atta-ur-Rahman Institute for Natural Product Discovery (AuRIns), Universiti Teknologi MARA Selangor Branch, Selangor, Malaysia

Che Puteh Osman Atta-ur-Rahman Institute for Natural Product Discovery (AuRIns), Universiti Teknologi MARA Selangor Branch, Selangor, Malaysia

Fatimah Salim Atta-ur-Rahman Institute for Natural Product Discovery (AuRIns), Universiti Teknologi MARA Selangor Branch, Selangor, Malaysia

Sharifah Aminah Syed Mohamad Atta-ur-Rahman Institute for Natural Product Discovery (AuRIns), Universiti Teknologi MARA Selangor Branch, Selangor, Malaysia

Biologically Active Compounds of Malaysian Medicinal and Aromatic Plants

Nurhuda Manshoor, Fatimah Salim, Che Puteh Osman, Norizan Ahmat, Aisyah Salihah Kamarozaman, Nurul 'Izzah Mohd Sarmin, and Sharifah Aminah Syed Mohamad

Contents

N. Manshoor · F. Salim (✉) · C. P. Osman · N. 'I. Mohd Sarmin · S. A. Syed Mohamad
Atta-ur-Rahman Institute for Natural Product Discovery (AuRIns), Universiti Teknologi MARA Selangor Branch, Selangor, Malaysia
e-mail: fatimah2940@uitm.edu.my

N. Manshoor
e-mail: nurhuda15@uitm.edu.my

C. P. Osman
e-mail: cheputeh@uitm.edu.my

N. 'I. Mohd Sarmin
e-mail: izzahsarmin@uitm.edu.my

S. A. Syed Mohamad
e-mail: sharifah459@uitm.edu.my

N. Ahmat · A. S. Kamarozaman
Centre of Foundation Studies, Universiti Teknologi MARA, Cawangan Selangor, Selangor, Malaysia
e-mail: noriz118@uitm.edu.my

A. S. Kamarozaman
e-mail: aisyah2910@uitm.edu.my

A. D. Kinghorn, H. Falk, S. Gibbons, Y. Asakawa, J.-K. Liu, V. M. Dirsch (eds.),
Biologically Active Compounds of Malaysian Medicinal and Aromatic Plants,
Progress in the Chemistry of Organic Natural Products 127,
https://doi.org/10.1007/978-3-031-86378-3_1

1 Introduction

Malaysia is a biodiversity hotspot, home to an exceptional variety of medicinal and aromatic plants that have played a crucial role in traditional healing practices for centuries. Its unique tropical climate, the rich ecosystems, and unparalleled floral diversity present support the growth of numerous plant species, many of which possess well-investigated phytochemical profiles and biological activities [1]. The Malaysian government has acknowledged the significance of these species, designating some as high-value plants under national economic development programs. In recognition of their potential, 17 species have been designated as high-value herbal products under the National Key Economic Area of Malaysia in the Eleventh Malaysia Plan, highlighting their importance not only for health benefits in local healthcare practices, but also for contributing to the national economy [2]. These initiatives aim to enhance the cultivation, research, and commercialization of these plants. The growing demand for plant-based healthcare solutions aligns with global trends toward natural therapies, thus placing Malaysian medicinal plants at the forefront of opportunities to integrate traditional remedies into contemporary medical frameworks. However, despite this growing interest, challenges persist such as the variability in bioactive compound profiles, their limited clinical validation, and knowledge gaps in their mechanisms of action. Addressing these issues through multidisciplinary approaches, including omics technologies and molecular analyses, is crucial for maximizing their potential.

In this contribution, the authors provide an overview of the therapeutic potential of these plants from Malaysia, focusing on the bioactive compounds that are responsible for their efficacy. These plants produce diverse phytochemical constituents, including alkaloids, terpenoids, flavonoids, and other phenolic compounds. Their phytochemical composition has been evaluated in combination with their biological activities, providing insights into how these compounds may contribute to specific health benefits, such as through their anti-inflammatory, antioxidant, antimicrobial, and cytotoxic effects. For instance, compounds derived from *Andrographis paniculata* exhibit anti-inflammatory and immunomodulatory properties [3], while *Eurycoma longifolia* has gained attention for its potential aphrodisiac and testosterone-boosting effects [4]. Additionally, *Centella asiatica* is of interest for its wound-healing and cognitive enhancement properties [5].

Included herein are not only 17 widely studied species but also the four lesser-known aromatic plants *Cynometra cauliflora*, *Cananga odorata*, *Citrus hystrix*, and *Polygonum minus*, for which their essential oils have been investigated phytochemically. These plants, renowned for their essential oils and aromatic properties, expand the scope of Malaysia's herbal legacy while offering emerging opportunities in the realms of skincare, wellness, and functional food applications. The inclusion of these species enriches the discussion by showcasing the breadth of therapeutic and aromatic possibilities from Malaysia's plant biodiversity.

The primary purpose of this contribution is to serve as a key reference for researchers and other interested parties, and it provides an overview of the scientific progress made and on the available literature on the 17 high-value herbs and the four additional aromatic species indicated above. By documenting systematically their phytochemical constituents and biological activities, the current authors aim to enhance an understanding of the roles these plants have in traditional medicine and their potential applications in contemporary healthcare. By delving into biochemical knowledge on these plants, it is hoped to pave the way for future research and development initiatives, ensuring the sustainable use of Malaysia's rich herbal heritage and fostering a greater appreciation for its contributions to global health. Through a systematic examination of these plants and their bioactive compounds, this contribution is intended to serve also as a valuable resource for researchers, healthcare professionals, and policymakers. It underscores the potential of Malaysian medicinal and aromatic plants as natural therapeutic agents as well as the importance of preserving the relevant biodiversity for both local communities and the national and international health sectors. By bridging traditional knowledge with cutting-edge research and fostering strategic collaborations, the future of medicinal plant research in Malaysia holds great promise for innovative healthcare solutions and economic growth.

2 Phytochemicals and Biological Activities of Malaysian Medicinal and Aromatic Plants

2.1 *Eurycoma longifolia*

Eurycoma longifolia Jack (Simaroubaceae) is an evergreen, slow-growing medicinal plant native to Southeast Asian countries such as Malaysia, Indonesia, and Vietnam. This plant species can also be found in isolated areas of Thailand, Laos, Myanmar, and Cambodia [6–10]. In Malaysia, *E. longifolia* is known traditionally as "Tongkat Ali" because of its long, twisted roots. It is mostly cultivated in Malaysia for better access for its therapeutic benefits and in order to preserve wild specimens [11–13]. This species is also referred to as "Ian-don" (Thailand), "Cay ba benh" (Vietnam), and "Tho Nan" (Laos) in addition to "Pasak Bumi" or "Bedara Pahit" in Indonesia.

The plant is a tall, thin, shrubby tree that thrives in soil that is sandy, well-drained, partially shaded, and receives sufficient water. Its limbs are covered in compound leaves that are pinnately shaped and green in hue. The many leaflets have smooth margins, and are oblong to ovate-lanceolate in shape, and opposite or sub-opposite (Plate 1). Tiny, reddish-purple, unisexual flowers are packed closely together. When ripe, the ovoid drupes develope a noticeable ridge that turns a dark reddish-brown color [14].

Plate 1 Leaf arrangement of *Eurycoma longifolia* (family Simaroubaceae)

2.1.1 Ethnobotany

Eurycoma longifolia has a long history of use in Malaysia as an aphrodisiac, as well as for its antipyretic, antimalarial, antidiabetic, and antimicrobial properties [15]. Traditionally, almost every part of the *E. longifolia* plant has been used for therapeutic purposes. The plant is one of the most widely used in folk medicine for its aphrodisiac properties and ability to treat malaria [16]. The roots are the most valuable part, and are used to treat lumbago and indigestion, fever, jaundice, cachexia, dropsy, and for various aches. They are also utilized as health supplements to improve blood flow, restore energy and vitality, and as an herbal ingredient for women after childbirth [17–19]. The roots are also employed to treat persistent fever, dysentery, and glandular swelling [20, 21]. Decoctions of *E. longifolia* leaves are used to wash itching skin, while the fruits are used to treat diarrhea and the bark is a vermifuge [11]. The leaves have also been employed by traditional healers to alleviate the symptoms of malaria and ulcers, to prevent gum diseases, and to treat sexually transmitted diseases like syphilis and gonorrhea [7]. The taproots are used to reduce high blood pressure, and the root bark for the treatment of diarrhea and fever [22].

This plant is a component of the traditional medicine systems of Malaysia, Indonesia, and Vietnam. In Malaysia and Indonesia, the roots of the plant are boiled in water, and the concoction is consumed for a variety of health issues, including fever, intestinal worms, diarrhea, indigestion, and jaundice, as well as postpartum recovery. A paste made from the plant is administered topically to treat headaches and stomachache in Malaysia [23]. In Vietnam, the roots are employed to treat fever and malaria, with the flowers and fruits serving as a treatment for dysentery [23, 24]. *Eurycoma longifolia* is reputed in traditional medicine to have aphrodisiac properties [23, 25]. Additional purported health advantages linked to this plant include its anti-inflammatory and anti-dengue [26], and antidiabetic, antibacterial, antipyretic, and immunomodulatory [27] properties.

The traditional method of consuming the roots of *E. longifolia* involves cutting them into small pieces, and boiling in water until becoming hot to the touch, and then drinking the resultant decoction as a tea. However, since the roots are very bitter-tasting, it is also advised to boil the roots vigorously in fresh water several times to eliminate the bitterness from the decoction. Some traditional healers also suggest combining the extractive of the chipped roots with dates, honey, or sugar syrup in order to reduce the bitter taste. Also, the powdered root may be combined with alcohol and then ingested after being allowed to stand for a predetermined time, which can last anywhere from six to eight weeks. The aqueous decoction of *E. longifolia* roots is consumed traditionally as a tonic beverage. More convenient formulas are now available, mostly as additives blended with tea or coffee, and various health supplement products are available on the market either in the form of a raw crude root powder or as capsules mixed with other plants [7].

In Malaysia and Indonesia, *E. longifolia* has been commercialized quite widely. Its bitter roots have been included in dietary supplement products and as additives for food and beverages [25]. This plant is marketed as a supplement with claims of enhancing sexual health, increasing energy and stamina, enhancing blood circulation, lowering stress levels [28], and reducing body fat [23]. It is found as an ingredient in certain coffee products that may be included in energy drinks promoted by the beverage industry. The demand for this species has increased considerably in recent years, and the health-food market now offers approximately 200 individual products inclusive of *E. longifolia*, with most of these promoting the aphrodisiac qualities of this plant [29]. The products are widely available and can be purchased in the form of raw root chips and flakes, a root powder, capsules combined with other aphrodisiac herbs, as an additive combined with coffee and tea, and a fortified juice in canned drinks, or as a substitute for ginseng.

2.1.2 Phytochemistry

Eurycoma longifolia has yielded a wide variety of bioactive compounds that have been isolated and characterized. These include quassinoids (simaroubolides), canthin-6-one alkaloids, β-carboline alkaloids, biphenyl neolignans, squalene derivatives, steroids, and miscellaneous lactones and triterpenoids. Among these, the bitter-tasting quassinoid constituents include eurycomanones (**1–9**), eurycomanols (**10–14**), hydroxyklaineanones (**15–23**), laurycolactones (**24** and **25**), eurycomalactones (**26–29**), eurycomadilactones (**30–32**), eurycomalides (**33–37**), hydroxyglaucarubol (**38**), longilactones (**39** and **40**), eurycolactones (**41–47**), eurylactones (**48–52**), longifolactones (**53–68**), and eurycomaosides (**69** and **70**) [18, 30–35], and represent the major constituents of *E. longifolia* roots. Quassinoids are a class of nortriterpenoids that effectively inhibit cellular proliferation, in some cases at nanomolar and subnanomolar concentrations [36, 37]. A canthin-6-one alkaloid (**71**) was isolated from different parts of the plant, and β-carboline alkaloids (**72** and **73**) were isolated from the roots [38–40]. Other components of *E. longifolia* include biphenyl neolignans (**74**), a squalene derivative (**75**), several miscellaneous triterpenoids (**76–82**), and triterpene lactones (**83–84**), and these have been isolated from various parts of the plant, including the bark, stem, and roots. The major constituents isolated and determined structurally from *E. longifolia* are summarized in the formula drawings.

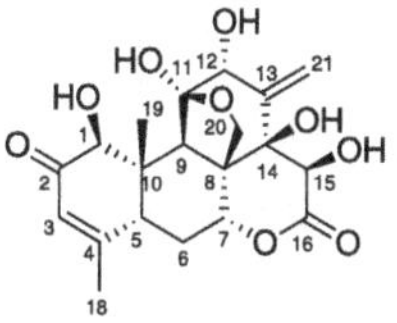

1 (eurycomanone = pasakbumin A)

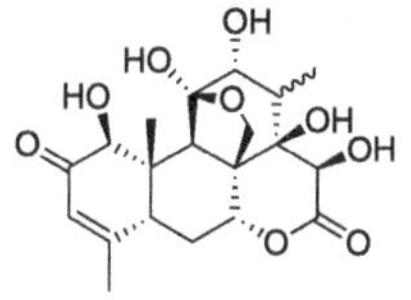

2 (13,21-dihydroeurycomanone = iandonone = pasakbumin C)
3 (13,21-dihydroeurycomanone)

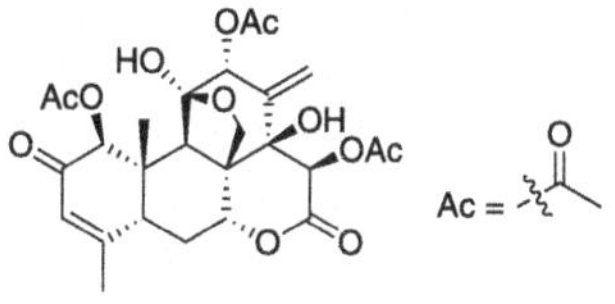

4 (1,12,15-triacetyleurycomanone)

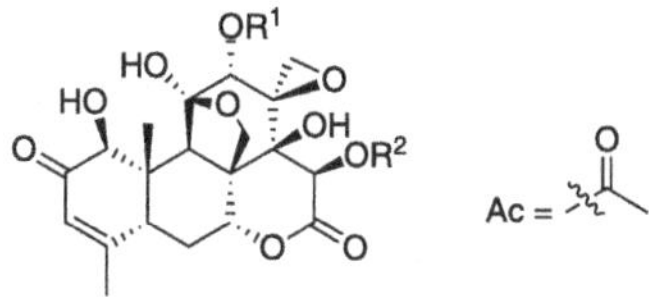

5 (13,21-epoxyeurycomanone = pasakbumin B) $R^1 = R^2 = H$
6 (15-acetyl-13(21)-epoxyeurycomanone) $R^1 = H$, $R^2 = Ac$
7 (12,15-diacetyl-13(21)-epoxyeurycomanone) $R^1 = R^2 = Ac$

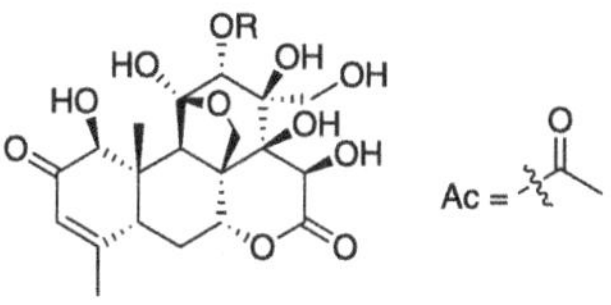

8 (13*β*,21-dihydroxyeurycomanone = pasakbumin D) R = H
9 (12-acetyl-13,21-dihydroeurycomanone) R = Ac

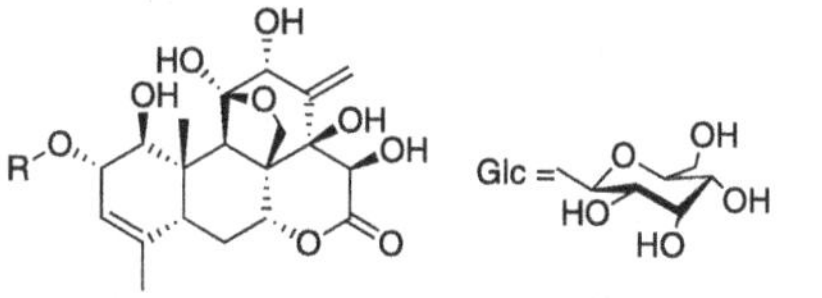

10 (eurycomanol) R = H
11 (eurycomanol-2-*O*-*β*-D-glucopyranoside) R = Glc

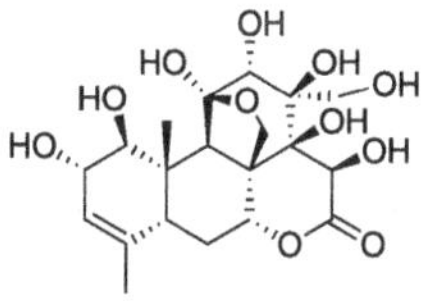

12 (13*β*-21-dihydroxyeurycomanol)

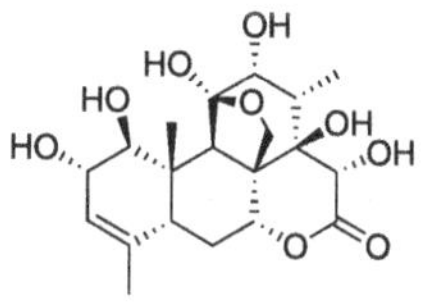

13 (iandonol)

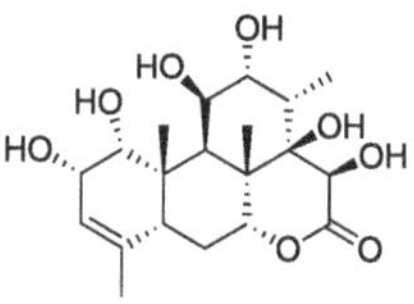

14 (1-dehydroxyklaineanol)

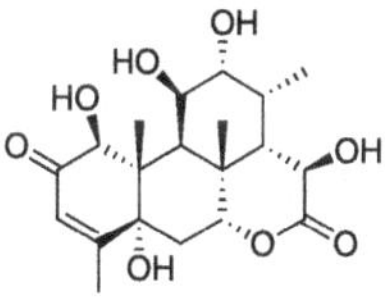

15 (hydroxyklaineanone)

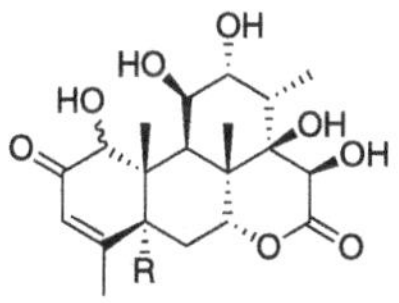

16 (5,14,15-trihydroxyklaineanone) R = OH
17 (1-*epi*-5,14*β*,15*β*-trihydroxyklaineanone) R = OH
18 (1-*epi*-14*β*,15*β*-dihydroxyklaineanone) R = H

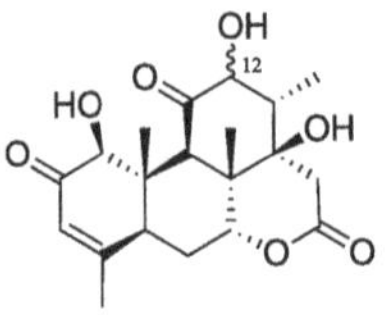

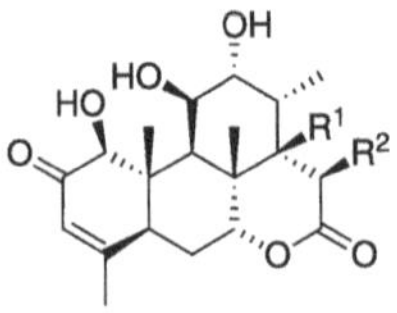

19 (11-dehydroklaineanone) 12β-OH
20 (12-*epi*-11-dehydroklaineanone) 12α-OH

21 (14,15-dihydroxyklaineanone) $R^1 = R^2 = OH$
22 (15-hydroxyklaineanone) $R^1 = H$, $R^2 = OH$
23 (15-*O*-acetyl-14-hydroxyklaineanone) $R^1 = OH$, $R^2 = OAc$

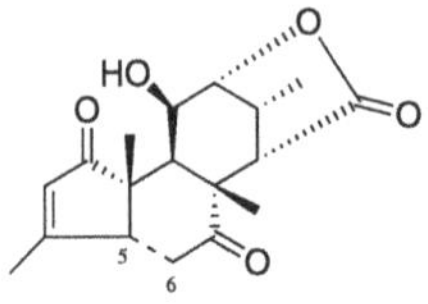

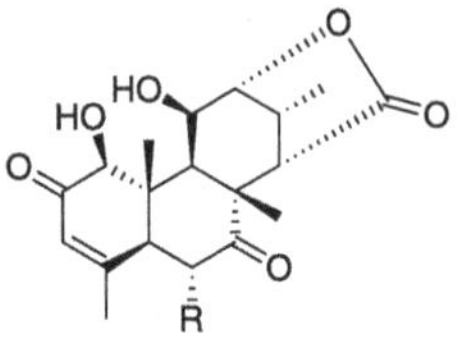

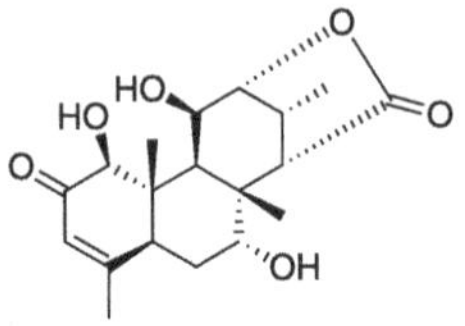

24 (laurycolactone A) 5–6
25 (laurycolactone B) 5=6

26 (eurycomalactone) R = H
27 (6-hydroxy-eurycomalactone) R = OH
28 (5,6-dehydroeurycomalactone) R=H

29 (7α-hydroxyeurycomalactone)

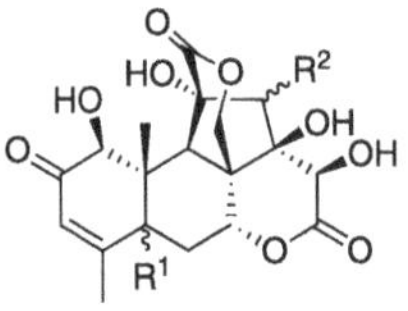

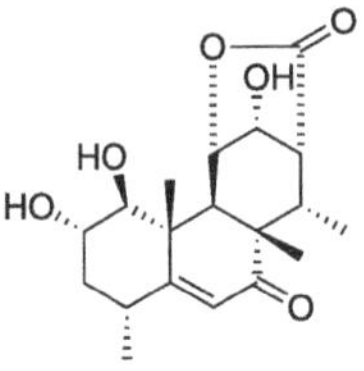

30 (eurycomadilactone) $R^1 = \alpha H$, $R^2 = \beta Me$
31 (5-*iso*-eurycomadilactone) $R^1 = \beta H$, $R^2 = \beta Me$
32 (13-*epi*-eurycomadilactone) $R^1 = \alpha H$, $R^2 = \alpha Me$

33 (eurycomalide A)

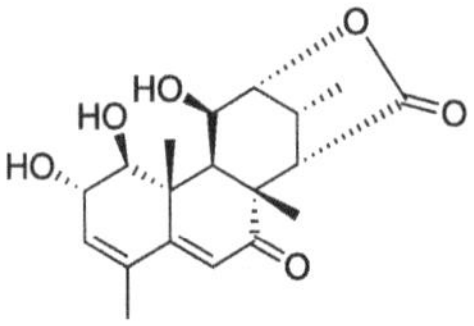

34 (eurycomalide B)

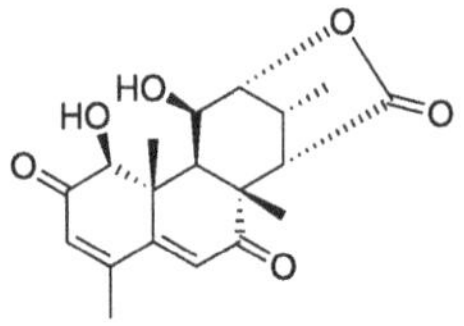

35 (eurycomalide C)

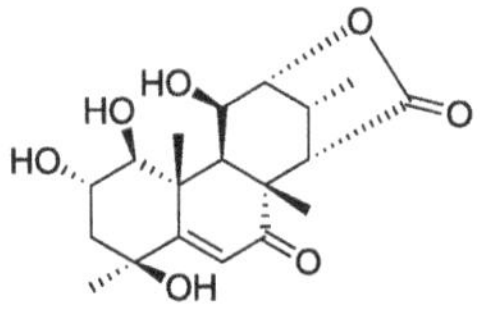

36 (eurycomalide D)

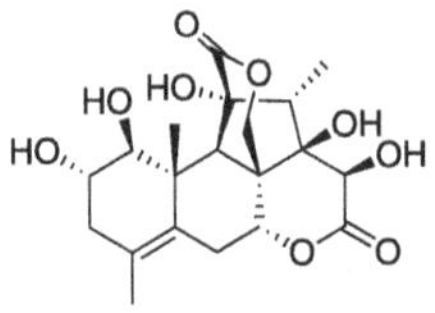

37 (eurycomalide E)

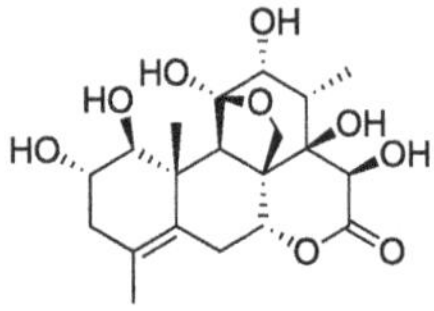

38 (hydroxyglaucarubol)

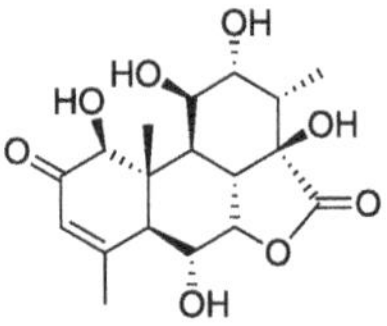

39 (longilactone)

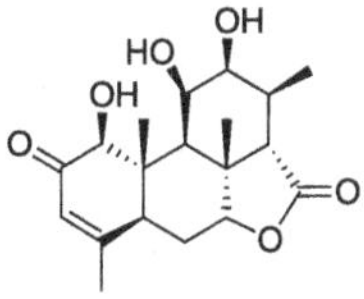

40 (6-dehydroxylongilactone)

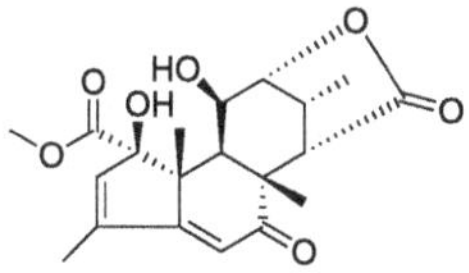

41 (eurycolactone A)

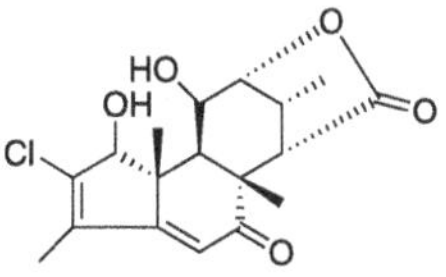

42 (eurycolactone B)

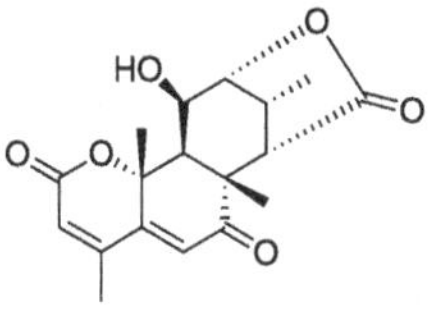

43 (eurycolactone C)

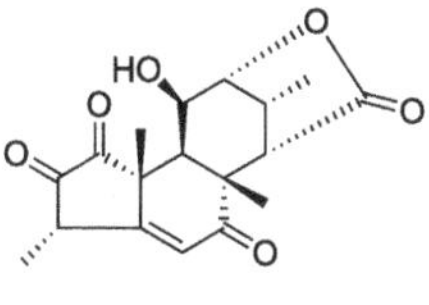

44 (eurycolactone D)

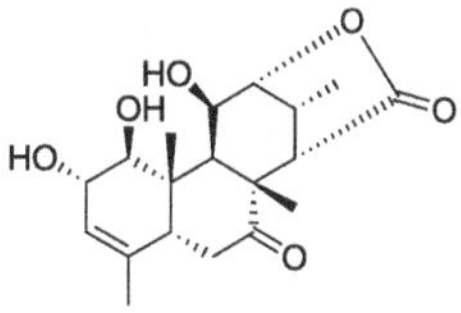

45 (eurycolactone E)

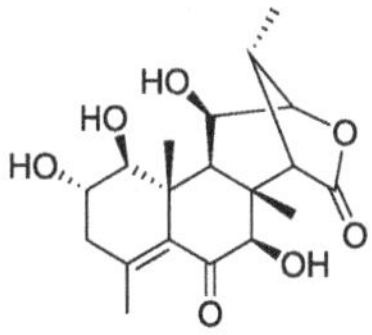

46 (5-dehydro-3-hydro-7-hydroxy-6-oxoeurycolactone E)

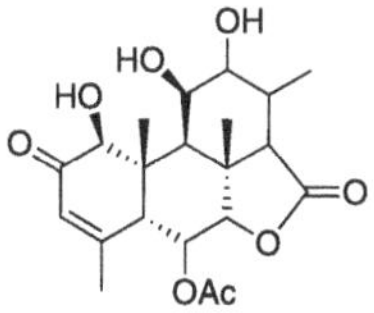

47 (eurycolactone F)

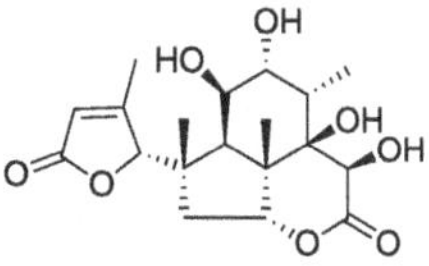

48 (eurylactone A)

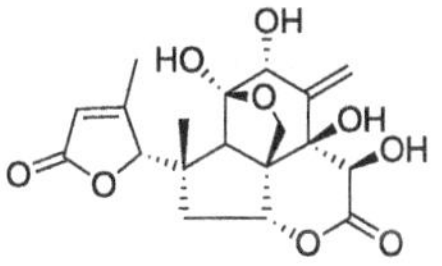

49 (eurylactone B)

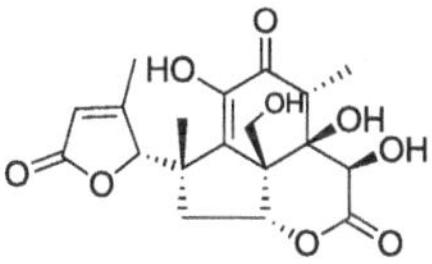

50 (eurylactone E)

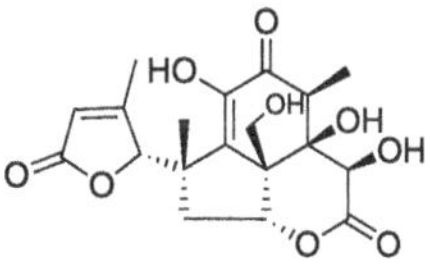

51 (eurylactone F)

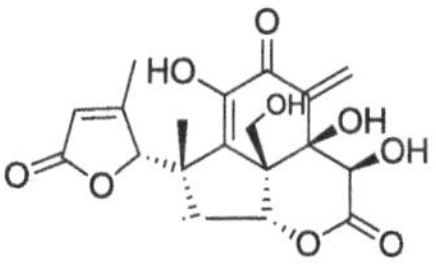

52 (eurylactone G)

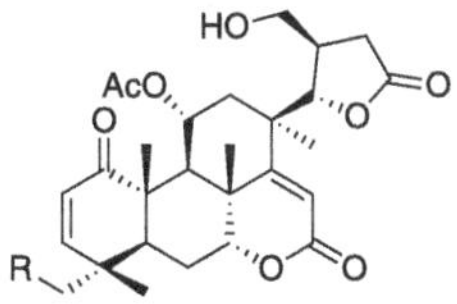

53 (longifolactone A) R = H
54 (longifolactone B) R = OH

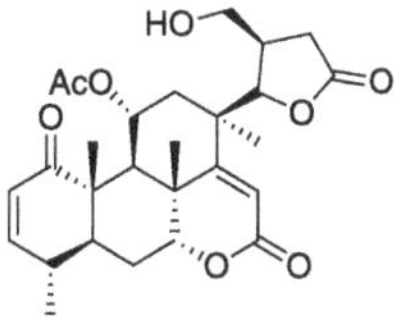

55 (longifolactone C)

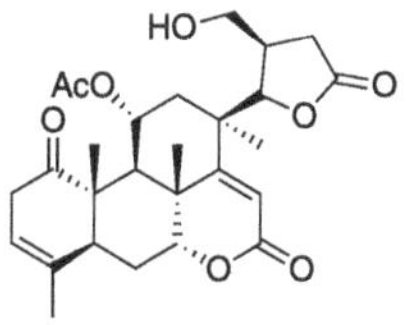

56 (longifolactone D)

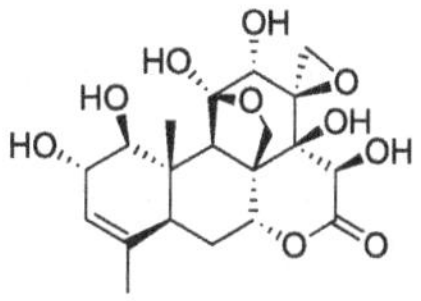

57 (longifolactone E)

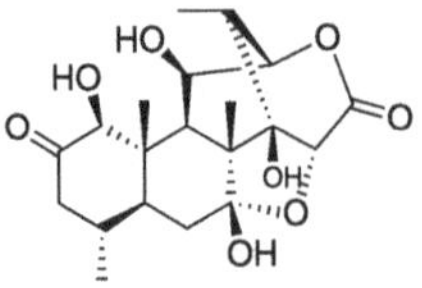

58 (longifolactone F)

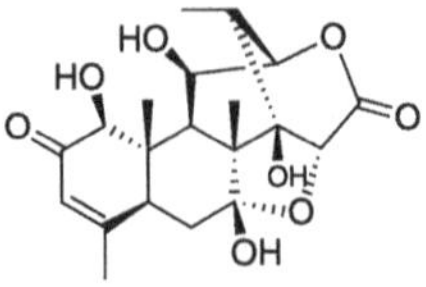

59 (longifolactone G)

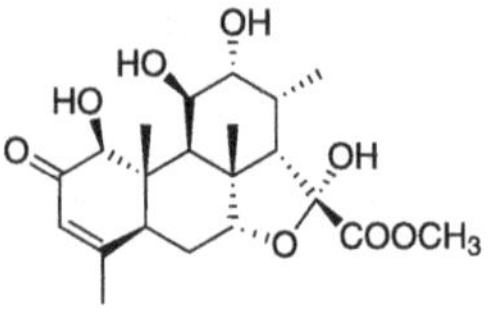

60 (longifolactone H)

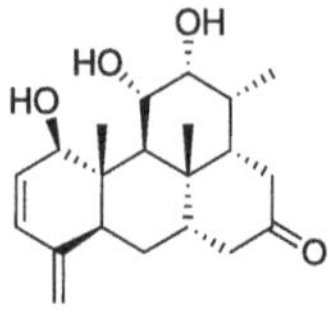

61 (longifolactone I)

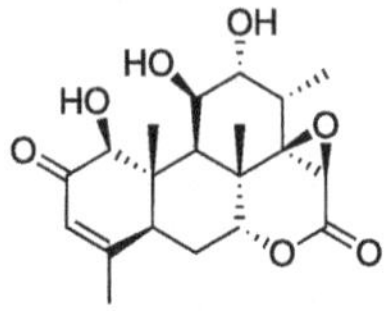

62 (longifolactone J)

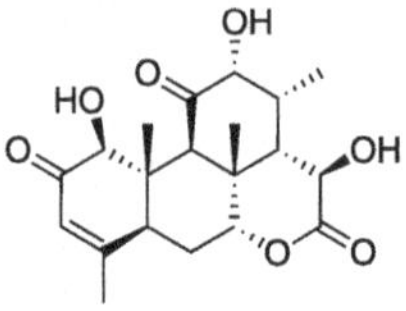

63 (longifolactone K)

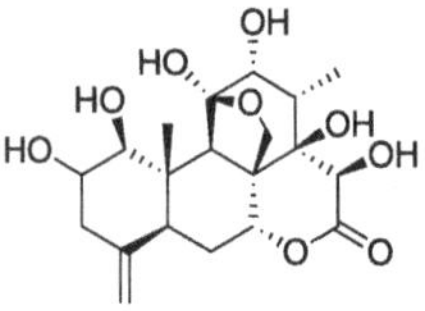

64 (longifolactone L)

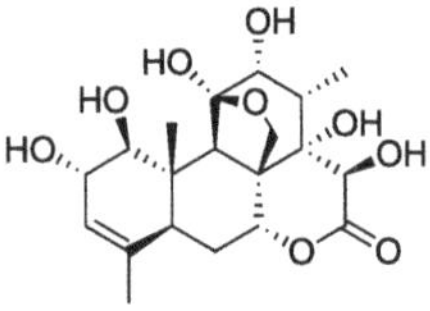

65 (longifolactone M)

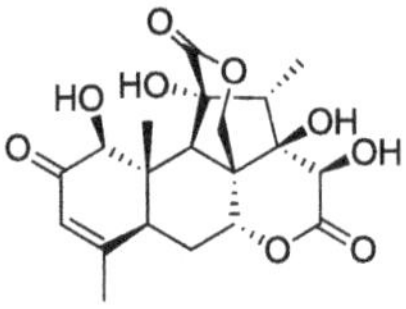

66 (longifolactone N)

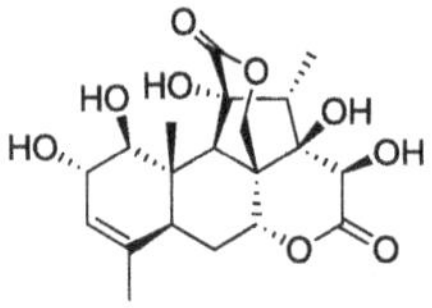

67 (longifolactone O) (5*S*)
68 (longifolactone P) (5*R*)

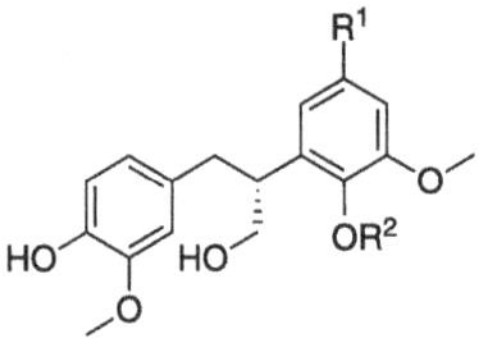

69 (eurylolignanoside A) R^1 = 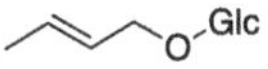 , R^2 = Glc
70 (eurylolignanoside B) R^1 = COOH, R^2 = Glc

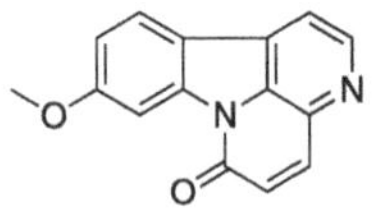

71 (9-methoxycanthin-6-one)

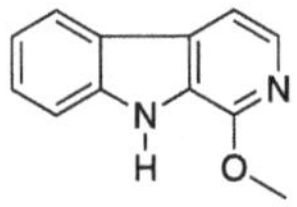

72 (1-methoxymethyl-β-carboline)

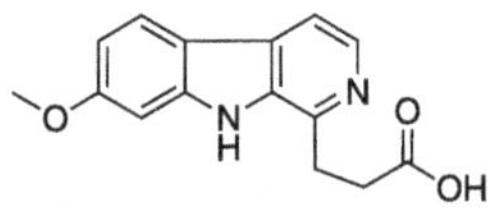

73 (7-methoxy-β-carboline-1-propionic acid)

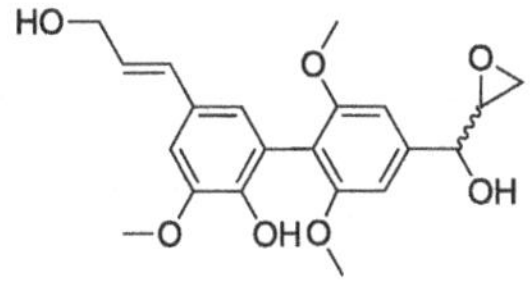

74 (biphenyl-neolignan)

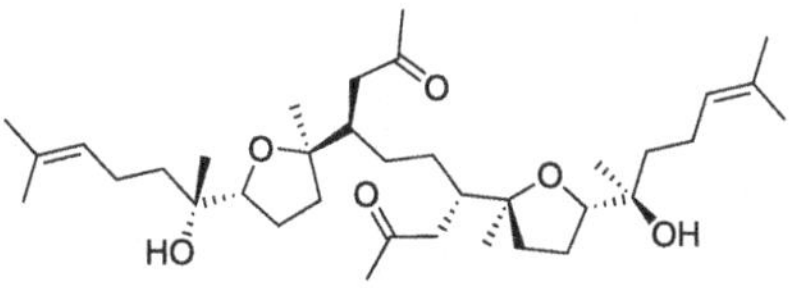

75 (eurylene)

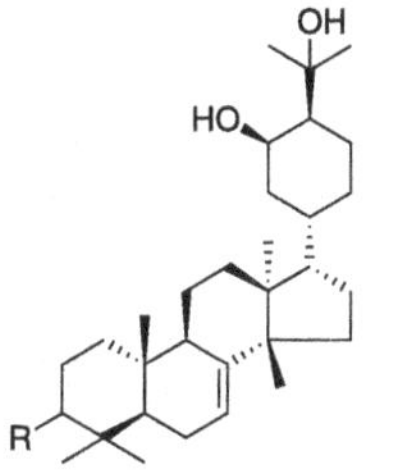

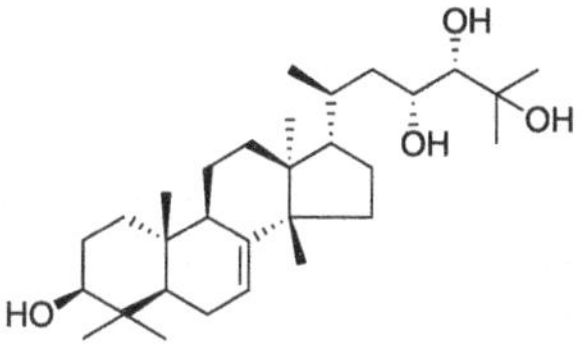

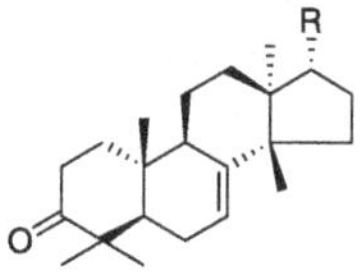

76 (3-*epi*-sapeline A) R = OH
77 (bourjotinolone A) R = =O

78 (hispidol)

79 (niloticin) R = A
80 (piscidinol A) R = B
81 (24-*epi*-piscidinol A) R = C
82 (bourjotinolone B) R = D

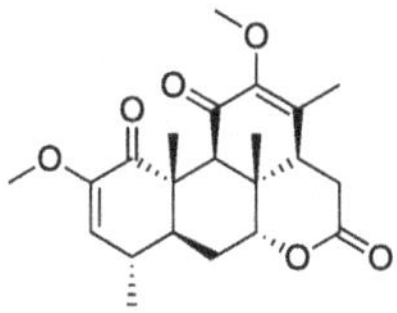

83 (quassin)

A = 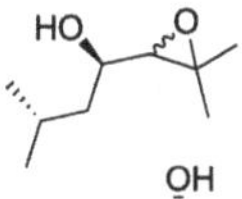

B = 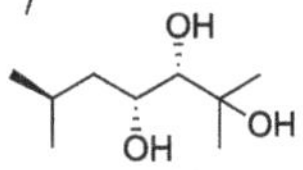

C = 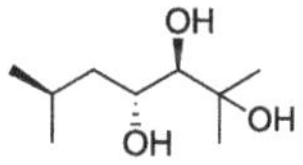

D = 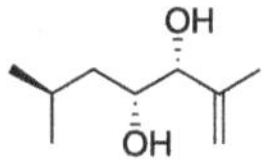

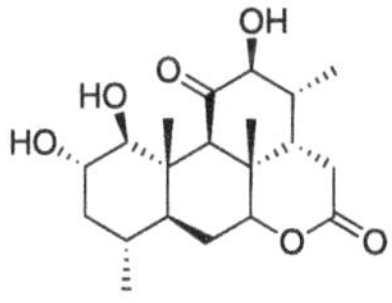

84 (chaparrolide)

2.1.3 Biological Activities

Eurycoma longifolia roots are popular as a traditional medicinal plant for their reputed aphrodisiac properties and as an alternative for enhancing libido, energy, sports performance, and weight loss [41], due to a perceived ability to stimulate the production or action of androgen hormones, especially testosterone. This plant is consumed also as a health supplement for hypertension, diarrhea, various aches, persistent fevers, dysentery, and glandular swelling, and it has been found to possess antibacterial [42], anti-inflammatory [43] anticoagulant (for potential use in complications during childbirth), aphrodisiac, antiplasmodial, cytotoxic, antipyretic, antidiabetic, antimalarial, hepatoprotective, anxiolytic, and antiulcer effects [44, 45].

Even though *E. longifolia* has several documented ethnomedical applications, enhancing its scientific validity and obtaining supporting clinical evidence are prerequisites for its further commercial exploitation. Besides the presently available data, more proof of its therapeutic effectiveness and safety is needed before *E. longifolia* can be regarded as a viable naturally occurring drug candidate. Moreover, it is very important to conserve this valuable medicinal plant for the health benefit of future generations. Several reports are now accessible on the pharmacological potential of the extracts, fractions, and isolated compounds of this plant. A summary of several significant studies on the biological and pharmacological characteristics, tests conducted, and their activity determined on *E. longifolia* is given in Table 1.

Table 1 Biological properties of *Eurycoma longifolia*

Activity	Assay	References
Antiangiogenic	Inhibited major angiogenesis steps such as proliferation, migration, and differentiation of human umbilical vein endothelial cells (HUVECs) in vitro Inhibited (63.1%) neovascularization in the chorioallantoic membrane of the chick embryo with an IC_{50} of 50 g/cm^3 in vivo Suppressed the sprouting of microvessels in the rat aorta with an IC_{50} of 11.5 g/cm^3 in vivo	[46]
Antidiabetic	Decreased blood glucose by 38% in streptozotocin-induced hyperglycemic adult rats after treatment of 150 mg/kg body weight using an aqueous extract	[47]
Anti-inflammation	Activated Nrf2 via a ROS-dependent p38 MAPK pathway and 7-MCPA by β-carboline alkaloids isolated from hairy-root cultures of *E. longifolia*	[48]
	Inhibited the LPS-induced protein expression of IL-6, NF-B, and iNOS in the NF-κB signaling pathway in a dose-dependent manner	[49]
	Stabilized human red blood cells in an in vitro anti-inflammatory test of the extract	[43]
Antimalarial	Inhibited a Malaysian chloroquine-resistant *Plasmodium falciparum* culture in vitro, after 1-day post-treatment	[50]
	Decreased glutathione (GSH) content of erythrocytes, caused by *P. falciparum* growth inhibition in vitro	[51]

(continued)

Table 1 (continued)

Activity	Assay	References
	Suppressed *Plasmodium yoelii* infection in an in vivo treatment of the experimental mice used	[52]
Antimicrobial	Antibacterial activity was shown by the leaf and stem extracts against *Staphylococcus aureus* and *Serratia marscesens*	[53]
	Inhibited the growth of *Staphylococcus aureus* and *Pseudomonas aeruginosa* with inhibition zone diameters ranging between 7 and 25 mm	[54]
	Improved the survival of *Staphylococcus aureus*-infected worms by 2.8-fold	[55]
Antiparasitic	Inhibition of *Toxoplasma gondii* activity by a crude extract and fractions from *E. longifolia*	[56]
	Reduced the cell count of *Blastocystis* sp. isolates in comparison to standard antiprotozoal allopathic drugs	[57]
	Increased knock-down activities of mosquitoes exposed to *E. longifolia*-containing smoke mosquito coils, but not increased mortality	[58]
Antistress	Improved stress hormone profiles and certain mood state parameters in 63 subjects (32 men and 31 women) supplemented with an *E. longifolia* hot-water extract	[41]
	Anti-anxiety effects in the mice tested, using various behavioral tests, were similar to those of the positive control diazepam	[59]
Aphrodisiac	Changed sexual behavior among middle-aged male rats towards receptive females after administering 800 mg/kg of a root extract	[60]
	Stimulated sexual arousal without genital sensation, in sexually vigorous non-copulator male rats treated with 0.5 g/kg root extract, using an electrical cage	[61]
	Increased the mounting frequency of the treated male rats with 400 mg/kg of fractions, with no erections, intromissions, ejaculations, or seminal emissions in a dose-dependent manner	[62]
	Enhanced sexual motivation activity in adult, middle-aged male mice and retired breeders that were administered with a root extract (500 mg/kg) daily for 10 days	[63]
	Increased aphrodisiac activities in sexually sluggish old male rats administered with 800 mg/kg of extract	[64]
	Improved sexual qualities by a decrease in their hesitation time of middle-aged male rats treated with 0.5 g/kg of a fraction, in an electrical copulation cage	[65]

(continued)

Table 1 (continued)

Activity	Assay	References
Cytotoxic and antiproliferative	Cytotoxic with IC_{50} values of 11 and 13 μg/cm^3, using the Hep2 and HFL1 cell lines, respectively	[66]
	Inhibited the growth of HL-60 leukemic cells with median inhibitory concentration (IC_{50}) values of 15.2–28.6 g/cm^3	[46]
	Inhibited growth fusing (sc) tumor in vitro using K-562 chronic myelogenous leukemia (CML) cells	[67]
	The cytotoxic activity of quassinoids from this plant was found not to be mediated through DNA cleaving properties	[35]
	Eurycomanone (**1**) and eurycomanol (**10**) isolated from *E. longifolia* inhibited Jurkat and K562 cell viability and proliferation without affecting healthy cells. Eurycomanone (**1**) inhibited the NF-κB signaling pathway through inhibition of IκBα phosphorylation and upstream mitogen-activated protein kinase (MAPK) signaling	[68]
	Cytotoxic effects of the root extract were shown against the KB, DU-145, RD, MCF-7, and CaOV-3 cell lines, with no significant cytotoxicity against on the MDBK (kidney) normal cell line	[69]
	Decreased the expression of Wnt signal target genes, *MITF* and zic2a, through the activation of GSK3 independent of CK1 in zebrafish embryos when treated with 9-hydroxycanthin-6-one, a compound isolated from *E. longifolia*	[70]
	Cytotoxic effects with IC_{50} values of 11 and 13 g/cm^3 for the Hep2 and HFL1 cell lines, respectively	[71]
	Fractions of an *E. longifolia* extract induced apoptosis in MCF-7 breast cancer cells, with the reduction of Bcl-2 protein	[72]
	Induced apoptosis in MCF-7 breast cancer cells via a caspase-9 and p53-independent manner in vitro	[73]
	Induced selective cytotoxicity in LNCaP human prostate cancer cells and inhibited growth in a mouse xenograft model using a standardized quassinoid mixture (SQ40) from *E. longifolia*	[74]
	Induced apoptosis in HepG2 cells via upregulation of p53	[75]
Male fertility enhancement	Enhanced and maintained a high level of all of the total number of successful crossovers, mountings, intromissions, and ejaculations during the 9–12th week observation period after treatment with an extract	[76]

(continued)

Table 1 (continued)

Activity	Assay	References
	Increase in the episodes of penile reflexes as evidenced by enhancements of quick flips, long flips, and erections of the treated male rats, during a 30-min observation period	[11]
	Enhanced the sexual behavior of the middle-aged male rats by decreasing their hesitation time, as compared to controls	[65]
	Increased the spermatozoal count, morphology, motility, plasma testosterone level, and Leydig cell count of rats when given a standardized methanol extract	[77]
	Elevated the serum testosterone:epitestosterone ratio, increased liver and renal functions, as well as showing increased muscle strength in male recreational athletes given a dose of 400 mg/day of a water extract for 6 weeks	[78]
	Affected male infertility by suppressing the α2-HS glycoprotein expression in rats infertile with diabetes, when treated with a standardized quassinoid-containing extract	[12]
	Improved penile erection index, increased libido (by 20%), and enhanced semen quality in boars when using a combination of *Eurycoma longifolia, Tribulus terrestri*s, and *Leuzea carthamoides*	[79]
	Enhanced testosterone production by the Leydig cells and freed up bound testosterone for use by the muscles	[15]
	Restored serum testosterone levels and significantly improved the physical condition and sexual health of human patients with ageing male symptoms	[80]
	Increased total and free testosterone concentrations and muscular force in men and women aged 57–72 years, when consuming *E. longifolia* extract 400 mg/day for 5 weeks	[81]
	Increased sperm motility by 44.4%, and semen volume at 18.2% after twelve weeks of study in 30–50 year-old men treated with 300 mg of a water extract	[4]
	Meta-analysis of randomized controlled clinidal trials showed improved erectile dysfunction performance using *E. longifolia* root extracts	[82, 95]
	Eurycomanone (**1**) enhanced testosterone steroidogenesis on the rat testicular Leydig's cells by inhibiting the conversion of testosterone to estrogen by aromatase	[83]
	Improved rat spermatogenesis shown by affecting the hypothalamic-pituitary gonadal axis	[83]
	Proandrogenic effects were exerted by *E. longifolia* that enhanced testosterone levels	[84]

(continued)

Table 1 (continued)

Activity	Assay	References
	Increased sperm quality, sperm count, motility, and viability occurred in male rats, in a dose-dependent manner with *E. longifolia*	[85]
	Enhanced erectile function in animal research, using a combination of *Panax quinquefolius, Eurycoma longifolia, Epimedium grandiflorum, Centella asiatica,* and flower pollen extracts	[86]
	Increased sperm motility and vitality, improved matrix metalloproteinase (MMP) levels, and increased testosterone concentrations along with enhanced muscle weight in male rats treated with an *E. longifolia* aqueous extract	[87]
	Improved all semen parameters evaluated in 350 men given 200 mg of an *E. longifolia* extract daily, allowing for 11 spontaneous pregnancies	[88]
	Improved the Ageing Male Symptom (AMS) score and the serum testosterone concentration in a group of male patients suffering from late-onset hypogonadism (LOH), after being treated with *E. longifolia* extract	[89]
	Enhanced sexual performance in healthy volunteer men aged 40–65 years, showing improvements in scores for their sexual intercourse attempt diary, erection hardness scale, sexual health inventory, and aging male symptom scale, all in a more effective manner than the placebo used	[90]
	Increased spermatogenesis and sperm count, and reversed the effects of estrogen in rats, after fourteen consecutive days of treatment	[44]
	Reduced ejaculation latencies and increased the percentage of mounting and ejaculating animals after an acute treatment of sexually sluggish rats Reduced the post-ejaculatory interval, increased the percentages of mounting and ejaculating, and enhanced serum testosterone levels in a subacute treatment of impotent rats	[91]
Hormonal effects	Normalized irregular estrous cycles and reduced the follicular morphological damage caused by chronic testosterone administration in female rats	[92]
	Antiestrogenic effects against 17-ethynylestradiol (EE)-induced uterotrophy of immature rats treated with a standardized extract containing eurycomanone (**1**) and 13,21-dihydroeurycomanone (**5**)	[93]

(continued)

Table 1 (continued)

Activity	Assay	References
Muscular effects	Increased weight of the levator ani muscle in castrated animals, but not in testosterone-treated animals and uncastrated animals	[11]
	Increased physiological responses and endurance exercise performance in a male rat model	[77]
Antiulcer effect	Treated ethanol-induced gastric lesions in rats, in as effective a manner to ranitidine	[94]
	Potent antiulcer activity shown for both pasakbumins A (eurocomanone) (**1**) and B (13,21-epoxyeurocomanone) (**5**), when purified from an *E. longifolia* root extract	[32]
Osteoporosis prevention	Enhanced testosterone levels, and stimulated osteoblast proliferation and osteoclast apoptosis. Maintained bone remodeling activity and reduced bone loss through nitric oxide generation and antioxidative properties	[84]
	Protected bone from androgen-deficient osteoporosis in experimental male rat models	[96]
	Protected against bone calcium loss in orchidectomized male rats, a model for androgen-deficient osteoporosis	[97–99]

Eurycoma longifolia is known for its extensive profile of presently known biologically active compounds, and these contribute to its therapeutic properties and long-standing use in traditional medicine. Compounds such as eurycomanone (**1**) and eurycomanol (**10**) are associated with diverse pharmacological effects, including anti-inflammatory, antimicrobial, antioxidant, and cancer-related properties, as shown in a laboratory setting. Also, *E. longifolia* is renowned for its reputed aphrodisiac and testosterone-boosting effects, which have been attributed to compounds like quassinoids that exhibit active biological characteristics that effectively inhibit cell proliferation in some cases at very low concentration levels, and also demonstrates potential antitumor and antimalarial properties. The occurrence of these bioactive compounds supports *E. longifolia* as a component of natural health products and for possible pharmaceutical development.

2.2 *Marantodes pumilum (Labisia pumila)*

Marantodes pumilum (Blume) Kuntze (Plate 2) is a small sub-herbaceous plant from the family of Primulaceae, although this was recognized formally as *Labisia pumila* (Blume) F. Vill-Naves (Primulaceae, formerly Myrsinaceae) [100]. *Marantodes pumilum* is referred to by several names in different areas of Malaysia, including "rumput Palis", "Tadah matahari", "Mata pelandok rimba", "Bunga belangkas hutan", "Kachip Patimah", "Selusuh Fatimah", "Rumput Siti Fatimah", "Akar Fatimah", "Kachit Fatimah", "Pokok pinggang" [101], and "Sangkoh (Iban)" [102]. *Marantodes* is a genus of small, leafy, woody plants distributed across the tropical

Plate 2 Whole plant of *Marantodes pumilum* (family Primulaceae). Photograph courtesy of Globinmed (https://www.globinmed.com/wp-content/uploads/2011/03/1_Kfatimah.jpg)

forests of Southeast Asia. In the rainforests of Malaysia, *M. pumilum* occurs in eight different varieties, of which, *M. pumilum* var. *pumila*, *M. pumilum* var. *alata*, and *M. pumilum* var. *lanceolata* have received the most research attention [103]. The petioles and leaf features of these three varieties can be used to identify each one. While *M. pumilum* var. *pumila* has an emarginate winged petiole and an oval leaf blade, *M. pumilum* var. *lanceolata* has a long, non-winged petiole, while *M. pumilum* var. *alata* has red veins and large winged petioles. The most prevalent variety in Malaysia is *M. pumilum* var. *alata*, which is highly valued for its ethnomedical qualities [104]. Since *M. pumilum* is mainly harvested from the wild, the supply of this plant from its native habitat is becoming scarce due to rising demand from the medicinal herb industry. To accommodate the future demand for this species, research teams in Malaysia have conducted experiments to domesticate and cultivate it [102].

2.2.1 Ethnobotany

Marantodes pumilum is a traditional medicinal plant in Malaysia that has been used by generations of Malay women for various purposes, including promoting childbirth, managing menstrual irregularities, and enhancing sexual function. Extractives of *M. pumilum*, obtained by boiling the roots, leaves, or whole plant in water, are reputed to help induce and facilitate childbirth, tighten the birth canal, tone the abdominal muscles, and regain body strength postpartum [105, 106]. The plant is also used to maintain a healthy female reproductive system free from infection, treat menstrual irregularities, and enhance sexual function. *M. pumilum* accordingly has been recognized as a popular female tonic and health product in Malaysia. The plants are usually boiled, and the water-soluble extract is taken as a beverage. In daily practice, it is used for the general well-being and vitality of women, helping to maintain their figure and appearance, and aiding in reducing symptoms related to hormonal imbalance and increasing libido [107].

The extensive use of *M. pumilum* by Malay women has led to the belief that it exhibits a phytoestrogenic activity, containing compounds with similar chemical structures to estrogens, which can relieve menopausal symptoms [104]. The water decoction of *M. pumilum* is typically taken as a healthy tonic drink for pre- and postpartum treatment, and other uses include treating dysentery, flatulence, dysmenorrhea, gonorrhea and to reduce the risk of osteoporosis [97, 108–110]. *Marantodes pumilum* is also used by males from several ethnic groups in Sarawak, Malaysia, to maintain and increase stamina. Other traditional uses include treating rheumatism [29].

Despite its long use in traditional medicine, the herb has only recently been used commercially as a health supplement. Due to its numerous perceived benefits, *M. pumilum* has been commercialized for women, leading to an increasing demand for its supply in the medicinal herb and food industries. Today, products in the form of tablets, capsules, health drinks, and tonics derived from extracts of *M. pumilum* are becoming more common in pharmacies and other retail outlets.

2.2.2 Phytochemistry

Marantodes pumilum is regarded as being valuable for disease prevention due to its presently known constituents, which have been shown to possess a range of biological activities. Thus, in earlier work, it has been reported that various relevant effects, including collagenase inhibitory, antioxidant antiasthmatic, and antimicrobial properties might be effected by flavonoid, saponin, and other constituents of *M. pumilum* [111–114]. These biological properties have raised interest from researchers for the further investigation of this plant.

Flavonoids, the main constituents reported to date in *M. pumilum*, include quercetin (**85**), rutin (**86**), apigenin (**87**), kaempferol (**88**), myricetin (**89**), catechin

(**90**), epigallocatechin (**91**), naringenin (**92**), daidzein (**93**), and genistein (**94**) [111, 115]. There are also reports on other phenolic compounds such as gallic acid (**95**), caffeic acid (**96**), methyl gallate (**97**), ascorbic acid (**98**), and pyrogallol (**99**) [116], and other constituents like tocopherols and carotenoids [101].

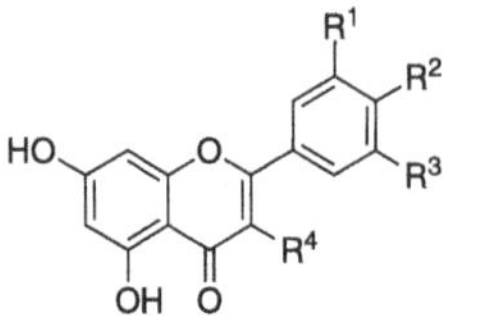

85 (quercetin) $R^1 = R^2 = R^3 = R^4 = OH$
86 (rutin) $R^1 = H$, $R^2 = R^3 = OH$, R^4 = Glc-*O*-Glc
87 (apigenin) $R^1 = R^3 = R^4 = H$, $R^2 = OH$
88 (kaempferol) $R^1 = R^3 = H$, $R^2 = R^4 = OH$
89 (myricetin) $R^1 = R^2 = R^4 = OH$, $R^3 = H$

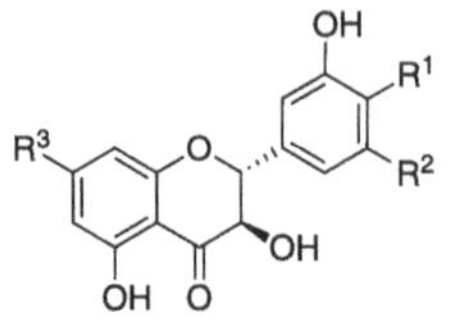

90 (catechin) $R^1 = R^3 = OH$, $R^2 = H$,
91 (epigallocatechin) $R^1 = R^2 = R^3 = OH$
92 (naringenin) $R^1 = OH$, $R^2 = H$, R^3 = Glc-*O*-Glc

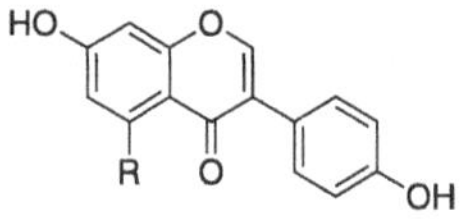

93 (daidzein) R = H
94 (genistein) R = OH

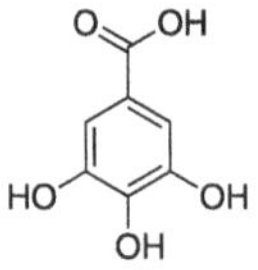

95 (gallic acid)

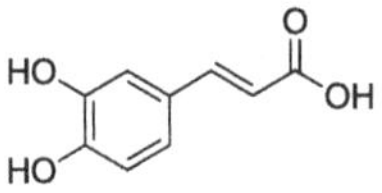

96 (caffeic acid)

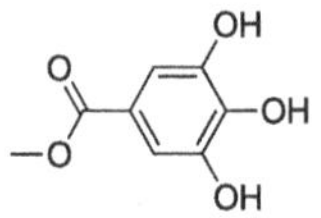

97 (methyl gallate)

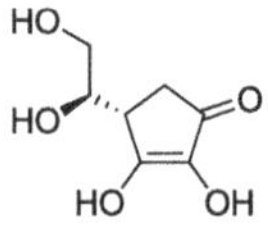

98 (ascorbic acid)

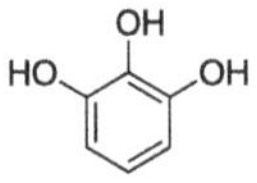

99 (pyrogallol)

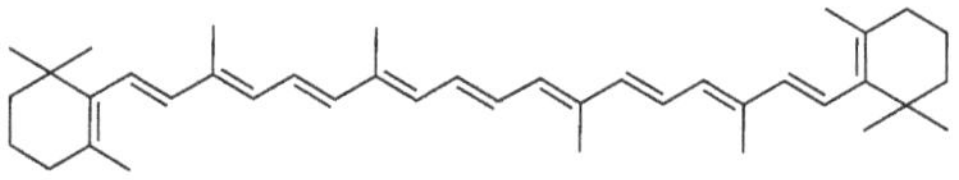

100 (β-carotene)

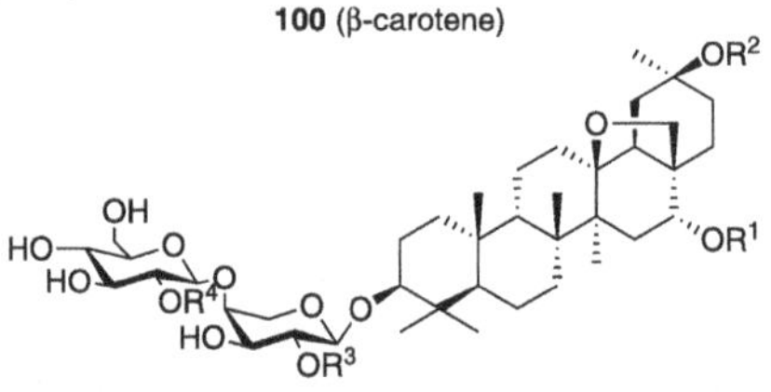

X = α-L-rhamnopyranose
Y = β-D-xylopyranose
Z = β-D-glucopyranose

101 (dexyloprimulanin) $R^1 = OH$, $R^2 = CHO$, $R^3 = R^4 = H$
102 (cyclamiretin A) $R^1 = OH$, $R^2 = CHO$, $R^3 = H$, $R^4 = X$
103 (primulanin O) $R^1 = R^3 = H$, $R^2 = CHO$, $R^4 = Y$
104 (ardisiacrispin A) $R^1 = OH$, $R^2 = CHO$ $R^3 = Z$, $R^4 = Y$
105 (ardisicrenoside B) $R^1 = OH$, $R^2 = CH_2OH$, $R^3 = Z$, $R^4 = Y$
106 (ardisicrenoside A) $R^1 = OH$, $R^2 = CH_2OH$, $R^3 = Z$, $R^4 = X$
107 (ardisimamilloside) $R^1 = {=}O$, $R^2 = CHO$, $R^3 = H$, $R^4 = X$
108 (3-*O*-β-D-xylopyranosyl-(1,2)-β-D-glucopyranosyl-(1,4)-α-L-arabinopyranosyl-3-hydroxy-13,28-epoxy-16-oxo-oleanan-30-al) $R^1 = {=}O$, $R^2 = CHO$, $R^3 = H$, $R^4 = Y$

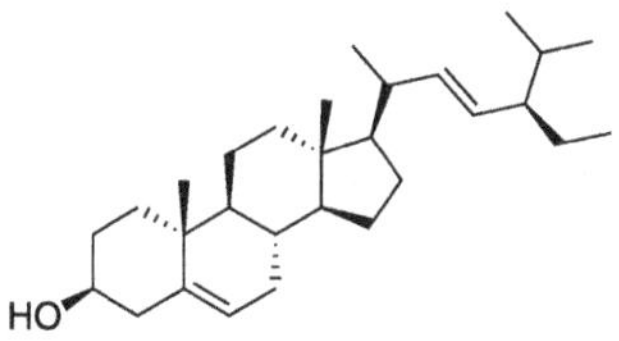

109 (stigmasterol)

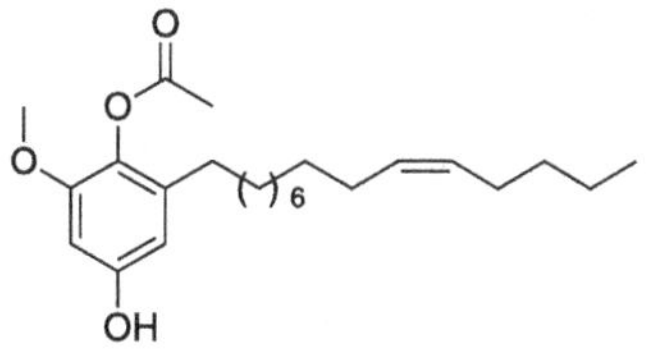

110 (1-*O*-methyl-6-acetoxy-5-(pentadec-(10*Z*)-enyl) resorcinol)

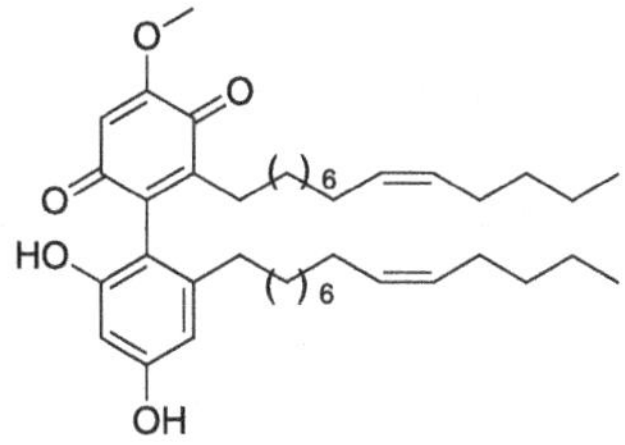

111 (demethylbelamcandaquinone B)

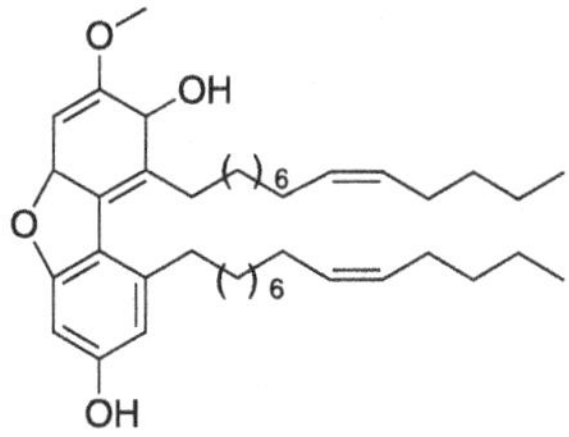

112 (fatimahol)

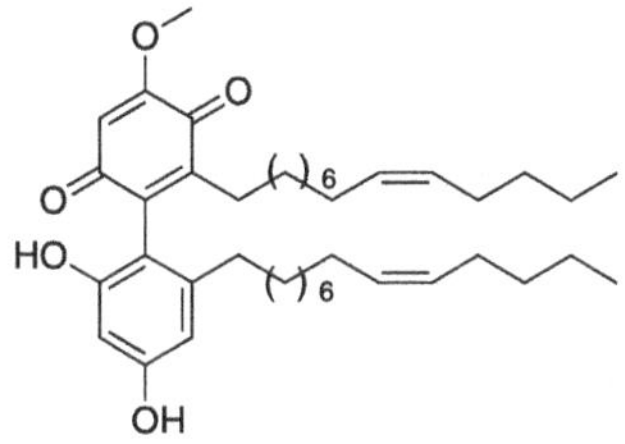

113 (labisiaquinone A)

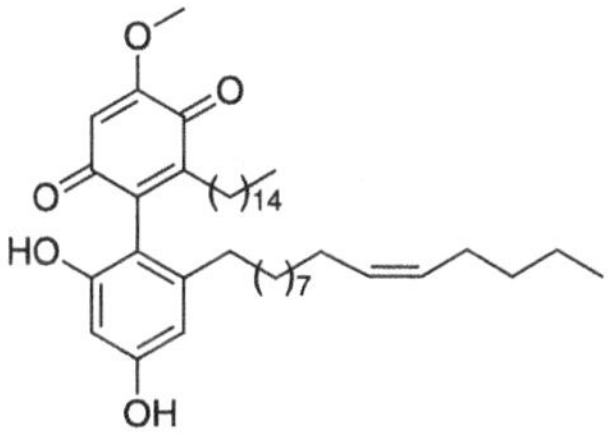

114 (labisiaquinone B)

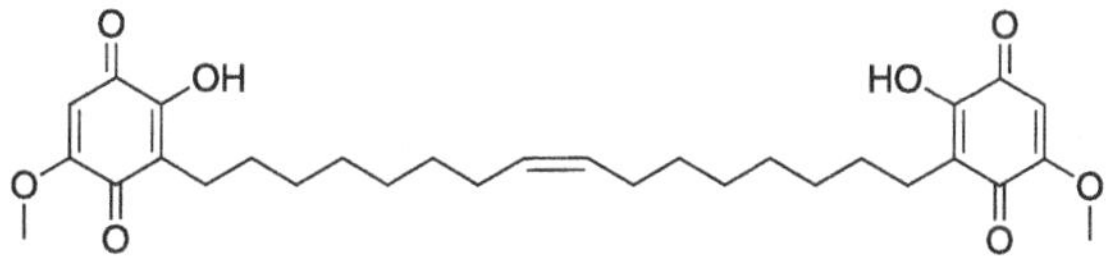

115 (irdisiaquinone A)

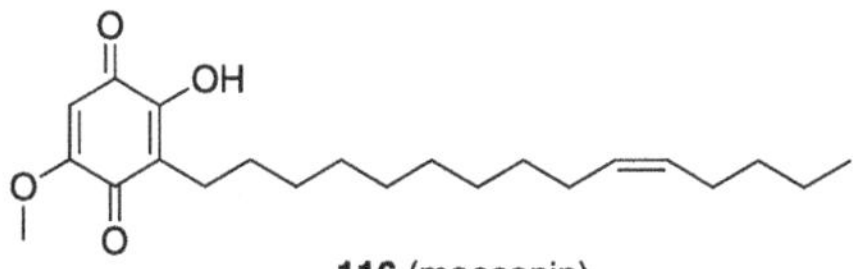

116 (maesanin)

The phenolic compounds of *M. pumilum* have exhibited antioxidant properties when measuring their free radical-scavenging, ferric-reducing antioxidant power, and β-carotene-bleaching activities using spectrophotometric measurements [111]. The role of phenolic compounds in modulating antioxidant mechanisms might contribute to the anticarcinogenic activity described for *M. pumilum* [117]. The carotenoid constituent, β-carotene (**100**) may help prevent sun-induced skin damage due to its high antioxidant properties [118].

Marantodes pumilum leaves also produce triterpene saponins (**101–108**) and stigmasterol (**109**) [111, 112, 119]. Other minor constituent content of this plant including an alkenylresorcinol (**110**), and isocoumarins (**111–116**), which have been investigated for their cytotoxic and xanthine oxidase inhibitory activities [113, 119, 120].

2.2.3 Biological Activities

Marantodes pumilum is used frequently by Malaysian women to maintain a healthy reproductive system and as a postpartum remedy [102]. This plant is well-known for its ability to treat ailments related to pregnancy, delivery, and those encountered postpartum, including flatulence, dysentery, dysmenorrhea, postpartum depression, rheumatism, and postmenopausal osteoporosis.

Apart from female reproductive health, many studies have shown that *M. pumilum* has a wide range of biological activities, including antibacterial [115], xanthine oxidase inhibitory [120], cytotoxic, antifungal [121], antioxidative [122], phytoestrogenic [123], anti-inflammatory, anticarcinogenic [124], antiaging [125], anti-hyperuricemic [126], anti-osteoporotic [127], anti-obesity [128], cardioprotective and hepatoprotective [129], and uterotonic [130] effects. The efficacy of this species in disease prevention and treatment is substantiated by its phytoestrogen, anti-inflammatory, and antioxidant properties (Table 2).

Marantodes pumilum, a traditional remedy used by Malaysian women, contains several known constituents that offer some support towards its ethnomedical uses described above. The main constituents found thus far including quercetin (**85**), rutin (**86**), apigenin (**87**), kaempferol (**88**), myricetin (**89**), and catechin (**90**), provide evidence that *M. pumilum* may be effective in treating conditions such as osteoporosis, obesity, and menopausal symptoms by modulating estrogenic pathways, improving bone health, and regulating metabolic parameters. Additionally, this plant exhibits antibacterial, anticarcinogenic, and hepatoprotective properties, supported by the presence of bioactive compounds that protect cells from oxidative stress, reduce inflammation, and modulate immune responses. Its currently known phytochemical profile to date supports *M. pumilum* as being as a promising ethnomedical plant for potential therapeutic development.

Table 2 Biological properties of *Marantodes pumilum*

Activity	Assay	References
Antibacterial	Active against Gram-positive (*Micrococcus luteus, Bacillus subtilis, Bacillus cereus, Staphylococcus aureus*) and Gram-negative organisms (*Enterobacter aerogenes, Klebsiella pneumoniae, Escherichia coli, Pseudomonas aeruginosa*) at low concentrations, when compared to kanamycin used as a control	[115, 131]
Anticarcinogenic	Inhibited skin tumor incidence, tumor burden, and tumor size formed in DMBA-croton oil-treated mice. when compared with a control group	[124]
Antifungal	Activity shown against *Fusarium* sp., *Candida* sp., and *Mucor* sp. using an agar disk diffusion assay, in comparison to streptomycin as the positive control	[115, 121]
Antihyperuricemic	Reduced serum uric acid levels in a hyperuricemic rat model induced by potassium oxonate	[126]
	Inhibited xanthine oxidase in a dose-dependent manner. with a root extract showing the most potent activity (IC_{50} value of 4.8 μg/cm^3), as compared to allopurinol (IC_{50} 0.2 μg/cm^3)	[132]
	Inhibited xanthine oxidase in vitro with an IC_{50} value of 0.66 ± 0.01 μg/cm^3	[120]
Anti-inflammatory	Inhibited NO release in RAW 264.7 cells stimulated with bacterial lipopolysaccharide (LPS) and interferon-gamma (IFN-γ)	[121]
	Inhibited the increased expression of TNF-α and IL-6 in the serum of treated animals	[117]
	Inhibited IL-1, IL-1, IL-6, IL-8, TNF-, and PGE2 secretion in monosodium urate-induced human peripheral blood mononuclear cells	[126]
	Reduced inflammation in a formalin-induced paw licking test using a rat model	[133]
	Suppressed the edema formation in a carrageenan-induced rat paw edema experiment	[133]
Antioxidative	Inhibited free radicals in ABTS radical scavenging (IC_{50} of 399.0 ± 1.1 μg/cm^3) and NO scavenging activity (IC_{50} of 411.5 ± 1.7 μg/cm^3) tests	[131]
	Comparative antioxidative activities were determined for two *L. pumila* varieties, var. *alata* and var. *pumila*, using DPPH, FRAP, and β-carotene bleaching methods	[122]

(continued)

Table 2 (continued)

Activity	Assay	References
	A root extract displayed DPPH free radical scavenging activity, with an IC_{50} value of 1.7 μg/cm^{3}that was compared to ascorbic acid (IC_{50} of 5.4 μg/cm^3)	[132]
	The reported antioxidant activities of the plant extract were attributed to the presence of carotenoids, ascorbic acid, anthocyanins, other flavonoids, and other phenolic compounds	[122]
Body weight management	An extract regulated body weight gain in ovariectomized rats by modulating the secretion of leptin and resistin, and the expression of the adipokines in adipose tissues	[128]
Cardiovascular effects	An extract modulated atherogenic parameters by improving serum lipid profiles, reducing hypercholesterolemic-triggered oxidative stress, and altering oxidized lipoprotein-related atherogenic risks	[129]
	An aqueous extract increased the size and vasculature of adipose tissue and thickened the adipocyte membrane of ovariectomized rats	[134]
	Modulated postmenopausal cardiovascular risk by maintaining the elasticity of the aorta lamellar architecture in ovariectomized rats	[135]
	Reduced triglyceride levels in postmenopausal women with no significant changes in terms of body weight, BMI, and waist circumference in a human trial	[136]
Cytotoxic	Cytotoxic against the MCF-7, MDA-MB-231, and Chang cell lines	[121]
	Two alkenylresorcinol derivatives showed cytotoxic activities against the PC3, HCT116, and MCF-7 cell lines (IC_{50} values $\leq$ 10 μM)	[113]
Toxicity	No significant reproductive toxicity, or complications in pregnancy, delivery, and early pup growth were observed in rats treated with 800 mg/kg/day of an aqueous extract	[105]
	An aqueous extract was devoid of toxicity at a low dose of 50 mg/kg in a sub-acute study of 28 days, whereas higher doses were associated with some toxicity concerns	[137]
	Decreased the polychromatic erythrocyte:normochromatic erythrocyte (PCE:NCE) ratio, at 2000 mg/kg of body weight at a 48 h harvest time point, in mammalian bone marrow cells using micronuclei formation to assess the genotoxicity	[138]
	Absence of external congenital malformations in rats observed with 1000 mg/kg/day of an aqueous extract, which suggested no teratogenic effects	[139]

(continued)

Table 2 (continued)

Activity	Assay	References
Hepatoprotective	Reduced the damage in the liver caused by fatty infiltration in hypercholesterolemia as well as attenuating the vascular endothelial permeability	[129]
Osteoporosis prevention	Protected bones against estrogen deficiency by regulating the gene expression of receptor activator of nuclear factor kappa-β ligand (RANKL), osteoprotegerin (OPG), and bone morphogenetic protein-2 (BMP-2)	[110]
	Changed the bone biochemical characteristics of osteoporosis in ovariectomized rats	[99]
	Time and dose-dependent effects of *L. pumila* were shown on bone strength using a postmenopausal osteoporosis rat model	[140]
	Demethylbelamcandaquinone B (**111**) increased osteocalcin expression that promoted matrix mineralization, as evidenced by an increase in calcium deposition	[127]
	Prevented changes in bone biochemical markers in ovariectomized rats, but did not inhibit bone calcium loss induced by ovariectomy	[99]
	Prevented osteoporotic changes, in as effective a manner as estrogen replacement therapy (ERT), in the bones of ovariectomized rats, as measured with histomorphometric parameters	[109]
	Improved femoral bone strength in ovariectomized rats, and was as effective as ERT in preventing fractures due to postmenopausal osteoporosis	[108]
Phytoestrogenic effects	An in vitro estrogenic effect was shown in human endometrial adenocarcinoma cells in estrogen-free basal medium, thus enhancing the secretion of alkaline phosphatase	[104]
	An aqueous extract modulated post-menopausal adiposity, thus reducing obesity in ovariectomized rats treated at a 17.5 mg/kg/day dose, using 120 μg/kg/day of ERT as a positive control	[134]
	A plant extract (10–50 mg/kg/day) was found to regulate the body weight gain of Sprague–Dawley rats through uterotrophic action in the rat model used	[128]
	Increased uterine weight up to 27% and improved insulin sensitivity up to 36% as well as improving the lipid profile in a postmenopausal osteoporosis (PCOS) rat model	[141]
	Prevented changes in bone biochemical markers from the administration of 17.5 mg/kg/day of a plant extract	[99]

(continued)

Table 2 (continued)

Activity	Assay	References
	Reduced triglyceride levels after 6 months of treatment in a randomized, double-blind placebo-controlled study on postmenopausal women	[136]
Skin protection	Stimulated the growth of human skin fibroblast sand keratinocyte cells in vitro after exposure to ultraviolet (UVB) radiation	[125]
	Protected human skin from the reactive oxygen species attacks generated by UVB exposure, mainly due to the presence of flavonoids and phenolic acids in the plant extract	[142]
	Restored the decreased collagen synthesis of human fibroblasts induced by UVB irradiation to a normal level, after treatment with an extract	[125]
Uterine contraction	Stimulated uterine smooth muscle contraction by decreasing estradiol levels while the progesterone levels remained low in blood serum at day-8 post-partum, in rats receiving 250–500 mg/kg/day of extract	[130]

2.3 *Andrographis paniculata*

Andrographis paniculata (Burm. f.) Wall. ex Nees (Plate 3) is a herbaceous plant from the family Acanthaceae. The genus *Andrographis* contains approximately 40 species. Only a few are used widely in folk medicine to treat various health conditions, with *A. paniculata* being the most significant of these. The plant is native to Taiwan, China, India [143], and Sri Lanka, and has been introduced to Thailand, Malaysia, Indonesia [144], and other countries [145]. In Malaysia, the plant is referred to as "Hempedu bumi" or "Pokok cerita". In India, it is known as "Mahatikta" (Sanskrit), "Kiryato" (Gujarati), "Mahatita" (Hindi), "Kalmegh" (Bengali), and "Sambiloto" in Indonesia, "Fah Talai Jone" in Thai [145], and "The Creator of Bitters" or "King of Bitters" in English.

This is an annual plant that grows to a height of about 1 m in damp, shady areas. The slender stem is dark green, branching, and has longitudinal furrows. The leaves of this species are small, green, simple, opposite, lanceolate, and glabrous, with a short petiole (Plate 3). The small flowers are borne in spreading racemes and have white petals with reddish-purple spots. The fruit is a capsule containing many seeds [146]. The seeds are tiny, plentiful, and yellowish brown. The aerial parts and roots are important in traditional medicine [147]. The ethnomedicinal value coupled with the ability of this plant to grow in most soil types has motivated farmers to cultivate *A. paniculata* in countries such as Malaysia, India, China, Thailand, Indonesia, the West Indies, and Mauritius [148, 149].

Plate 3 *Andrographis paniculata* (family Acanthaceae) leaves and flowers

2.3.1 Ethnobotany

Several Asian traditional medicine systems have used *Andrographis paniculata* for generations to treat various ailments. The herb may be administered orally or applied topically. The plant is used by traditional medical practitioners from Malaysia, Indonesia, Thailand, the Philippines, India, Bangladesh, China, and other countries to treat illnesses of the local people [143, 144, 147, 150–152]. The treatments include those for a wide range of chronic and diseases, from sore throats, dysentery, and snakebites, to diabetes, liver disease, herpes, and cancer [149, 153–155].

According to the Unani medical system, *A. paniculata* has uses in cases of leprosy, gonorrhea, scabies, boils, skin eruptions, chronic and seasonal fevers, pharyngolaryngitis, coughs with thick sputum, carbuncles, and sores. The plant is also regarded as an aperient, emollient, astringent, diuretic, emmenagogue, gastric tonic, and carminative [156]. Traditional Chinese medicine practitioners have used this species to treat inflammation, fevers, colds, and laryngitis. The plant has been attributed with a cooling effect and an an ability to help the body expel toxins and heat [148, 149, 157–160]. In addition, *A. paniculata* is a key component of herbal formulations used in one of the Indian traditional medical systems to treat a variety of ailments, including fever, liver diseases, vitiligo, gallbladder disorders, worm infections, and anti-inflammatory conditions [148, 149, 157].

Many other countries acknowledge *A. paniculata* as an important plant in traditional medicine practices. In Thailand, *A. paniculata* has been prioritized by the Thai Food and Drug Administration as an important plant as an anthelmintic, anti-inflammatory, and antipyretic agent, and for digestive problems, boils, chronic or seasonal fevers, general debility, gonorrhea, loss of appetite and skin diseases such as leprosy [151, 161]. The Chinese Pharmacopeia and the United States Pharmacopeia officially have each recorded *A. paniculata* as an herbal medicine for the treatment of upper respiratory tract infections, sore throat, and influenza [162]. In India, an environmental information system center focusing on medicinal plants has listed *A. paniculata* as a traditional remedy for the treatment of fever, sprue, edema, purification of breast milk, intrinsic hemorrhage, and vomiting [163].

To treat illnesses, the various plant parts of *A. paniculata* can be used singly or in combination. The leaves are used most commonly, followed by the aerial parts, the roots, and a combination of the leaves and roots. The herbal formulations are prepared in the form of dried powders, pastes made from ground plant materials, decoctions, infusions with boiling water, or pressed into a juice. In addition, the plant leaves can be consumed when raw. In Malaysia, a decoction of the aerial parts is used to treat malaria, diabetes, hypertension, cancer, snakebites, and the common cold [17].

2.3.2 Phytochemistry

The phytochemical constituents of *A. paniculata* have been isolated and identified from both the aerial parts and the roots of the plant. The main biologically active principles, which are also the major components of the plant, are a group of *ent*-labdane-type diterpene lactones (**117–166**). The composition of these compounds may fluctuate depending on various factors, including the plant processing technique, the harvesting period, and the geographic location of growth [164]. The highest concentration levels of diterpene lactones are found in the aerial parts (leaves and stems), with andrographolide (**117**) being the main constituent that has been evaluated for its pharmacological activity. Other constituents of this plant reported to date include flavonoids, xanthones, and noriridoids [147].

Andrographolide (**117**) is the most abundant diterpenoid found in *A. paniculata*. This diterpenoid lactone was first isolated in pure form in 1911, and has a highly bitter taste and is colorless [145].

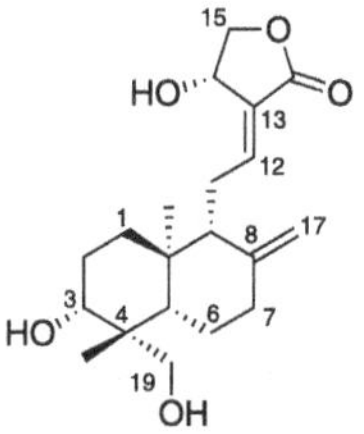

117 (andrographolide)

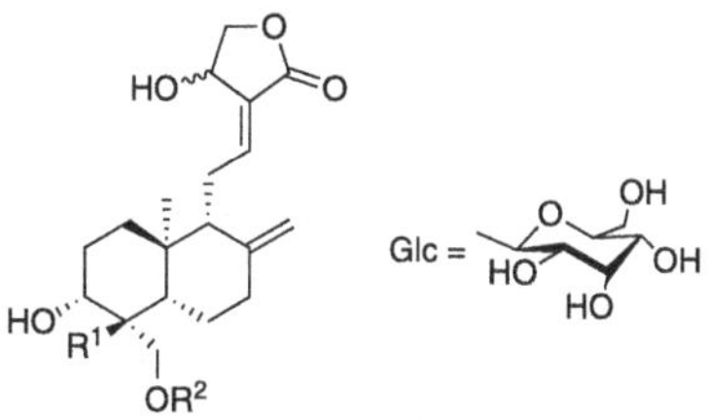

118 (andrographiside) R^1 = Me, R^2 = Glc
119 (14-*epi*-andrographolide) R^1 = Me, R^2 = H
120 (andropanolide) R^1 = R^2 = H

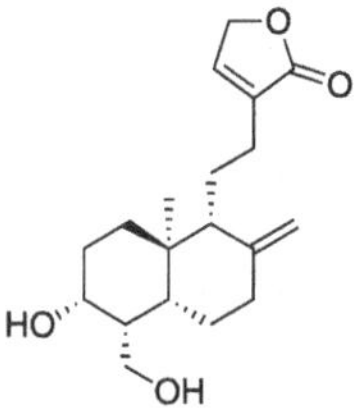

121 (3,18,19-trihydroxy-*ent*-labda-8(17),13-dien-16,15-olide)

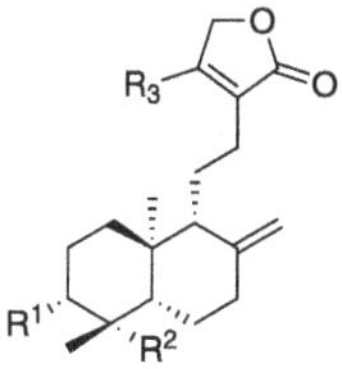

122 (andrograpanin) R^1 = OH, R^2 = Et, R^3 = H
123 (6′-acetylneoandrographolide) R^1 = R^3 = H, R^2 = *O*-Glc-6-Ac
124 (3-*O*-β-D-glucopyranosyl-14,19-dideoxyandrographolide) R^1 = *O*-Glc, R^2 = CH_2OH, R^3 = H
125 (3-*O*-β-D-glucosyl-14-deoxyandrographiside) R^1 = *O*-Glc, R^2 = CH2OH, R^3 = H
126 (19-hydroxy-*ent*-labda-8(17),13-dien15,16-olide) R^1 = R^3 = H, R^2 = CH_2OH
127 (neoandrographolide) R^1 = H, R^2 = *O*-Glc, R^3 = OH
128 (14-dideoxyandrographolide) R^1 = OH, R^2 = CH_2OH, R^3 = OH

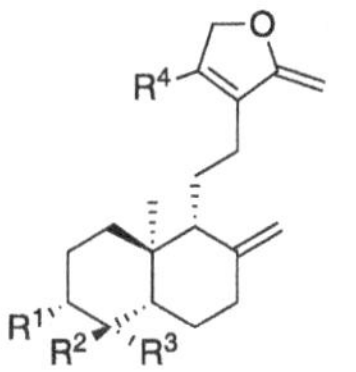

129 (7-hydroxy-14-deoxyandrographolide) R^1 = OH, R^2 = OH, R^3 = CH_2OH, R^4 = H
130 (19-*O*-[β-D-apiofuranosyl-β-D-glucopyranosyl]-3,14-dideoxyandrographolide) R^1 = R^4 = H, R^2 = CH_3, R^3= *O*-Api-Glc
131 (andropanoside) R^1 = OH, R^2 = CH_3, R^3 = *O*-Glc, R^4 = H
132 (12-hydroxyandrographolide) R^1 = H, R^2 = CH_3, R^3 = CH_2OH, R^4 = OH

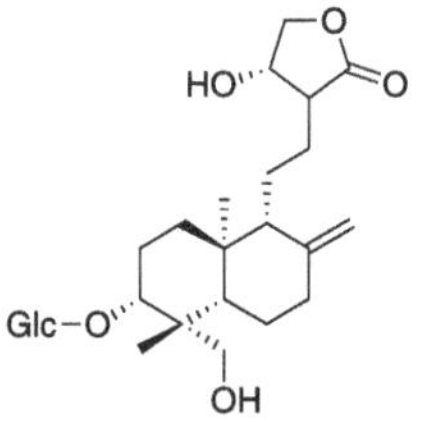

133 (3-*O*-β-D-glucopyranosyl-andrographolide)

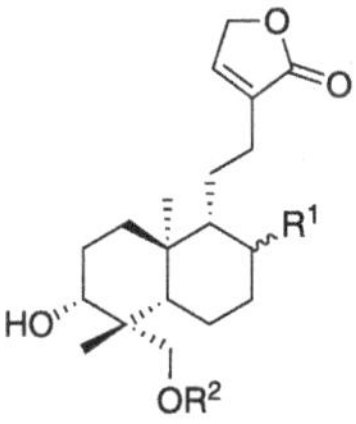

134 (14-deoxyandrographiside) R^1 = H, R^2 = *O*-Glc
135 (14-deoxy-17-β-hydroxy andrographolide) R^1 = R^2 = CH_2OH

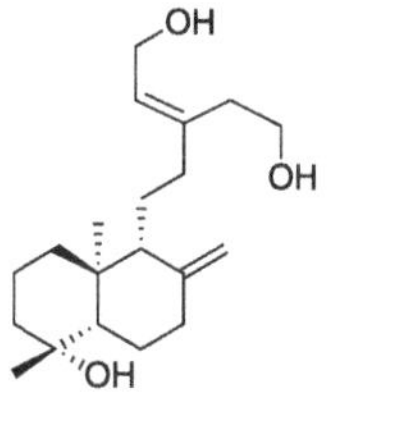

136 (8(17),13-*ent*-labda-diene-15,16,19-triol)

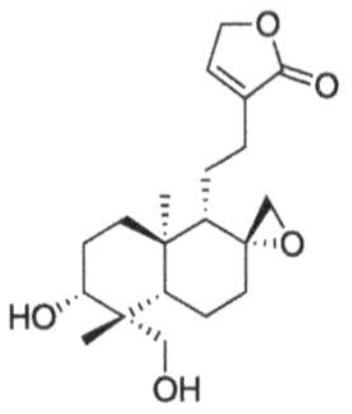

137 (8,17-epoxy-14-deoxy-andrographolide)

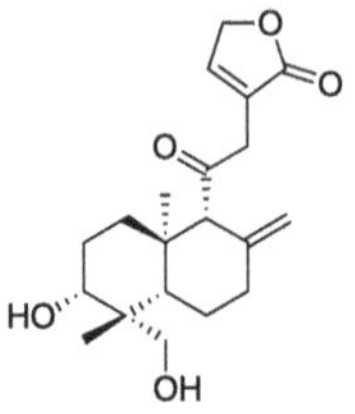

138 (14-deoxy-11-oxo-andrographolide)

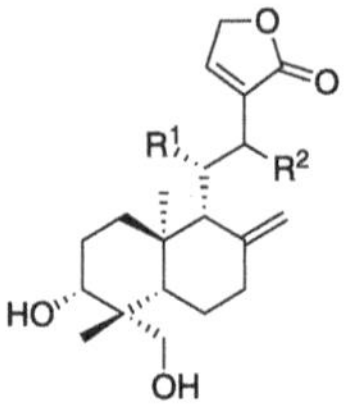

139 (14-deoxy-11-hydroxy-andrographolide)
R^1 = OH, R^2 = H
140 (14-deoxy-12-methoxy-andrographolide)
R^1 = H, R^2 = OMe

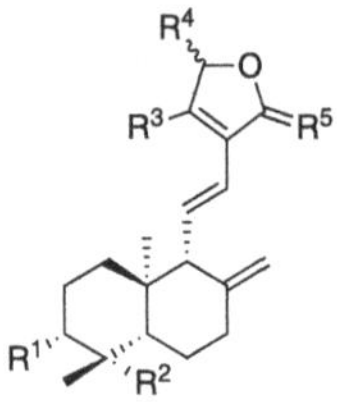

141 (14-deoxy-11,12-didehydroandrographolide)
R^1 = OH, R^2 = CH_2OH, R^3 = OH, R^4 = H, R^5 = O
142 (15-methoxy-3,19-dihydroxy-8(17)11,13-*ent*-labda-trien-16,15-olide)
R^1 = H, R^2 = OH, R^3 = H, R^4 = OMe, R^5 = CH_2
143 (3-*O*-β-D-glucosyl-14-deoxy-11,12-didehydroandrographiside)
R^1 = *O*-Glc, R^2 = CH_2OH, R^3 = R^4 = H, R^5 = O
144 (14-deoxy-15-isopropylidene-11,12-didehydroandrographolide)
R^1 = OH, R^2 = CH_2OH, R^3 = H, R^4 = 2-methylpropenyl, R^5 = O
145 (14-deoxy-11,12-didehydroandrographiside)
R^1 = OH, R^2 = CH_2OH, R^3 = R^4 = H, R^5 = O

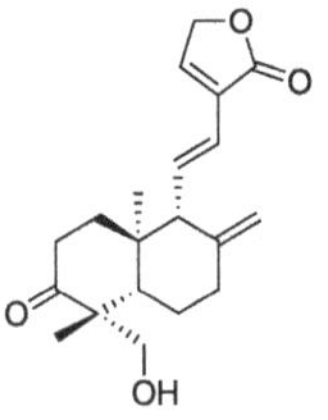

146 (3-oxo-14-deoxy-11,12-didehydroandrographolide)

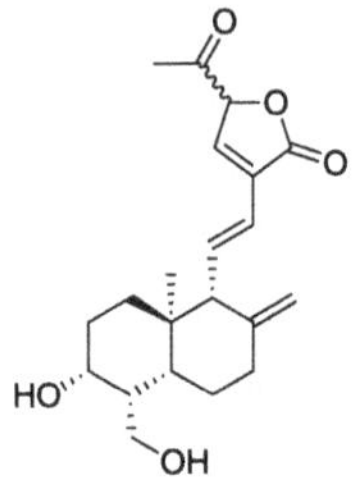

147 (3,19-dihydroxy-15-methoxy-*ent*-labda-8(17),11,13-trien-16,15-olide)

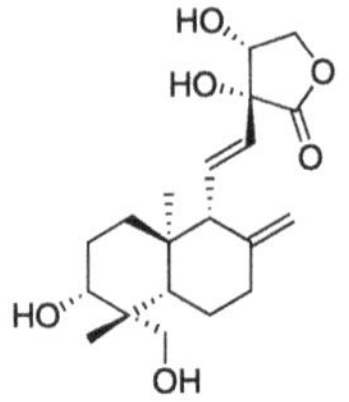

148 (3,13,14,19-tetrahydroxy-*ent*-labda-8(17),11-dien-16,15-olide)

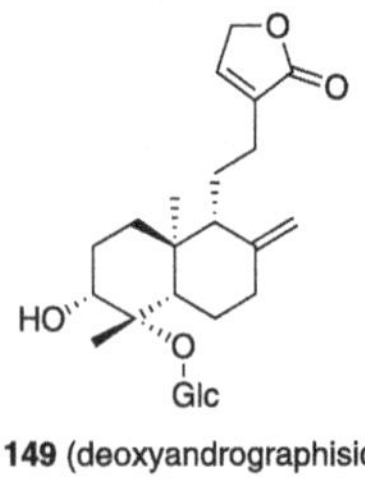

149 (deoxyandrographiside)

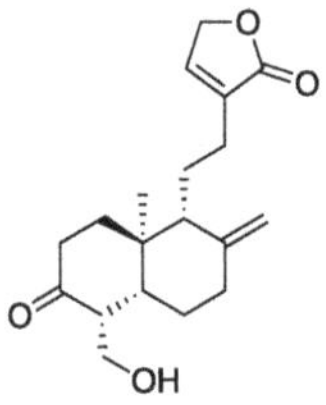

150 (19-hydroxy-3-oxo-*ent*-labda-8(17),11,13-trien-16,15-olide)

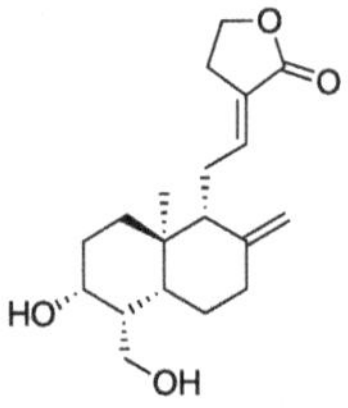

151 (3,19-dihydroxy-*ent*-labda-8(17),12-dien-16,15-olide)

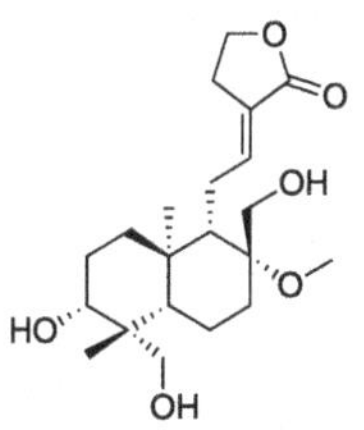

152 (8-methoxy-14-deoxy-17-β-hydroxyandrographolide)

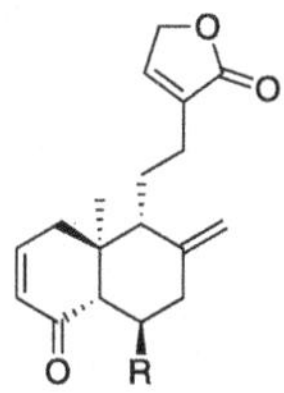

153 (19-norandrographolide B) R = H
154 (19-norandrographolide C) R = OH

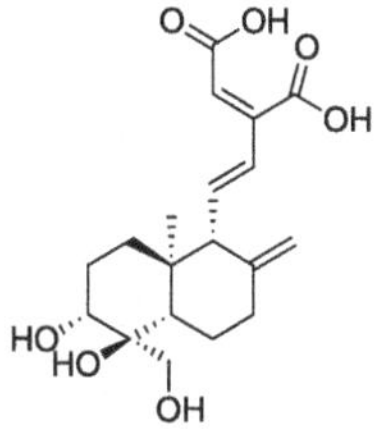

155 (andrographic acid)

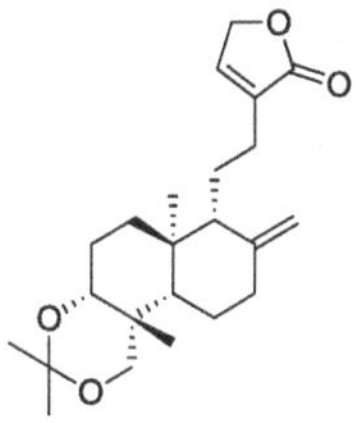

156 (3,19-isopropylidene-14-deoxy-*ent*-labda-8(17),13-dien-16,15-olide)

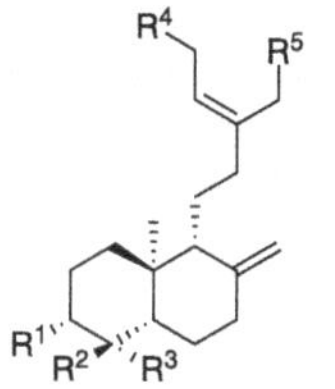

157 (andrographatoside) R^1 = H, R^2 = OH, R^3 = COO-Glc, R^4 = R^5 = OH

158 (19-*O*-β-D-glucopyranosyl-*ent*-labda-8(17),13-dien-15,16,19-triol) R^1 = H, R^2 = CH_3, R^3 = *O*-Glc, R^4 = R^5 = OH

159 (3,15,19-trihydroxy-*ent*-labda-8(17),13-dien-16-oic acid) R^1 =OH, R^2 = H, R^3 = CH_2OH, R^4 = OH, R^5 = COOH

160 (*ent*-labda-8(17),13-dien-15,16,19-triol) R^1 = R^2 = H, R^3 = CH_2OH, R^4 = R^5 = OH

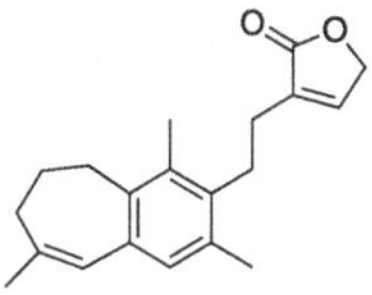

161 (andrographolactone)

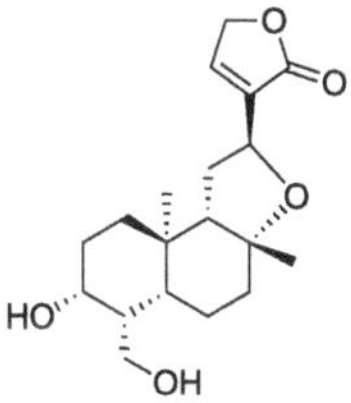

162 (isoandrographolide)

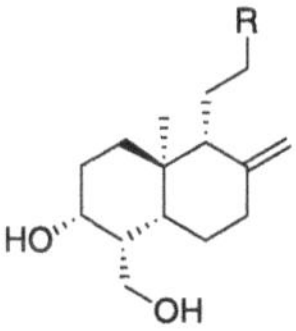

163 (13,14,15,16-tetra-*nor*-*ent*-labda-8(17)-ene-3,12,19-triol) R = OH
164 (3,19-dihydroxy-14,15,16-trinor-*ent*-labda-8(17),11-dien-13-oic acid) R = COOH

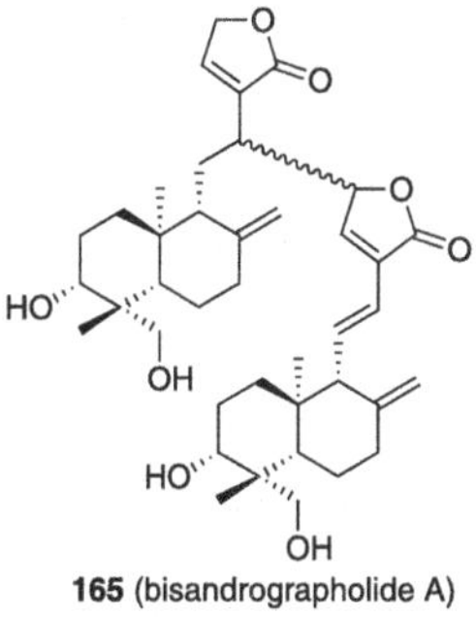

165 (bisandrographolide A)

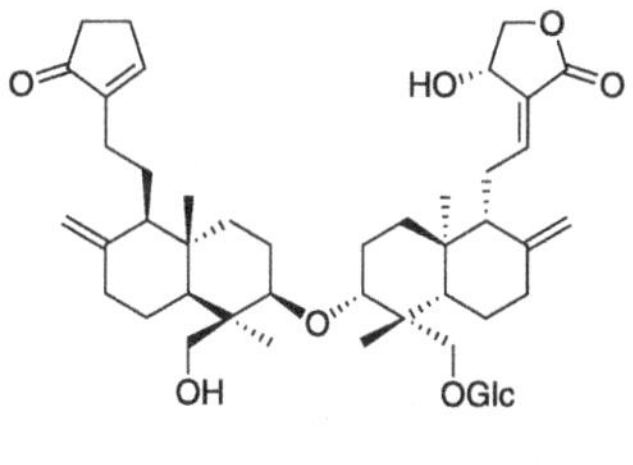

166 (bisandrographolide ether)

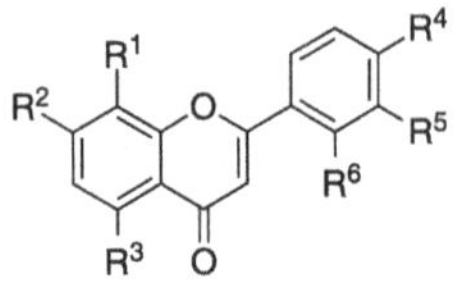

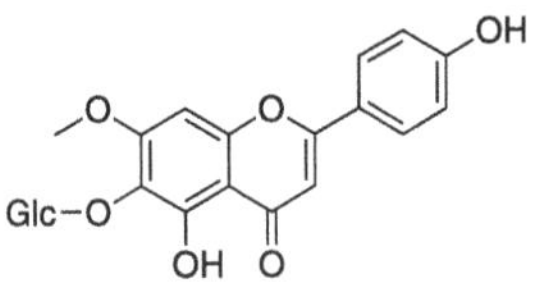

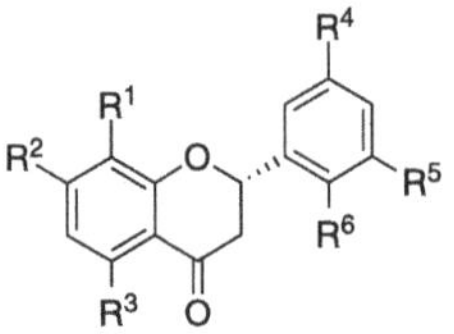

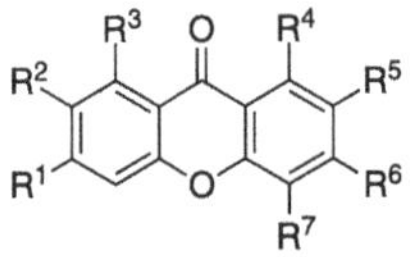

167 (7-*O*-methylwogonin) $R^1 = R^2 = OMe$, $R^3 = OH$, $R^4 = R^5 = R^6 = H$

168 (5,7,2',3'-tetramethoxyflavanone) $R^1 = R^4 = H$, $R^2 = R^3 = R^5 = R^6 = OMe$

169 (5-hydroxy-7,8-dimethoxyflavone) $R^1 = R^2 = OMe$, $R^3 = OH$, $R^4 = R^5 = R^6 = H$

170 (skullcapflavone I) $R^1 = R^2 = OMe$, $R^3 = OH$, $R^4 = R^5 = H$, $R^6 = OH$

171 (skullcapflavone-12'-methylether) $R^1 = R^2 = OMe$, $R^3 = OH$, $R^4 = R^5 = H$, $R^6 = OMe$

172 (skullcapflavone-12'-glucoside) $R^1 = R^2 = OMe$, $R^3 = OH$, $R^4 = R^5 = H$, $R^6 =$ *O*-Glc

173 (apigenin-7-*O*-*β*-D-methylglucuronide) $R^1 = H$, $R^2 =$ *O*-Glu-OMe, $R^3 = OH$, $R^4 = OH$ $R^5 = R^6 = H$

174 (cosmosiin) $R^1 = H$, $R^2 =$ *O*-Glc, $R^3 = R^4 = OH$, $R^5 = R^6 = H$

175 (andropaniculoside A) $R^1 = R^2 = OMe$, $R^3 =$ *O*-Glc, $R^4 = R^5 = H$, $R^6 = OH$

176 (isoswertisin) $R^1 =$ *O*-Glc, $R^2 = OMe$, $R^3 = R^4 = OH$, $R^5 = R^6 = H$

177 (scutellarin-6-*O*-*β*-D-glucoside-7-methylether)

178 (5-hydroxy-7,8-dimethoxyflavanone) $R^1 = H$, $R^2 = Me$, $R^3 =$ *O*-Glc, $R^4 = R^5 = R^6 = H$

179 (7-*O*-methyldihydrowogonin) $R^1 = H$, $R^2 = Me$, $R^3 =$ *O*-Glc, $R^4 = R^5 = R^6 = H$

180 (7-*O*-methyldihydrowogonin-5-glucoside) $R^1 = H$, $R^2 = Me$, $R^3 =$ *O*-Glc, $R^4 = R^5 = R^6 = H$

181 (andropaniculosin A) $R^1 = R^2 = OMe$, $R^3 = R^4 = OH$, $R^5 = H$, $R^6 = OMe$

182 (andrographidine A) $R^1 = R^2 = OMe$, $R^3 =$ *O*-Glc, $R^4 = R^5 = R^6 = H$

183 (dihydroskullcapflavone I) $R^1 = R^2 = OMe$, $R^3 = OH$, $R^4 = R^5 = H$, $R^6 = OH$

184 (5-hydroxy-7,2',3'-trimethoxyflavone) $R^1 = H$, $R^2 = OMe$, $R^3 = OH$ $R^4 = H$ $R^5 = R^6 = OMe$

185 (1,2-dihydroxy-6,8-dimethoxyxanthone) $R^1 = R^3 = OMe$, $R^2 = H$, $R^4 = R^5 = OH$, $R^6 = R^7 = H$

186 (1,8-dihydroxy-3,7-dimethoxyxanthone) $R^1 = H$, $R^2 = OMe$, $R^3 = R^4 = OH$, $R^5 = H$, $R^6 = OMe$, $R^7 = H$

187 (3,7,8-trimethoxy-1-hydroxyxanthone) $R^1 = H$, $R^2 = R^3 = OMe$, $R^4 = OH$, $R^5 = H$, $R^6 = OMe$, $R^7 = H$

188 (4,8-dihydroxy-2,7-dimethoxyxanthone) $R^1 = H$, $R^2 = OMe$, $R^3 = OH$, $R^4 = H$, $R^5 = OMe$, $R^6 = H$, $R^7 = OH$

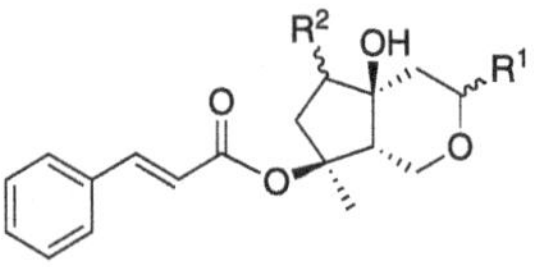

189 (andrographidoid A) $R^1 = OMe$, $R^2 = OH$
190 (andrographidoid B) $R^1 = OMe$, $R^2 = OH$
191 (andrographidoid C) $R^1 = OMe$, $R^2 = OH$
192 (andrographidoid D) $R^1 = R^2 = OH$

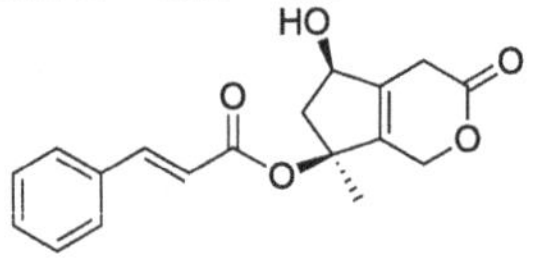

193 (andrographidoid E)

The major flavonoids (**167–184**) of *A. paniculata* are flavones (**167–177**), which have been isolated from the aerial parts, roots, and the whole plant [165, 166]. Some flavanones and flavonoid glycosides were also isolated and identified from this plant [160, 165, 167]. Several miscellaneous compound types have also been purified, especially from the roots of *A. paniculata*. These compounds include xanthones (**185–188**) [168], quinic acid [169], and some rare noriridoids (**189–193**) [170]. (*E*)-Cinnamic acid (**670**), caffeic acid (**490**), ferulic acid (**489**), and chlorogenic acid (**392**) were also reported as minor isolates [165, 171]. The major isolated compounds from *A. paniculata* are listed in the formula drawings provided.

2.3.3 Biological Activities

The extensive use of the different parts of *A. paniculata* plant in folk medicine to treat various disorders has raised interest among scientists to study its pharmacological activities to validate its purported therapeutic properties. Numerous investigations have demonstrated various diverse biological properties of this plant, including antimicrobial, cytotoxic, antiprotozoal, anti-inflammatory, antioxidant, immunostimulant, antidiabetic, anti-infective, antiangiogenic, antipyretic, antihypertensive, antiallergic, antidiarrheal, hepatorenal protective, sex hormone modulatory, liver enzyme modulatory, and insecticidal activities [145, 147, 172, 173] (Table 3). Previous biological assays and screening methods have included those on cancer cell growth inhibition [174–177], and on the antidiabetic [177–180], anti-inflammatory [181–183], antimalarial [184], antiretroviral [185], antioxidative [186–188], antihypertensive [189], antipsychotic [190], antithrombotic [191], anti-atherosclerotic [192], cholestatic [177], respiratory condition modulation [193], analgesic, antipyretic, antiulcerogenic [194], and immunomodulatory [195] effects.

Andrographis paniculata contains a range of bioactive compounds, primarily labdane diterpenoid lactones, which are mainly responsible for its currently investigated pharmacological activities. Known for its use in traditional medicine, this plant

Table 3 Pharmacological properties of *Andrographis paniculata*

Activity	Assay	References
Analgesic	Analgesic effects were shown for andrographolide (**117**) using hot plate and acetic acid-induced writhing tests in mice and the Randall Selitto's test in rats	[194]
Antiallergic	Andrographolide (**117**) and neoandrographolide (**127**) both showed mast cell stabilization activity in a rat model, in a similar manner to disodium chromoglycate	[196]
Antidiabetic	An andrographolide (**117**)-lipoic acid conjugate lowered blood glucose and stimulated GLUT4 membrane translocation in alloxan-treated diabetic mice	[200]

(continued)

Table 3 (continued)

Activity	Assay	References
	Andrographolide (**117**) decreased the levels of blood glucose and improved islet and beta-cell functions	[201]
	An aqueous extract prevented glucose-induced hyperglycemia (orally administered) in nondiabetic rabbits without affecting epinephrine-induced hyperglycemia	[202]
	An ethanol extract of *A. paniculata* lowered fasting serum glucose in diabetic rats treated at doses of 0.1, 0.2, and 0.4 g/kg, when compared to a control	[203]
	A dose of 125 mg/kg of an extract of *A. paniculata* possessed antidiabetic effects in alloxan-induced diabetic rats	[178]
Antidiarrheal	Andrographolide (**117**) was found to act against *Escherichia coli* endotoxin in an in vivo antidiarrheal model	[236]
Antifertility	A combination of curcumin and andrographolide (**117**) produced antifertility effects in female rats	[209, 210]
	Andrographolide (**117**) was found to arrest spermatogenesis when administered to male Wistar rats for 48 days	[210]
	A powder of *A. paniculata* showed antispermatogenic and antiandrogenic effectsin albino rats	[211]
Anti-HIV	Neoandrographolide (**127**) inhibited furin with an IC_{50} value of 53.5 μM	[185]
	The three diterpenoid lectones andrographolide (**117**), neoandrographolide (**127**), and 14-deoxy-11,12-didehydroandrographolide (**141**) all showed viricidal activity against herpes simplex virus 1 (HSV-1), and no cytotoxicity at their virucidal concentrations	[227]
	A phase I trial of andrographolide (**117**) was conduced in HIV-positive patients and normal volunteers, and this compound was proposed to inhibit HIV-induced cell cycle dysregulation	[238]
Antihypertensive	14-Deoxyandrographolide was found to exhibit a vasorelaxation effect in a rat thoracic aorta model	[189]
Anti-inflammation	Andrographolide (**117**) produced an inhibitory effect on the release of tumor necrosis factor alpha (TNF-α) in the THP-1 human monocytic leukemia cell line	[214]
	Andrographis paniculata exhibited an antipyretic action by inhibiting free radical production involving the modulation of a protein kinase C (PKC)-dependent pathway	[164]

(continued)

Table 3 (continued)

Activity	Assay	References
	Andrographolide (**117**) exerted in vitro and in vivo antiinflammatory effects by showing NF-κB-dependent luciferase activity	[215]
	14-Deoxy-11,12-didehydroandrographolide (**141**), andrographolide (**117**), and neoandrographolide (**127**) exhibited an antiinflammatory effect in an in vitro study, with compound **117** having multiple effects on gene expression	[216]
	Neoandrographolide (**127**) inhibited NO production both in vitro and ex vivo in murine models, and was postulated to play an important role in the use of *A. paniculata* as an potential anti-inflammatory crude drug	[235]
Antileishmanial	An extract showed cytostatic and growth inhibition of *Leishmania donovani* amastigotes and reduced the splenic parasitic burden	[190]
	A liposomal formulation of andrographolide (**117**) was found to target *L. donovani*-infected macrophages	[212]
Antimalarial	An ethanol extract of *A. paniculata* reduced the level of parasitemia against *Plasmodium berghei* NK65 in *Mastomys natalensis*, in a dose-dependent manner	[205]
	A chloroform extract showed complete inhibition of *Plasmodium falciparum* within a 24-h incubation period, at a dose of 0.05 mg/cm^3	[218]
Antimicrobial	A controlled delivery system of an *A. paniculata* extract inhibited *Staphylococcus aureus* ATCC 25,923, *Escherichia coli* wild type, *Candida tropicalis* ATCC 750, and *Candida albicans* ATCC 10,231	[219]
	A methanol extract inhibited *E. coli* and *S. typhi*, but was less effective for *Staphylococcus aureus*, using a 30–120 mg/cm^3 dose range	[163]
	A leaf ethanol extract inhibited bacterial growth at doses of 300, 400, 500 μg/cm^3	[220]
	Potent inhibition occurred of the growth of both Gram-negative and Gram-positive bacteria at 0, 10, 25, 50, 100, 200 μg/cm^3, using an ethanol extract of *A. paniculata*	[221]
	Activity was shown against methicillin-resistant *Staphylococcus aureus*, *Enterococcus faecalis*, and *Mycobacterium tuberculosis* with a dose range of 1.0–3.0 mg, using a methanol extract of the plant leaves	[222]

(continued)

Table 3 (continued)

Activity	Assay	References
	An *A. paniculata* extract exhibited antibacterial activities against several Gram-positive and Gram-negative bacterial strains that were tested	[223]
	Activity against *Microsporum nanum*, *Curvularia* sp. and *Fusarium dimerum* was shown by an aqueous extract of *A. paniculata*	[224]
	A leaf protein from *A. paniculata* showed antifungal effects against *Aspergillus flavus* and *Macrophomina phaseolina*	[225]
	A methanol extract of *A. paniculata* inhibited the activity of DENV-1 using an in vitro anti-dengue assay at doses of 1–3 mg/g	[226]
	It was postulated that the antimicrobial activity of aqueous extracts of *A. paniculata* is due to a combination of andrographolide (**117**), related diterpenoid lactones, and arabinogalactan proteins	[237]
Antioxidative	The antioxidant activity on lymphoma was shown of an aqueous extract of *Andrographis paniculata*	[179]
	A leaf extract reduced MDA levels and elevated the total antioxidant status in rat urine at a dose of 1 g/kg body weight	[217, 228]
	Andrographolide (**117**) inhibited the proliferation of 3T3-L1 preadipocytes at 10 and 20 μg/cm^3, through the abrogation of a glutathione antioxidant system	[229]
	Andrographis pancilata extracts showed free radical-scavenging activities when evaluated by the CUPRAC, FRAP, and DPPH methods	[230, 231]
	It was proposed that neoandrographolide (**127**) may scavenge free radicals by hydrogen abstracton from C-15 in the lactone ring	[204]
	Andrographolide (**117**) suppressed rat neutrophil ROS production and adhesion	[233]
Antistress	Andrographolide (**117**) altered body weight, gastric ulcer, adrenal and spleen weights, the depressive state, and sexual behavior in stress-induced male rats	[232, 234]
Cholerectic	Andrographolide (**117**) showed a cholerectic action in rats by increasing bile flow	[213]
Cytotoxic	Inhibited the adhesion of gastic cancer cells to endothelial cells by blocking E-selectin expression	[162, 176]
	Inhibited the proliferation of HT-29 colon cancer cells and augmented the proliferation of human peripheral blood lymphocytes (HPBLs) at low concentrations	[175]

(continued)

Table 3 (continued)

Activity	Assay	References
	Inhibited OVCAR-5 ovarian cancer cells by 51.1% with 100 μg/cm^3 of a hydroalcohol extract	[197]
	Induced apoptosis in the TD-47 human breast cancer cell line in a time and concentration-dependent manner	[198]
	Andrographolide (**117**) induced apoptosis in human cancer cells via activation of caspase 8, involving the participation of the mitochondria	[206]
	A polysaccharide from *A. paniculata* suppressed the proliferation of HepG2 cells via inducing mitochondrial-mediated apoptosis	[239]
	A stem and leaf extract of *A. paniculata* showed a weak antiproliferative effect in Hela cells using a MTT assay	[240]
	A methanol extract of *A. paniculata* aerial parts and several diterpenoid constituents showed cell-differentiating activities in the M1 mouse myeloid leukemia cell line	[241]
	Andrographolide (**117**) and isoandrographolide (**162**) were cytotoxic for KB cells, with IC_{50} values of 6.5 and 5.1 μM, respectively	[242]
Effects on the respiratory sstem	In a clinical study, school pupils were administered tablets prepared from *A. paniculata*, and this led to a decrease in the incidence of the common cold in the third month of treatment	[243]
Hepatoprotective	Andrographolide (**117**) and *A. panculata* extracts have been found to inhibit hepatic cytochrome P450 enzymes	[244–246]
	An ethanol extract of *A. paniculata* (25 mg/kg) and andrographolide (**117**) and neoandrographolide (**127**) (6 mg/kg/day for two weeks) showed significant antihepatotoxic action in *Plasmodium berghei* K173-induced hepatic damage in *Mastodys natalensis*	[247]
	An ethanol *A. paniculata* extract, and purified andrographolide (**117**), andrographiside (**118**), and neoandrographolide (**127**) all showed antihepatotoxic effects against carbon tetrachloride-induced liver damage in rodent models	[248, 249]
	A 500 mg/kg dose of an ethanol leaf extract showed antihepatotoxic activity against paracetamol-induced liver damage in rats	[250]

(continued)

Table 3 (continued)

Activity	Assay	References
	Andrographolide (**117**) displayed a hepatoprotective effect against hexachlorocyclohexane-induced oxidative injury	[251]
	An *A. paniculata* extract protected against hepatotoxicity induced by paracetamol when treated with 100–200 mg/kg doses	[252]
	A leaf powder extract protected against hepatic cell damage in rats caused by paracetamol induction	[253]
	Doses of 200 and 800 mg/kg andrographolide (**117**) led to a hepatoprotective activity against galactosamine or paracetamol toxicity in a rat model	[254]
	Significant hepatoprotective effects were shown against CCl_4-induced liver damage when treated with 300 mg/kg of an *A. paniculata* ethanol extract	[255]
	Andrographis paniculata enhanced the recovery from CCl_4-induced hepatic damage due to its antioxidant and hepatoprotective properties	[256]
Immunomodulation	An *A. paniculata* extract and andrographolide (**117**) modulated the immune response in pretreated mice with an inactivated *Salmonella typhimurium* vaccine	[257]
	An *A. paniculata* extract showed anaphylactoid potential using a popliteal lymph node assay and P815 cell degranulation in vitro	[258]
	An *A. paniculata* extract and andrographolide (**117**) affected antigen-specific and nonspecific immune, humoral, and cellular adaptive immune responses	[151, 212]
	Andrographolide (**117**) modulated innate and adaptive immune responses by regulating macrophage phenotypic polarization	[259]
Insecticidal	Andrographolide (**117**) was found to be an effective insect growth regulator against *Tribolium confusum*	[208]
	Four flavonoids from *A. paniculata* showed inhibitory effects against *Anopheles minimus* CYP6AA3 and CYP6P7 enzymes	[260]
Muscle protection	Inhibition of aberrant vascular smooth muscle cells and cerebral endothelial cells in a dose-dependent manner using andrographolide (**117**)	[199]
Platelet aggregation	Andrographolide (**117**) has an inhibitory effect on PAF-induced platelet aggregation	[207]

exhibits antimicrobial, anti-inflammatory, cytotoxic, and antioxidant effects. Key compounds like andrographolide (**117**) contribute to its anti-inflammatory effects by inhibiting pathways like NF-κB, while others enhance immune responses, making it of potential use as an immunomodulatory agent. Additionally, *A. paniculata* has demonstrated cancer-related properties by promoting apoptosis in cancer cells and has shown hepatoprotective effects against drug-induced liver damage. These findings position *A. paniculata* as being well worthy of further laboratory and clinical investigation for a potential wide range of therapeutic applications related to various health conditions.

2.4 *Morinda citrifolia*

Morinda citrifolia L. (Rubiaceae), with the common name, noni, is known locally as "mengkudu besar" in Malaysia. This species is a shrub or small tree, typically ranging from 3 to 10 m in height, and is characterized by having large, broad elliptical leaves (measuring 5–17 cm in length and 10–40 cm in width). Its small, tubular white flowers are clustered and attached to the peduncles. The petioles have distinct ring-like scars on the stems, and the corollas are a pale greenish-white color [261] (Plate 4).

Plate 4 Fruits and leaves of *Morinda citrifolia* (family Rubiaceae). Photograph courtesy Ninetechno (https://pixabay.com/photos/tree-plant-leaves-fruit-noni-6705882/)

2.4.1 Ethnobotany

Morinda citrifolia has many uses in traditional medicine, including as a treatment for dysentery, heartburn, and liver disease, and as an antidiabetic [262]. The leaves are used for postpartum care among women of the Malay community [263]. In turn, the fruits and leaves of *M. citrifolia* have been used as an abortifacient, with the leaves and shoots being of use to treat eye injuries and infections. The fruits are employed for the treatment of diabetes and tuberculosis [264]. In the Jah Hut Orang Asli community, a root and leaf decoction of *M. citrifolia* is utilized for treating boils on the head. Temuan villagers consume its raw fruits to purify the blood. The young shoots are consumed in Malaysia to help tolerate the effects of high blood pressure [265]. The juice from the fruits of noni is available as a health tonic in Malaysia, Australia, the Philippines, the United States, and elsewhere. Other reported uses of the various parts of *M. citrifolia* are to alleviate various symptoms related to cardiovascular diseases such as dyslipidemia, hypertension, in addition to type II diabetes, and non-alcoholic fatty liver disease.

2.4.2 Phytochemistry

The roots and stem bark of *M. citrifolia* are reported to contain mainly anthraquinones [266–269], whereas the fruits biosynthesize flavonoid glycosides, iridoids, and phenolic acids as well as anthraquinones [270–275]. In turn, the leaves accumulate iridoids, flavonoids, and anthraquinones [276–279]. This contribution will discuss compounds from noni of the anthraquinone type only.

Anthraquinones from *M. citrifolia* are of the *Rubia* type, which are derived biosynthetically from the shikimate and MEP/mevalonate pathways. They are typically substituted in ring C, with the substitution in ring A being usually unsymmetrical. *Rubia* anthraquinones are sometimes present in the form of glycosides [280]. The common sugars attached to the anthraquinone aglycones are primeverose and gentibiose (two units of D-glucose). Most anthraquinones from *M. citrifolia* contain substituents at C-1, C-2, and C-3 on ring C [281]. Several anthraquinones contain functionalities on both rings A and C. Substituents are commonly located at C-5 and C-6 on ring A and more rarely located at C-4, C-7, and C-8. To date, 63 anthraquinones have been reported from the various parts of *M. citrifolia*, especially the roots. Table 4 lists the anthraquinones reported from *M. citrifolia* and the formula drawings of **194–247** display their structures.

Table 4 Anthraquinones from *Morinda citrifolia*

Plant collection location	Part	Compounds	References
Vietnam	Stems	Nordamnacanthal (**194**) and rubiadin (**195**)	[282]
Hawaii, United States	Roots	Nordamnacanthal (**194**), damnacanthal (**196**), rubiadin (**195**), 1-hydroxy-3-methyl-anthraquinone (**197**), 2-methylanthraquinone (**198**), 2-formyl-1-hydroxyanthraquinone (**199**), 2-formylanthraquinone (**200**), alizarin-1-methyl ether (**201**), 1-methoxy-3-hydroxyanthraquinone (**202**), 1-methyl-3-hydroxyanthraquinone (**203**), 1,7-dihydroxy-8-methoxy-2-methyl-anthraquinone (**204**), 1,3-dimethoxy-2-methoxymethylanthraquinone (**205**), 2-(ethoxymethyl)-1,3-dihydroxy-ibericine (**206**)	[283]
Okinawa, Japan	Roots	Digiferruginol-1-methylether-11-*O*-*β*-gentiobioside (**207**), digiferruginol-11-*O*-*β*-primeveroside (**208**), damnacanthol-11-*O*-*β*-primeveroside (**209**), 1-methoxy-2-primeverosyloxymethyl-anthraquinone-3-olate (**210**), 1-hydroxy-2-primeverosyloxymethyl-anthraquinone-3-olate (**211**), 1-hydroxy-5,6-dimethoxy-2-methyl-7-primeverosyloxy-anthraquinone (**212**)	[270]
Okinawa, Japan	Roots	5,15-Dimethylmorindol (**213**), alizarin-1-methylether (**201**), anthragallol-1,3-dimethylether (**214**), anthragallol-2-methylether (**215**), 6-hydroxy-anthragallol-1,3-dimethylether (**216**), morindone-5-methylether (**217**)	[270]
Karachi, Pakistan	Fruits	1,3-Dimethoxyanthraquinone (**218**), 1,2-dihydroxyanthraquinone (**219**)	[273]
Karachi, Pakistan	Stems	Morindicinone (2-hydroxy-1,8-dimethoxy-7-methoxymethylanthraquinone) (**220**), morindicininone (4-hydroxymethyl-1,3-dimethoxyanthraquinone) (**221**), 2-hydroxy-anthraquinone (**222**), 2-methoxy-anthraquinone (**223**)	[268]
Okinawa, Japan	Leaves	Damnacanthol-3-*O*-*β*-D-primeveroside (**224**), lucidin 3-*O*-*β*-D-primeveroside (**225**), morindone-6-*O*-*β*-D-primeveroside (**226**)	[284]
Fiji Islands	Fruits	1,6-Dihydroxy-5-methoxy-2-methoxymethylanthraquinone (**227**), 1,5,7-trihydroxy-6-methoxy-2-methoxymethylanthraquinone (**228**)	[285]

(continued)

Table 4 (continued)

Plant collection location	Part	Compounds	References
Okinawa, Japan	Roots	Digiferruginol-1-methylether-11-*O*-*β*-gentiobioside (**207**), digiferruginol-11-*O*-*β*-primeveroside (**208**), damnacanthol-11-*O*-*β*-primeveroside (**209**), 1-methoxy-2-primeveryosyloxymethylanthaquinone-3-olate (**210**), 1-hydroxy-2-primeverosyl-oxymethylanthraquinone-3-olate (**211**), 1-hydroxy-5,6-dimethoxy-2-methyl-7-primeverosyloxyanthraquinone (**212**), lucidin-3-*O*-*β*-D-glucoside (**229**), damnacanthol-3-*O*-*β*-D-glucoside (**230**), rubiadin-1-methylether-3-*O*-*β*-primeveroside (**231**), soranjidiol-6-*O*-*β*-primeveroside (**232**)	[267]
Xisha Islands, China	Bark	1,3-Dihydroxy-5-methoxy-6-methoxymethyl-2-methylanthraquinone (**233**), 1,3-dihydroxy-5-methoxy-2,6-bismethoxymethylanthraquinone (**234**), 1,5,15-trimethylmorindol (**235**), morindone-6-methyl-ether (**236**), 1-hydroxy-5-methoxy-anthraquinone (**237**), 1,4-dimethoxyl-2-hydroxyanthraquinone (**238**), 2,3-dihydroxy-1-methoxylanthraquinone (**239**), rubiadin (**195**), lucidin 3-methyl ether (**240**), lucidin 1,3-dimethyl ether (**241**), damnacanthol (**242**), digiferruginol (**243**)	[269]
Okinawa, Japan	Leaves	5,15-Dimethylmorindol (**213**), 1,5,15-trimethyl-morindol (**236**)	[276]
Malaysia	Roots	Morindone-6-methylether (**236**), 2-ethoxy-1-hydroxyanthraquinone (**244**), 2-(ethoxy-ethyl)-1-hydroxyanthraquinone (**245**), 1-hydroxy-3-methylanthraquinone (**197**), damnacanthal (**196**), nordamnacanthal (**194**), 2-formyl-1-hydroxyanthraquinone (**199**)	[266]
Information not available	Roots	1,2-Dihydroxyanthraquinone (**219**), 1,3-dihydroxy-2-methylanthraquinone (**197**), 2-hydroxy-3-(hydroxymethyl) anthraquinone (**246**), 1,3,6-trihydroxy-2-methylanthraquinone (**247**)	[286]

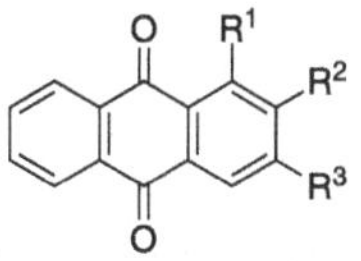

194 (nordamnacanthal) $R^1 = R^3 = OH$, $R^2 = CHO$
195 (rubiadin) $R^1 = R^3 = OH$, $R^2 = CH_3$
196 (damnacanthal) $R^1 = OMe$, $R^2 = CHO$, $R_3 = OH$
202 (1-methoxy-3-hydroxyanthraquinone) $R^1 = OMe$, $R^2 = H$, $R^3 = OH$
203 (1-methyl-3-hydroxyanthraquinone) $R^1 = CH_3$, $R^2 = H$, $R^3 = OH$
205 (1,3-dimethoxy-2-methoxymethylanthraquinone) $R^1 = R^3 = OMe$, $R^2 = CH_2OCH_3$
206 (2-ethoxymethyl)-1,3-dihydroxy-ibericine) $R^1 = R^3 = OH$, $R^2 = CH_2OCH_2CH_3$
209 (damnacanthol-11-*O*-β-primeveroside) $R^1 = OMe$, $R^2 =$ *O*-β-D-prim, $R^3 = OH$
210 (1-methoxy-2-primeverosyloxymethyl-anthraquinone-3-olate) $R^1 = OMe$, $R^2 = CH_2$-*O*-β-D-prim, $R^3 = OH$
211 (1-hydroxy-2-primeverosyloxymethyl-anthraquinone-3-olate) $R^1 = OH$, $R^2 = CH_2$-*O*-β-D-prim, $R^3 = OH$
214 (anthragallol-1,3-dimethylether) $R^1 = R^3 = OMe$, $R^2 = OH$
215 (anthragallol-3-methylether) $R^1 = R^3 = OH$, $R^2 = OMe$
218 (1,3-dimethoxyanthraquinone) $R^1 = R^3 = OMe$, $R^2 = H$
224 (damnacanthol-3-*O*-β-D-primeveroside) $R^1 = OMe$, $R^2 = CH_2OH$, $R^3 =$ *O*-β-D-prim
225 (lucidin 3-*O*-β-D-primeveroside) $R^1 = OH$, $R^2 = CH_2OH$, $R^3 =$ *O*-β-D-prim
229 (lucidin-3-*O*-β-D-glucoside) $R^1 = OH$, $R^2 = CH_2OH$, $R^3 =$ *O*-β-D-Glc
230 (damnacanthol-3-*O*-β-D-glucoside) $R^1 = OMe$, $R^2 = OH$, $R^3 =$ *O*-β-D-Glc
231 (rubiadin-1-methylether-3-*O*-β-primeveroside) $R^1 = OMe$, $R^2 = CH_3$, $R^3 =$ *O*-β-D-prim
239 (2,3-dihydroxy-1-methoxylanthraquinone) $R^1 = OMe$, $R^2 = R^3 = OH$
240 (lucidin 3-methyl ether) $R^1 = OH$, $R^2 = CH_2OH$, $R^3 = OMe$
241 (lucidin 1,3-dimethyl ether) $R^1 = R^3 = OCMe$, $R^2 = CH_2OH$
242 (damnacanthol) $R^1 = OMe$, $R^2 = R^3 = OH$
246 (2-hydroxy-3-(hydroxymethyl)anthraquinone) $R^1 = H$, $R^2 = OH$, $R^3 = CH_2OH$

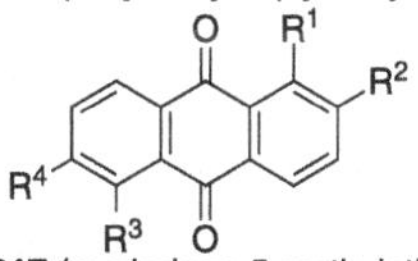

217 (morindone-5-methylether) $R^1 = R^4 = OH$, $R^2 = R^3 = OMe$
220 (morindicinone) $R^1 = R^4 = OH$, $R^2 = CH_2OCH_3$, $R^3 = OMe$
226 (morindone-6-*O*-β-D-primeveroside) $R^1 = R^3 = OH$, $R^2 = CH_3$, $R^4 =$ *O*-β-D-prim
227 (1,6-dihydroxy-5-methoxy-2-methoxymethylanthraquinone) $R^1 = R^4 = OH$, $R^2 = CH_2OCH_3$, $R^3 = OMe$
232 (soranjidiol-6-*O*-β-primeveroside) $R^1 = OH$, $R^2 = CH_3$, $R^3 = H$, $R^4 =$ *O*-β-D-prim
235 (1,5,15-trimethylmorindol) $R^1 = R^3 = OMe$, $R^2 = OH$, $R^4 = CH_2OCH_3$
236 (morindone-6-methyl-ether) $R^1 = R^3 = OH$, $R^2 = CH_3$, $R^4 = OMe$
237 (1-hydroxy-5-methoxyanthraquinone) $R^1 = OH$, $R^2 = R^4 = H$, $R^3 = OMe$

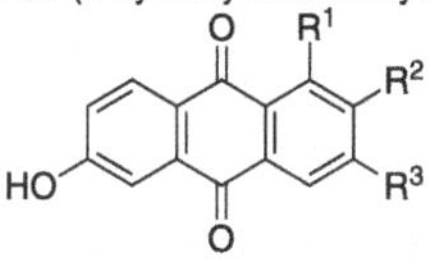

216 (6-hydroxy-anthragallol-1,3-dimethylether) $R^1 = R^3 = OMe$, $R^2 = R^6 = OH$
247 (1,3,6-trihydroxy-2-methylanthraquinone) $R^1 = R^3 = OH$, $R^2 = CH_3$)

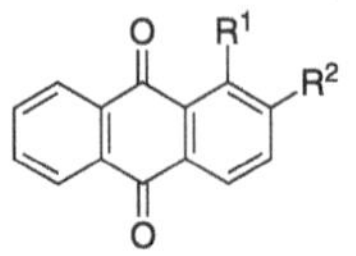

197 (1-hydroxy-3-methylanthraquinone) $R^1 = OH$, $R^2 = CH_3$
198 (2-methylanthraquinone) $R^1 = H$, $R^2 = CH_3$
199 (2-formyl-1-hydroxyanthraquinone) $R^1 = OH$, $R^2 = CHO$
200 (2-formylanthraquinone) $R^1 = H$, $R^2 = CHO$
201 (alizarin-1-methyl ether) $R^1 = OMe$, $R^2 = OH$
207 (digiferruginol-1-methylether-11-*O*-*β*-gentiobioside) $R^1 = OMe$, $R^2 = CH_2$-O-gen
208 (digiferruginol-11-*O*-*β*-primeveroside) $R^1 = OH$, $R^2 = CH_2$-*O*-*β*-D-prim
219 (1,2-dihydroxyanthraquinone) $R^1 = R^2 = OH$
222 (2-hydroxyanthraquinone) $R^1 = OH$, $R^2 = OH$
223 (2-methoxyanthraquinone) $R^1 = H$, $R^2 = OMe$
243 (digiferruginol) $R^1 = OH$, $R^2 = CH_2OH$
244 (2-ethoxy-1-hydroxyanthraquinone) $R^1 = OH$, $R^2 = OCH_2CH_3$
245 (2-(ethoxyethyl)-1-hydroxyanthraquinone) $R^1 = OH$, $R^2 = CH_2CH_2OCH_2CH_3$

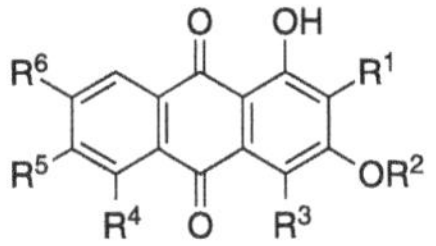

228 (1,5,7-trihydroxy-6-methoxy-2-methoxymethylanthraquinone) $R^1 = CH_2OCH_3$, $R^2 = R^5 = H$, $R^3 = OMe$, $R^3 = R^4 = OH$
212 (1-hydroxy-5,6-dimethoxy-2-methyl-7-primeverosyloxyanthraquinone) $R^2 = CH_3$, $R^2 = H$, $R^3 = OMe$, $R^4 = $ *O*-prim
233 (1,3-dihydroxy-5-methoxy-6-methoxymethyl-2-methylanthraquinone) $R^1 = R^3 = OH$, $R^2 = CH_3$, $R^4 = H$, $R^5 = OMe$, $R^6 = CH_2OCH_3$
234 (1,3-dihydroxy-5-methoxy-2,6-bismethoxymethylanthraquinone) $R^1 = R^5 = CH_3$, $R^2 = R^3 = H$, $R^4 = OMe$

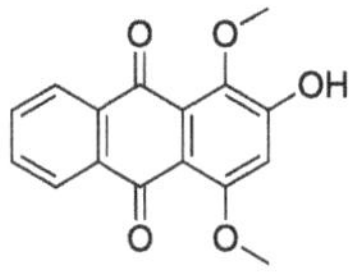

238 (1,4-dimethoxy-2-hydroxy-anthraquinone)

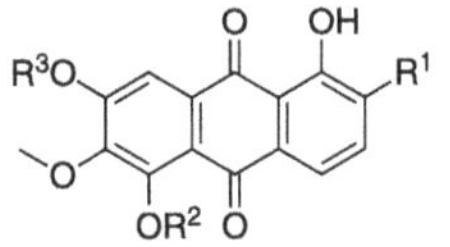

215 (1-hydroxy-5,6-dimethoxy-2-methyl-7-primeverosyl-oxyanthraquinone) $R^1 = R^2 = CH_3$, $R^3 = $ prim

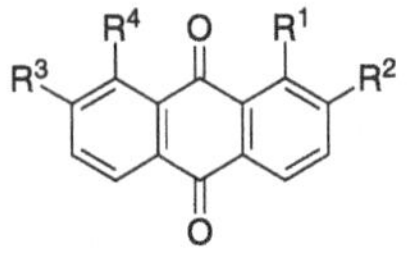

204 (1,7-dihydroxy-8-methoxy-2-methyl-anthraquinone) $R^1 = R^3 = OH$, $R^2 = CH_3$, $R^4 = OMe$
221 (morindicininone) $R^1 = CH_3$, $R^2 = OH$, $R^3 = CH_2OCH_3$, $R^4 = OMe$

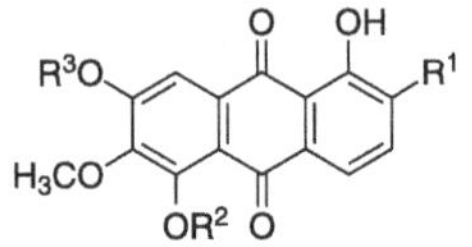

213 (5,15-dimethylmorindol) $R^1 = R^6 = OH$, $R^2 = OCH_2OCH_3$, $R^3 = R^4 = H$, $R^5 = OMe$

2.4.3 Biological Activities

Several constituents of *Morinda citrifolia* have been investigated for their cancer-related biological activities in both in vivo and in vitro models. Anthraquinones, inclusive of nordamnacanthal (**194**), and damnacanthal (**196**), as found in the stem-bark and roots of noni, have shown relevant biological activities. Their mechanism of action includes inducing apoptosis, arresting the cell cycle, inhibiting angiogenesis, and modulating the immune system. They are lead compound candidates for both cancer chemotherapeutic and chemopreventive treatments [287]. The next part of this subsection will focus on the potential of *M. citrifolia* in ameliorating cardiovascular diseases.

Several parts of *M. citrifolia* have been found to display antidiabetic, antihypertensive, and antihyperlipidemic properties. A key mechanism of action includes inhibiting angiotensin-converting enzyme, thus lowering the blood pressure in rat models [288]. A noni root extract significantly lowered blood glucose levels in diabetic rats by modulating calcium and glucose metabolism, with certain iridoids and anthraquinones identified as active the compounds [284].

Morinda citrifolia extracts have shown antidyslipidemic activity by reducing cholesterol, triglycerides, and low-density lipoprotein (LDL) cholesterol through inhibition of key enzymes, such as 3-hydroxy-3-methylglutaryl coenzyme A reductase (HMG-CoA reductase) and acyl-CoA cholesterol acyl transferase (ACAT), while some extracts also increased high-density lipoprotein (HDL) cholesterol and reduced body weights [289]. A root extract acted as a vasodilator by blocking receptor and voltage-operated calcium channels and by inhibiting calcium release from intracellular stores, which contributed to its hypotensive effects [290].

Fermented noni juice has demonstrated a reduction of liver, heart, and visceral fat in high-fat diet hamsters by upregulating lipid metabolism genes like PPAR-alpha and UCP2, increasing fecal lipid excretion, and reducing lipid peroxidation [291]. Clinical trials showed that the consumption of noni fruit consumption reduced cholesterol and LDL levels by inhibiting HMG-CoA reductase and ACAT [292].

Even though a noni extract did not reduce body fat and cholesterol in a high-fat diet mice model, a noni fruit preparation was found to prevent glucose intolerance by inhibiting glucose absorption and modulating insulin sensitivity [293]. Noni leaf and fruit extracts also lowered blood pressure in hypertensive rats by targeting calcium channels [294]. However, chronic use of *M. citrifolia* caused nephrotoxicity in diabetic rats [295]. Table 5 summarizes the key findings on the antidiabetic, antihypertensive, and antihyperlipidemic effects of *M. citrifolia.*

Table 5 Antidiabetic, antihypertensive, and antihyperlipidemic activities of *Morinda citrifolia*

Part(s)	Extraction method/ compounds	Findings	References
Fruits	Fruit juice	*Morinda citrifolia* fruit juice inhibited ACE activity. A single oral administration reduced the systolic blood pressure of male spontaneously hypertensive rats	[288]
Roots	The dried roots, were macerated in MeOH, and then, partitioned with $CHCl_3$, EtOAc, and *n*-BuOH	Streptozocin-induced diabetic rats were fed with the *n*-butanol soluble partition at 3 g/kg (oral route, p.o) and showed a significant reduction of blood glucose levels at 5 h after administration (3039 ± 25.6 mg/dm^3) relative to those at 0 h after administration (4559 ± 17 mg/dm^3)	[284]
Fruits, leaves, and roots	The separate dried plant materials were macerated in 70% EtOH	All three extracts reduced total cholesterol (TC) and lowered triglyceride levels in both triton- and high-fat diet-induced dyslipidemia. In high-fat diet-induced dyslipidemia, all extracts reduced also LDL-C, the atherogenic index, and the TC/HDL ratio. The root extract also increased HDL-C levels. The leaf and root extracts reduced body weight and food intake, while the fruit extract did not affect these factors	[289]
Roots	The dried ground roots were macerated in 70% ethanol	The root extract acted as a vasodilator by blocking calcium channels and inhibiting muscle contractions in the gut, heart, and blood vessels. It resembled verapamil in its action, specifically targeting smooth muscle tissues more than cardiac tissue, and worked by directly blocking calcium channels without involving nitric oxide or other receptor pathways	[290]
Fruits	The blended juice was stored in a steel bottle for one year	Six weeks of treatment with fermented noni juice reduced fat levels, improved blood lipid profiles, and provided cardiovascular protection in hamsters on a high-fat diet, mainly due to its phenolic constituents	[291]
Fruits	The fruits were dried, ground, and encapsulated	In a double-blind randomized control trial conducted in Indonesia, consumption of 1 g of the dried fruits three times daily for 14 days reduced TC and LDL-C by 13.8 and 15.5%, respectively	[296]
Fruits	Commercial noni fruit juice at 50 g/dm^3	In hyperlipidemic Wistar rats, noni fruit juice (50 and 100 mg/kg) reduced total cholesterol, triglycerides, and very low-density lipoprotein cholesterol substantially, when compared to untreated controls. However, at the lower dose (50 mg/kg), reductions in the total cholesterol and LDL-C were not statistically comparable to those of atorvastatin	[297]

(continued)

Table 5 (continued)

Part(s)	Extraction method/ compounds	Findings	References
Fruits and leaves	The dried powder of fruits and leaves were macerated in 96% ethanol	The ethanol extracts of the fruits and leaves, alone and when combined (500 mg/kg), reduced blood pressure in dexamethasone-induced hypertensive rats. The combined extract showed the most potent effect, lowering systolic pressure by 16.7%, diastolic pressure by 21.5%, and mean arterial pressure by 19.6%	[294]
Fruits	Fruits blended with 400 cm^3 deionized water	Seven weeks of treatment wth the extract (250 and 500 mg/kg) did not reduce body weight, visceral fat, cholesterol, triglycerides, insulin levels, insulin resistance, or liver fat in mice on a high-fat/fructose diet. However, the 500 mg/ kg dose improved glucose tolerance	[293]
Leaves	Extraction using hot water at 80°C	The catechin-rich water extract prepared increased LDL-C uptake up to 1.11 ± 0.03-fold, but this was insignificant relative to the control. Purified catechin gallate and epigallocatechin gallate from the extract significantly increased LDL-C uptake up to 1.37 ± 0.19-fold and 1.59 ± 0.19-fold, respectively, in an in vitro test protocol	[298]
Fruits	Tahitian Noni Original® juice	In a pilot trial, ten non-smoking participants (age 40+) with normal or borderline cholesterol levels consumed two ounces of the juice twice daily for 30 days. This reduced total cholesterol by 2.0 times, increased HDL by 0.3, lowered LDL by 0.8, and improved the TC/HDL ratio by 0.1. The juice, at increasing concentrations (1, 5, and 10 mg/cm^3), inhibited HMG-CoA reductase by up to 83% and also reduced ACAT enzyme activity in the liver and intestine by up to 17 and 31%, respectively	[292]
Fruits	Blended fruits	Treatment with blended noni juice (2 cm^3/kg) daily for 60 days in streptozotocin-induced diabetic rats reduced glycated hemoglobin, and increased VO_2 max values after exercise. However, prolonged intake with insulin treatment caused kidney toxicity through increasing Bowman's space	[295]

2.5 *Centella asiatica*

Centella asiatica (L.) Urb. (Apiaceae) is an economically important medicinal plant. The plant is native to India and widespread in subtropical regions including Malaysia, where is it known as "gotu kala" and "pegaga" [299]. This species thrives in relatively shaded and moist environments, such as wetlands, riverbanks, ponds, wet meadows, and forests, and occurs typically at altitudes ranging from 300 to 1,800 m. This species is a perennial, flowering, mildly aromatic herb that grows low to the ground and spreads through creeping stems. The stems are either prostrate or semi-erect, with roots developing at the nodes, and the plant reaches a height of 10 to 45 cm. The leaves are reniform (kidney-shaped) or shovel-like, growing alternately in rosettes at the nodes [300]. Plate 5 illustrates the leaves of *C. asiatica.*

Plate 5 Leaves of *Centella asiatica* (family Apiaceae). Photograph courtesy Bishnu Sarangi (https://pixabay.com/photos/plant-herb-medicinal-1646382/)

2.5.1 Ethnobotany

Centella asiatica is widely used in traditional medicine to treat symptoms related to metabolic disorders like diabetes as well as neurodegenerative diseases such as Alzheimer's disease and dementia. In Malaysia, *C. asiatica* is a popular salad item, often eaten raw. It is utilized as a nutritional food as well as for medicinal purposes. Also, *C. asiatica* is used in traditional medicine for the treatment of leprosy, and for skin tuberculosis, wound healing, stomachache, arthritis, varicose veins, high blood pressure, and as a memory enhancer [299]. Several laboratory studies have shown that *C. asiatica* extracts exhibit antioxidant properties, and show antiproliferative effects in tumor cells, and also improve venous wall conditions in chronic venous hypertension including protecting the venous endothelium. The triterpenoids of *C. asiatica* are components of some medicinal agents and also occur in cosmetic preparations for skin care.

2.5.2 Phytochemistry

Triterpene saponins are prominent chemical constituents produced by *C. asiatica*. Among these triterpene saponins, asiatic acid (**248**), madecassic acid (**249**), asiaticoside (**253**), and madecassoside (**254**) are the most important biologically active compounds. These are widely studied in in vivo models and certain clinical trials. Variations in the triterpenoid saponin content have been noted depending on geographic region, phenotype, and genotype. The present section will focus on triterpene saponins from *C. asiatica*, and Table 6 summarizes the saponins isolated from *C. asiatica* and the formula drawings shown contain the structures of this class of phytochemicals isolated from *Centella asiatica*, namely, ursane-type saponins (**248–251**), an oleanane-type triterpene (**252**), ursane-type triterpene glycosides (**253–274**), oleanane-type triterpene glycosides (**275–287**), and dammarane-type triterpene glycosides (**288–289**).

Table 6 Saponins from *Centella asiatica*

Plant material collection location	Compounds	References
Information not available	Asiaticoside (**253**), asiatic acid (**248**)	[301]
India	Madecassoside (**254**), asiaticoside (**253**)	[302]
India	Asiaticoside (**253**), madecassoside (**254**), asiatic acid (**248**), madecassic acid (**249**)	[303]
Sri Lanka	Centellasaponin A (**276**), centellasaponin B (**255**), centellasaponin C (**256**), centellasaponin D (**277**), centellasapogenol A (**252**), madecassoside (**254**), asiaticoside (**253**), scheffoleoside A (**286**)	[304]
Vietnam	Centellasapogenol A (**252**), centellasaponin A (**276**)	[304]
Yunnan province, China	Asiaticoside C (**257**), asiaticoside D (**258**), asiaticoside E (**259**), asiaticoside F (**260**), madecassoside (**254**)	[305]
Hebei province, China	2*α*,3*β*,20,23-Tetrahydroxyurs-28-oic acid (**251**), asiatic acid (**248**), madecassic acid (**249**), 3*β*,6*β*,23-trihydroxyolean-12-en-28-oic acid (isothankuniside) (**261**)	[306]

(continued)

Table 6 (continued)

Plant material collection location	Compounds	References
Hebei province, China	2*α*,3*β*,23-Trihydroxyurs-20-en-28-oic acid (**250**), 2*α*,3*β*,23-trihydroxyurs-20-en-28-oic acid *O*-*α*-L-rhamnopyranosyl-(1⟶4)-*O*-*β*-D-glucopyranosyl-(1⟶6)-*O*-*β*-D-glucopyranosyl ester (isoasiatic acid) (**262**)	[306]
University of Missisippi, USA	Asiatic acid (**248**), madecassic acid (**249**), asiaticoside C (**257**), asiaticoside F (**260**), asiaticoside (**253**), madecassoside (**254**), 23-*O*-acetylasiaticoside B (**278**)	[307]
Guangxi Zhuang Autonomous, Region, China	Centelloside A (**288**), centelloside B (**289**)	[308]
Guangxi Zhuang Autonomous Region, China	Centelloside D (**279**), centelloside E (**280**), centellasaponin B (**255**), asiaticoside E (**259**), scheffoleoside A (**286**), (2*α*,3*β*,6*β*)-2,3,6-trihydroxyolean-12-en-28-oic acid 28-[*O*-*α*-L-rhamnopyranosyl-(1⟶4)-*O*-*β*-D-glucopyranosyl-(1⟶6)-*β*-D-glucopyranosyl] ester (**281**), asiaticoside F (**260**)	[308]
Guangxi Zhuang Autonomous Region, China	Centellasaponin I (**265**), centellasaponin J (**266**), centellasaponin E (**267**), centelloside D (**279**), centellasaponin B (**255**)	[309]
Australia	Asiatic acid (**248**), madecassic acid (**249**), asiaticoside (**253**), madecassoside (**254**)	[310]
Sichuan Province, China	Centelloside G (**268**), 11-oxo-asiaticoside B (**285**), 11-oxo-madecassoside (**273**), 11*β*-methoxy asiaticoside B (**287**), 11*β*-methoxy madecassoside (**274**), asiaticoside (**253**), asiaticoside B (**275**), madecassoside (**254**), centellasaponin A (**276**), scheffoleoside A (**286**), centelloside E (**280**)	[311]
Sichuan Province, China	Asiaticoside H (**269**), asiaticoside I (**270**)	[312]
Hebei Province, China	Madecassic acid (**249**), centelloside C (**271**), centellasaponin B (**255**), madecassoside (**254**), 11-oxo-madecassoside (**273**), asiaticoside B (**275**), 11-oxo-asiaticoside B (**285**), asiatic acid (**248**), asiaticoside E (**259**), asiaticoside (**253**), asiaticoside C (**257**), isomadecassoside (**272**), chebuloside II (**283**)	[313]

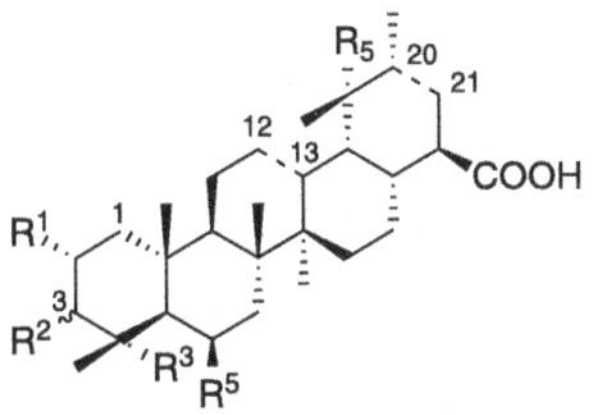

248 (asiatic acid) R^1 = OH, R^2 = β-OH, R_3 = CH_2OH, R^4 = R^5 = H, $\Delta^{12,13}$
249 (madecassic acid) R^1 = R^4 = OH, R^2 = β-OH, R^3 = CH_2OH, R^5 = H, $\Delta^{12,13}$
250 (2α,3β,23-trihydroxyurs-20-en-28-oic acid) R^1= OH, R^2 = β-OH, R^3 = CH_2OH, R^4 = R^5 = H, $\Delta^{20,21}$

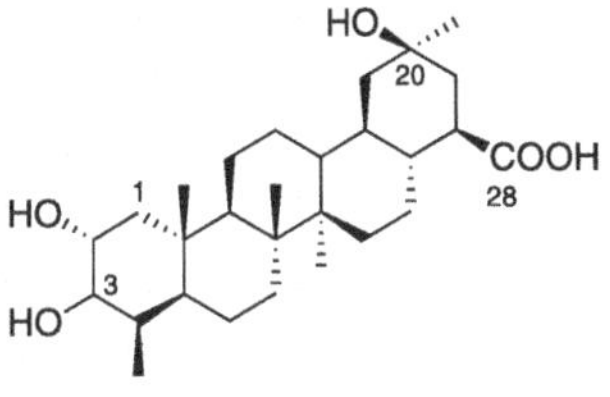

251 (2α,3β,20,23-tetrahydroxyurs-28-oic acid)

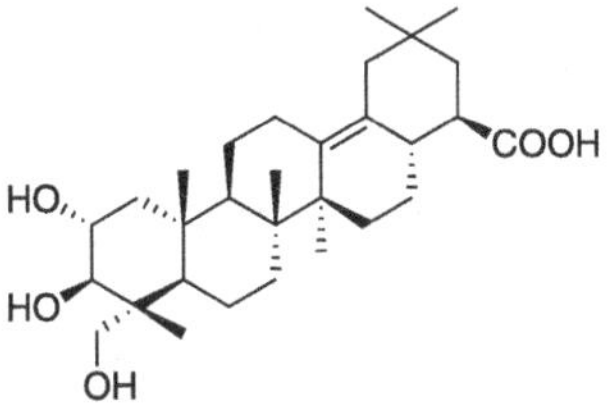

252 (centellasapogenol A)

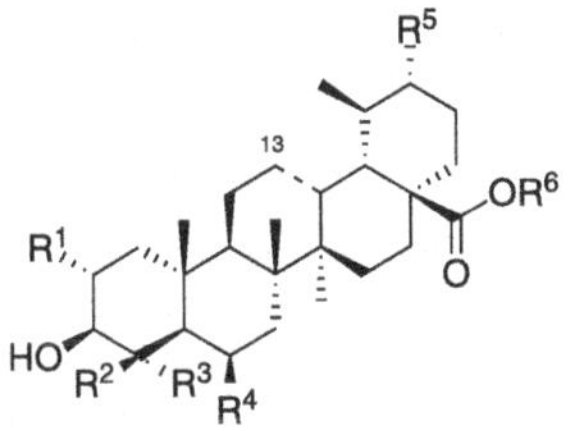

253 (asiaticoside) R^1 = OH, R ^{2}= R^5 = CH_3, R^3 = CH_2OH, R^4 = H, R^6 = Glc-Glc-Rha, $\Delta^{12,13}$
254 (madecassoside) R^1 = R^4 = OH, R^2 = R^5 = CH_3, R^3 = CH_2OH, R^6 = Glc-Glc-Rha, $\Delta^{12,13}$
255 (centellasaponin B) R^1 = R^4 = OH, R^2 = R^5 = CH_3, R^6 = Glc-Glc, $\Delta^{12,13}$
256 (centellasaponin C) R^1 = R^4 = OH, R^2 = R^3 = R^5 = CH_3, Glc-Glc, $\Delta^{12,13}$
257 (asiaticoside C) R^1 = OH, R^2 = R^5 = CH_3, R^3 = CH_2OaAc, R^6 = Glc-Glc-Rha, $\Delta^{12,13}$
258 (asiaticoside D) R^1 = OH, R^2 = R^3 = R^5= CH_3, R^4 = H, R^6 = Glc-Glc-Rha, $\Delta^{12,13}$
259 (asiaticoside E) R^1 = OH, R^2 = R^5 = CH_3, R^3 = CH_2OH, R^4 = H, R^6 = Glc-Glc-Rha, $\Delta^{12,13}$
260 (asiaticoside F) R^1 = R^4 = H, R^2 = R^5 = CH_3, R^3 = CH_2OH, R^6 = Glc-Glc-Rha, $\Delta^{12,13}$
261 (isothankuniside) R^1 = H, R^2 = R^5= CH_3, R^3 = CH_2OH, R^4 = OH, R^5 = Glc-Glc-Rha, $\Delta^{12,13}$
262 (isoasiatic acid) R^1 = OH, R^2 = R^5 = CH_3, R^3 = CH_2OH, R^4 = H, R^6 = Glc-Glc-Rha, $\Delta^{12,13}$
263 (23-*O*-acetyl-madecassoside) R^1 = R^4 = OH, R^2 = R^5 = CH_3, R^3 = CH_2OAc, R^6 = Glc-Glc-Rha, $\Delta^{12,13}$
264 (centelloside E) R^1 = OH, R^2 = R^5 = CH_3, R^3 = CH_2OH, R^4 = H, R^6 = Glc-Glc-Rh, $\Delta^{12,13}$
265 (centellasaponin I) R^1 = R^4 = OH, R^2 = R^5 = CH_3, R^3 = CH_2OGlc, R^6 = Glc-Glc-Rha, $\Delta^{12,13}$
266 (centellasaponin J) R^1 = OH, R^2 = R^3 = CH_2OH, R^4 = H, R^5 = Glc-Glc-Rha, $\Delta^{12,13}$
267 (centellasaponin E) R^1 = OH, R^2 = R^3 = CH_2OH, R^4 = H, R^5 = CH_3, R^6 = Glc-Glc-Rha, $\Delta^{12,13}$
268 (centelloside G) R^1 = R^4 = OH, R^2 = R^5 = CH_3, R^3 = CH_2OH, R^6 = Glc-Glc-Rha, $\Delta_{9,11}$, $\Delta^{12,13}$
269 (asiaticoside H) R^1 = OH, R^2 = R^5 = CH_3, R^3 = CH_2OGlc, R^4 = H, R^6 = Glc-Glc-Rha, $\Delta^{12,13}$
270 (asiaticoside I) R^1 = R^4 = OH, R^2 = R^5 = CH_3, R^3 = CH_2OH, R^6 = Glc-Glc
271 (centelloside C) R^1 = R^4 = OH, R^2 = R^4 = CH_3, R^3 = CH_2OH, R^6 = Glc, $\Delta^{12,13}$
272 (isomadecassoside) R^1 = R^4 = OH, R^2 = R^5 = CH_3, R^3 = CH_2OH, R^6 = Glc-Glc-Rha, $\Delta^{12,13}$

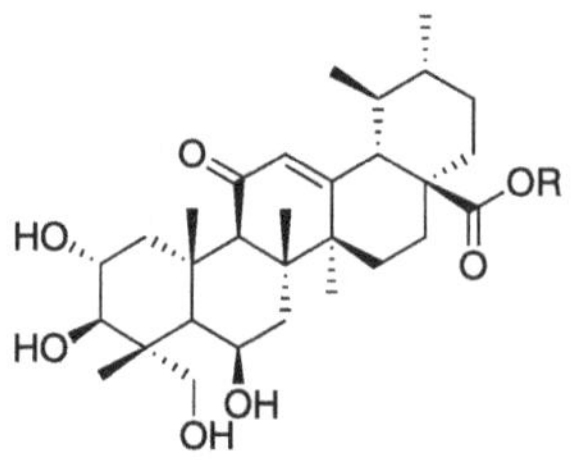

273 (11-oxo-madecassoside)
R = Glc-Glc-Rha

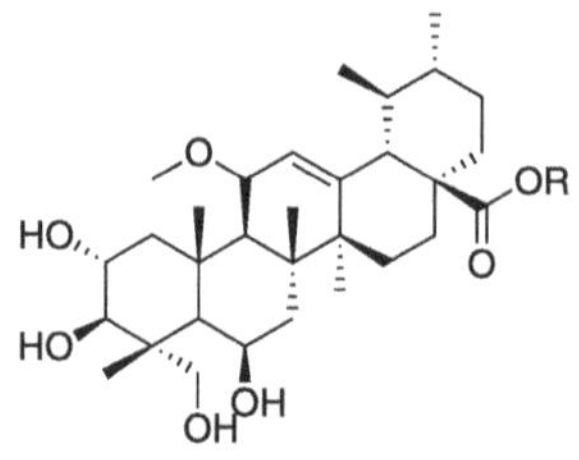

274 (11β-methoxy madecassoside)
R = Glc-Glc-Rha

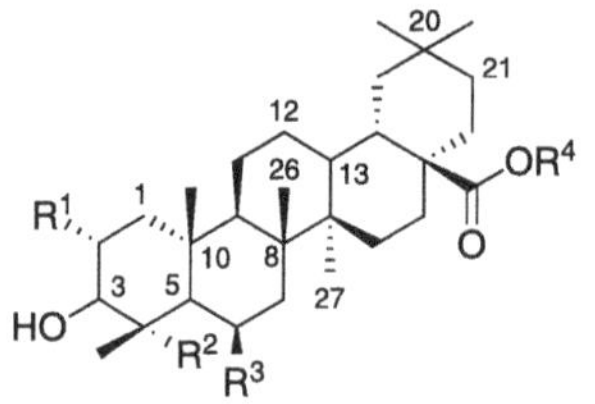

275 (asiaticoside B) $R^1 = R^3 = OH$, $R^2 = CH_2OH$, R^4 = Glc-Glc-Rha, $\Delta^{12,13}$

276 (centellasaponin A) $R^1 = OH$, $R^2 = CH_2OH$, $R^3 = H$, R^4 = Glc-Glc-Rha, $\Delta^{13,18}$

277 (centellasaponin D) $R^1 = H$, $R^2 = CH_2OH$, $R^3 = OH$, R^4 = Glc-Glc-Rha, $\Delta^{12,13}$

278 (23-*O*-acetylasiaticoside B) $R^1 = R^3 = OH$, $R^2 = CH_2OAc$, R^4 = Glc-Glc-Rha, $\Delta^{12,13}$

279 (centelloside D) $R^1 = R^3 = OH$, $R^1 = CH_2OH$, R^4 = Glc-Glc, $\Delta^{12,13}$

280 (centelloside E) $R^1 = R^3 = OH$, $R^2 = CH_2OH$, R^4 = Glc-Glc-Rha, $\Delta^{9,11}$, $\Delta^{12,13}$

281 ((2α,3β,6β)-2,3,6-trihydroxyolean-12-en-28-oic acid 28-[*O*-α-L-rhamnopyranosyl-(1→4)-*O*-β-D-glucopyranosyl-(1→6)-β-D-glucopyranosyl] ester) $R^1 = R^3 = OH$, $R^2 = CH_3$, Glc-Glc-Rha, $\Delta^{12,13}$

282 (centellsaponin J) $R^1 = R^3 = OH$, $R^2 = CH_2Glc$, R^4 = Glc-Glc-Rha, $\Delta^{12,13}$

283 (chebuloside II) $R^1 = R^3 = OH$, $R^2 = CH_2OH$, R^4= Glc, $\Delta^{12,13}$

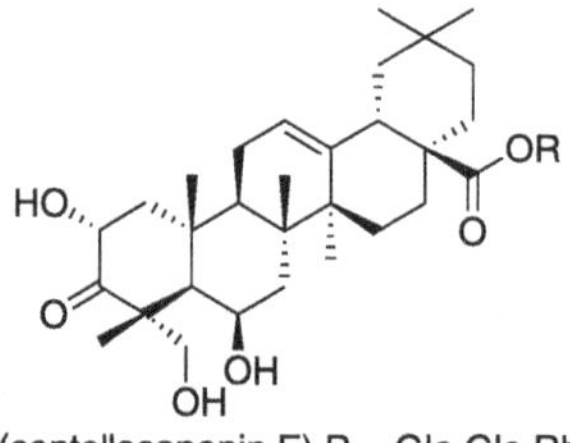

284 (centellasaponin E) R = Glc-Glc-Rha

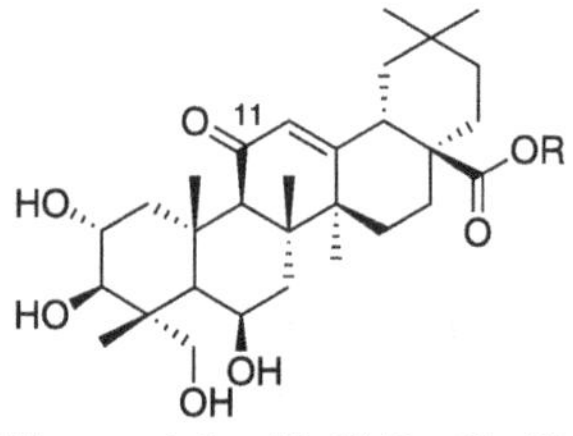

285 (11-oxo-asiaticoside B) R = Glc-Glc-Rha

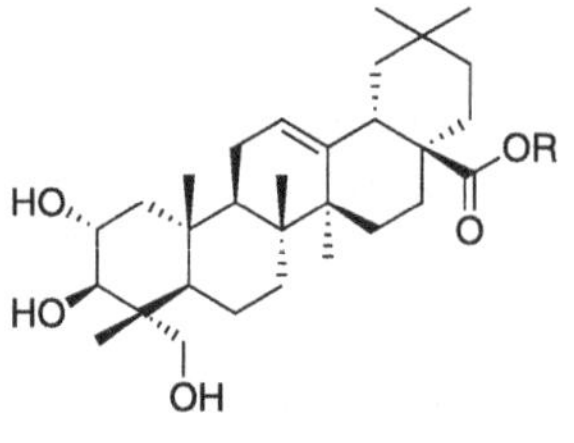

286 (scheffoleoside A) R = Man-Glu-Glu Ester

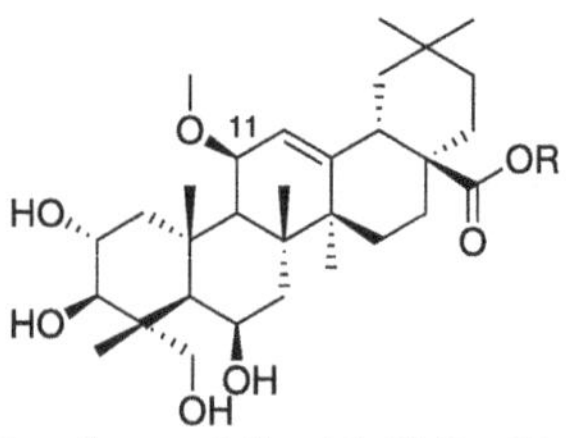

287 (11β-methoxy asiaticoside B) R = Glc-Glc-Rha

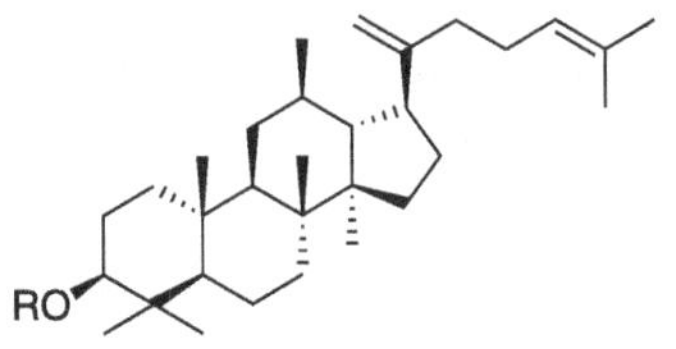

288 (centelloside A) R = Xyl-Glc-Glc

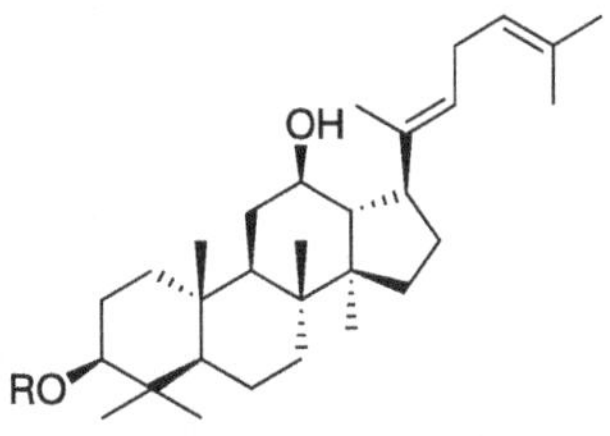

289 (centelloside B) R = Xyl-Glc-Glc

2.5.3 Biological Activities

Centella asiatica is widely used ethnomedically for the treatment of neurodegenerative diseases and for diabetes, such as in Ayurvedic medicine. Various studies have demonstrated that *C. asiatica* possesses antihyperglycemic activity by reducing blood glucose levels in diabetic rat models [314–319]. *Centella asiatica* demonstrated antihyperglycemic activity by diminishing glucose homeostasis and glucose absorption [314], improving muscle glucose and glycogen metabolism [317], preserving β-cell structure and function due to the recovery of islet structural degeneration and an increased area of islets [320], modulating glucose metabolism through the inhibition of gluconeogenesis and activation of glucose utilization pathways, and stimulation of insulin synthesis and the repair of pancreatic β-cells [319].

The marker triterpene saponins, asiatic acid (**248**), madecasoside (**254**), and asiaticoside (**253**) are responsible for the antihyperglycemic property of *C. asiatica* [319–322]. Treatment by asiatic acid (**248**) at 20 mg/kg for 45 days significantly reduced blood glucose and increased serum insulin glucose levels in diabetic rats [319].

Supplementation by asiatic acid (**248**) in streptozotocin-niacinamide-induced diabetic rats over 45 days significantly improved certain pathological abnormalities, such as by normalizing the activity and expression of HMG-CoA reductase reductase in the plasma and tissue. This compound improved the activity of plasma lipoprotein lipase and lecithin cholesterol acyl transferase, consequently improving dyslipidemia [321].

In addition, *C. asiatica* was found to reduce symptoms related to chronic diabetes. *Centella asiatica* may protect diabetic tissues from disease-related complications through its antioxidant and anti-inflammatory activities. *Centella asiatica* reduced the renal levels of malondialdehyde, TNF-α, and IFN-γ, while increasing IL-10 when compared to diabetic control rats. Additionally, *C. asiatica* reduced DA, TNF-α, and IFN-γ with increases in IL-4 and IL-10 in the brain [323]. *Centella asiatica* ameliorated oxidative injury and improved antioxidant capacity in diabetic rat testes [324].

Asiatic acid (**248**) is the key component of *C. asiatica* that is responsible for its hypoglycemic effects. Treatment with asiatic acid (**248**) in an animal model was shown to lower blood glucose levels and body weight, demonstrating hypoglycemic effects through enhanced peripheral glucose utilization, improved hepatic glycolysis, and reduced gluconeogenesis. Asiatic acid (**248**) also was found to affect lipid metabolism, reducing levels of total cholesterol, triglycerides, LDL cholesterol, and very LDL cholesterol, while increasing high-density lipoprotein cholesterol, thereby lowering cardiovascular risks associated with diabetes. Furthermore, asiatic acid (**248**) decreased the activity of HMG-CoA reductase involved in cholesterol synthesis, and enhanced lecithin cholesterol acyltransferase and lipoprotein lipase activities, promoting better cholesterol and fatty acid metabolism. Histopathological analysis indicated that asiatic acid (**248**) reduced fat buildup and hepatic steatosis in diabetic rats, corroborating its liver-protective properties [321]. Asiatic acid (**248**) was also found to ameliorate diabetic cardiomyopathy by activation of the AMPK/Nrf2 pathway in a diabetic mouse model [325]. Table 7 summarizes the key findings on antidiabetic, antihypertensive, and antihyperlipidemic properties of *C. asiatica*.

Table 7 Biological activities of *Centella asiatica*

Extraction method/compound	Key findings	References
Fresh *C. asiatica* plants were shade-dried, powdered, and made into a juice	A 39.4% decrease in glucose levels, a 37.4% reduction in triglycerides, and a 29.1% decrease in total cholesterol occurred in a test group treated with 50 mg/kg of *C. asiatica* extract	[318]
Whole plants were macerated in ethanol	Supplementation of *C. asiatica* at 1000 mg/kg did not reduce fasting blood sugar in a diabetic rat model. Nevertheless, the treatment improved glucose homeostasis when loaded with glucose after 60 and 120 min. The *C. asiatica* extract reduced glucose absorption, increased gastrointestinal motility, and inhibited disaccharide enzyme activity	[314]
Asiatic acid (**248**)	Treatment with asiatic acid (**248**) at 20 mg/kg for 45 days significantly reduced blood glucose and increased serum insulin glucose levels in diabetic rats. The activities of the enzymes glucose-6-phosphatase and fructose-1,6-bisphosphatase were significantly higher in diabetic rats compared to normal rats	[319]
Leaves were macerated in methanol	Oral administration of the extract to diabetic rats significantly increased enzyme activities compared to untreated diabetic rats. In diabetic rats, G6PDH activity was significantly decreased, while LDH activity was increased. Daily oral administration for 30 days significantly restored both G6PDH and LDH activities, similar to normal rats. The effects were similar to those of glibenclamide	[326]
Leaves were macerated in ethanol	Treatment with 300 mg/kg of the extract reduced serum glucose levels and increased high-density lipids significantly. The serum insulin levels in diabetic rats were normalized	[315]
Leaves were macerated in methanol	Treatment with *C. asiatica* raised the body weight of diabetic rats by 93%. Liver weights and liver weight/body weight ratios were reduced in diabetic rats. Decreased hepatic MDA, increased FRAP, and increased GSH, GST, GPX were evident, and the hepatic concentrations of IL-1β, MCP-1, and TNF-α were reduced to 68, 75, and 63%, respectively. Serum TG concentrations decreased when compared to diabetic controls	[316]

(continued)

Table 7 (continued)

Extraction method/ compound	Key findings	References
Leaves were macerated in methanol	Treatment with *C. asiatica* reduced the renal levels of malondialdehyde (MDA) by 33%, TNF-α by 78%, and IFN-γ by 42%, while increasing IL-10 by 18% compared to diabetic control rats. In the brain, treatment with the extract also led to significant reductions in MDA (37%), TNF-α (30%), and IFN-γ (37%), with increases in IL-4 (94%) and IL-10 (20%)	[326]
Asiatic acid (**248**)	Supplementation with asiatic acid (**248**) for 45 days in diabetic rats normalized plasma glucose, reduced insulin by 65.4%, and decreased body weight by 51.9%. It restored lipid profiles to near-normal levels and reduced liver and kidney lipid levels. Asiatic acid treatment also lowered the activity of HMGR, and increased liver LCAT mRNA levels, helping restore these metabolic parameters levels to near normal	[321]
Leaves were macerated in absolute methanol	Treatment with 0.5 g/kg of the extract increased phosphofructokinase activity (sevenfold) and fructose 1,6-bisphosphatase activity (23%), while treatment with 1 g/kg increased glycogen synthase activity (27%) and glycogen phosphorylase activity (50%)	[317]
Leaves macerated in ethanol	The extract reduced blood glucose and increased liver weight ($p = 0.03$) in treated rats, indicating a protective effect	[327]
Leaves macerated in ethanol	Treatment with the extract reduced omega-6 (16%) and MUFA (33%) in the testes of diabetic rats. The extract ameliorated oxidative injury and improved antioxidant capacity by increasing glutathione (49%), glutathione *S*-transferase (16%), and glutathione peroxidase (23%) when compared to diabetic controls	[324]
Leaves prepared in nanoemulsion form	*Centella asiatica* significantly reduced blood glucose by 37.54% and 51%, respectively. The extract in a SNEDDS formulation significantly lowered glucose in zebrafish ($p < 0.05$)	[328]
Madecassoside (**264**)	Administration of madecassoside (**254**) at 50 mg/kg in streptozocin-niacinamide-induced diabetic rats improved fasting blood sugar, increased serum insulin sensitivity, improved enzymatic antioxidants, and reduced lipid peroxidation	[320]
Ethanol extraction of *C. asiatica* using a Soxhlet apparatus, and dissolution in 96% ethanol	A significant decrease in blood glucose levels was observed, and the lipid profiles improved with higher HDL and lower LDL levels in treated groups ($p < 0.05$). The ethanol extract showed improvements in lipid profiles and reduction of blood glucose in treated rats	[329]

2.6 *Orthosiphon stamineus*

Orthosiphon stamineus Benth., locally known in Malaysia as "misai kucing", belongs to the family Lamiaceae. The plant is distributed also in other Asian countries such as Vietnam, Thailand, Indonesia, and Malaysia. It is also known as *Orthosiphon aristatus* (Blume) Miq. The plant comes in two varieties, one with white flowers and the other with light purple flowers, with the purple variety noted for producing a

Plate 6 The aerial part of *Orthosiphon stamineus* (family Lamiaceae). Photograph courtesy Latupeirissa (https://pixabay.com/photos/kidney-tea-flower-4982286/)

greater quantity of bioactive compounds. The plant is also known as "kumis kucing" or cat's whiskers due to its wispy appearance. Other names include "moustaches de chat" in French, "yaa nuat maeo" and "rau meo" in Thailand,"cay bac" in Vietnam, and "remujung" in Indonesia [330]. The herb is the source of Java tea and is consumed in the form of a herbal beverage and was likely introduced to western countries in the early twentieth century.

Orthosiphon stamineus is a perennial shrub or small tree with four-angled stems, reaching a height of up to 1.5 m. The leaves, which are utilized for medicinal purposes, are arranged in opposite pairs, measuring approximately 0.3 cm in length, and have reddish-purple petioles. The local name of the plant is inspired by the flower's resemblance to a cat's whiskers. The flowers are white to bluish in color, with verticils measuring around 16 cm in length, and feature long, protruding filaments [331]. Plate 6 illustrates the aerial parts of *O. stamineus.*

2.6.1 Ethnobotany

Orthosiphon stamineus is employed ethnomedically as a diuretic [262]. It is used traditionally also for the treatment of rheumatoid diseases, diabetes, hypertension, tonsillitis, epilepsy, menstrual disorders, gonorrhea, syphilis, renal calculus, gallstones lithiasis, edema, eruptive fever, influenza, hepatitis, and jaundice [330]. Additionally, *O. stamineus* is taken for the treatment of diabetes, hypertension [332], atherosclerosis, nephritis, kidney stones [333], inflammatory disorders (fever, cold, rheumatism), gonorrhea, syphilis, and hepatic disorders [330].

2.6.2 Phytochemistry

Orthosiphon stamineus contains a variety of bioactive compounds that contribute to its medicinal properties. The plant contains mainly diterpenoids, triterpenoids, flavones, flavonols, and other phenolic compounds. Some of the compounds isolated have been reported to exhibit antioxidant, anti-inflammatory, antidiabetic, antihyperlipidemic, and cytotoxic activities. Among these, diterpenes such as orthosiphols A–Y (**290–313**) and flavonoids, such as eupatorin (**347**) and sinensetin (**348**), were investigated for their anti-inflammatory and antioxidant activities and exhibited activities supportive of further studies on inflammation and the inhibition of tumor growth. Together, these bioactive compounds contribute to the therapeutic potential of *O. stamineus*, providing some support for its traditional use for a range of health conditions. This contribution will only discuss the diterpenoids and flavonoids from *O. stamineus*. Table 8 summarizes different compounds from different classes of compounds isolated from *O. stamineus,* while the formula drawings of **290–360** show their detailed structures.

Table 8 Classes of compounds isolated from *Orthosiphon stamineus*

Class of compounds	Compounds	References
Diterpenes	Orthosiphol A (**290**), B (**291**), C (**292**), D (**293**), E (**294**), F (**295**), G (**296**), H (**297**), I (**298**), J (**299**), K (**300**), L (**301**), M (**302**), N (**303**), O (**304**), P (**305**), Q (**306**), R (**307**), T (**308**), U (**309**), V (**310**), W (**311**), X (**312**), Y (**313**), 3-*O*-deacetylorthosiphol I (**314**), 2-*O*-deacetylorthosiphol J (**315**), 13-*epi*-orthosiphol N (**316**), siphonol A (**317**), B (**318**), C (**319**), D (**320**), E (**321**), 7-*O*-deacetylorthosiphol B (**322**), 6-hydroxyorthosiphol B (**323**), neoorthosiphol A (**324**) B (**325**), staminol A (**326**), B (**327**), C (**328**), and D (**329**), 14-deoxo-14-*O*-acetylorthosiphol Y (**330**), nororthosiphonolide A (**331**), norstaminone A (**332**), *seco*-orthosiphol A (**333**), B (**334**), and C (**335**), neoorthosiphonone A (**336**), stamilactone A (**337**) and B (**338**), norstaminol A (**339**), B (**340**), C (**341**), ortosiphonone A (**342**), B (**343**), C (**344**), D (**345**), norstaminolactone A (**346**)	[334–340]
Flavonoids	Eupatorin (**347**), sinensetin (**348**), 5-hydroxy-6,7,3′,4-tetramethoxyflavone (**349**), salvigenin (**350**), 6-hydroxy-5,7,3′-trimethoxyflavone (**351**), tetramethylscutellarein (**352**), ladanein (**353**), 5,6,7,3′-tetramethoxy-4′-hydroxy-8-*C*-prenylflavone (**354**), 3′,4′,6,7,8-pentamethoxyflavone (**355**), 5,6-dihydroxy-4′,7-dimethoxyflavone (**356**), scutellarein tetramethyl ether (**357**), eupatoretin (**358**), 7, 3′,4′-tri-*O*-methylluteolin (**359**), 3′-hydroxy-5,6,7,4′-tetramethoxyflavone (**360**)	[339, 341]

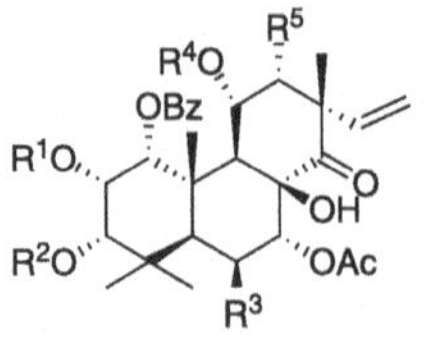

290 (orthosiphol A) R^1 = Ac, R^2 = R^3 = H, R^4 = Bz
291 (orthosiphol B) R^1 = R^3 = H,R^2 = Ac, R^4 = Bz
292 (orthosiphol C) R^1 = R^2 = R^3 = H, R^4 = Bz
300 (orthosiphol K) R^1 = R^2 = R^3 = H, R^4 = Bz
304 (orthosiphol O) R^1 = Bz, R^2 = Ac, R^3 = R^4 = H
305 (orthosiphol P) R^1 = H, R^2 = Ac, R^3 = R^4 = Bz

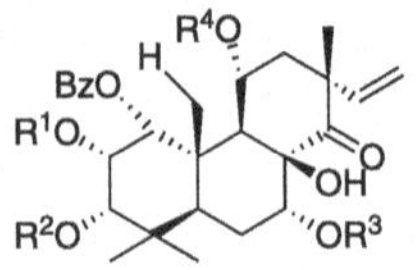

308 (orthosiphol T) R^1 = Ac, R^2 = R^3 = H, R^4 = Bz
309 (orthosiphol U) R^1 = R^2 = R^3 = Ac, R^4 = H
310 (orthosiphol V) R^1 = R^4 = H, R^2 = R^3 = Ac
311 (orthosiphol W) R^1 = R^3 = Ac, R^2 = R^4 = H
312 (orthosiphol X) R^1 = Ac, R^2 = Bz, R^3 = R^4 = H

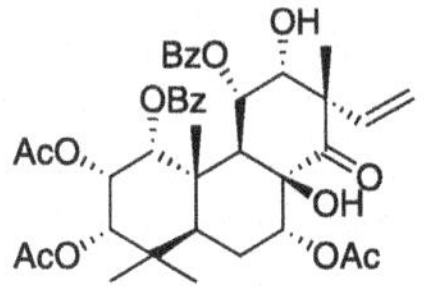

301 (orthosiphol L)

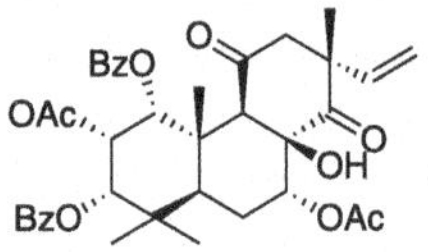

302 (orthosiphol M)

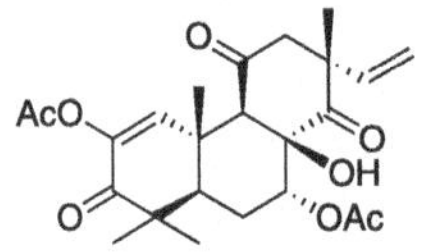

305 (orthosiphol Q)

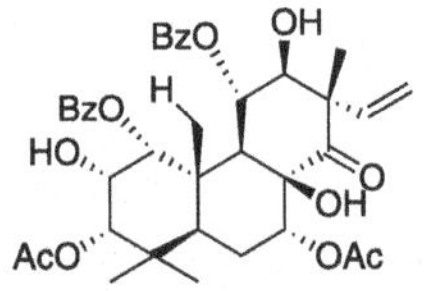

307 (orthosiphol R)

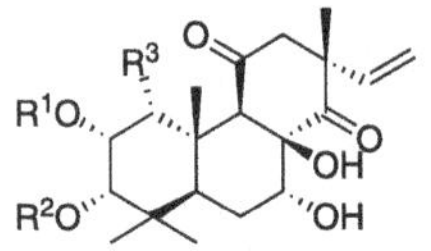

298 (orthosiphol I) R^1 = Ac, R^2 = H, R^3 = COOBz
299 (orthosiphol J) R^1 = $COCH_3$, R^2 = H, R^3 = COOBz
303 (orthosiphol N) R^1 = Ac, R^2 = Ph, R^3 = OBz
314 (3-*O*-deacetyl orthosiphol I) R^1 = Ac, R^2 = H, R^3 = COOBz
315 (2-*O*-deacetyl orthosiphol J) R^1 = H, R^2 = Ac, R^3 = COOBz
316 (13-*epi*-orthosiphol N) R^1 = Bz, R^2 = Ac, R^3 = OBz

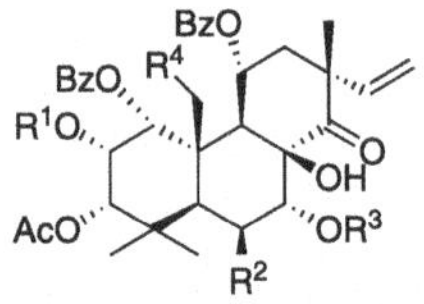

317 (siphonol A) R^1 = R^3 = H, R^2 = H, R^4 = OH
318 (siphonol B) R^1 = Ac, R^2 = R^3 = H, R^4 = OH
319 (siphonol C) R^1 = R^2 = H, R^3 = Ac, R^4 = OH
320 (siphonol D) R^1 = Ac, R^2 = R^3 = H, R^4 = OAc
322 (7-*O*-deacetyl orthosiphol B) R^1 = R^2 = R^3 = R^4 = H
323 (6-hydroxy orthosiphol B) R^1 = R^4 = H, R^2 = OH, R^3 = Ac

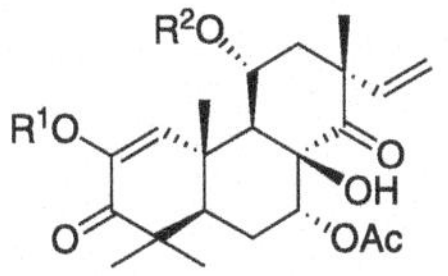

293 (orthosiphol D) R^1 = H, R^2 = Ac
294 (orthosiphol E) R^1 = H, R^2 = COBz
313 (orthosiphol Y) R^1 = Ac, R^2 = H

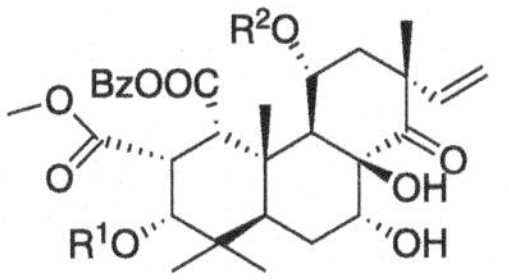

295 (orthosiphol F) R^1 = H, R^2 = COBz
296 (orthosiphol G) R^1 = $COCH_3$, R^3 = H
297 (orthosiphol H) R^1 = $COCH_3$, R^2 = H

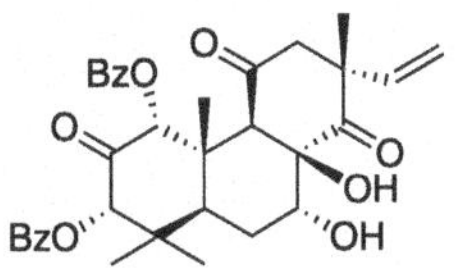

308 (orthosiphol S)

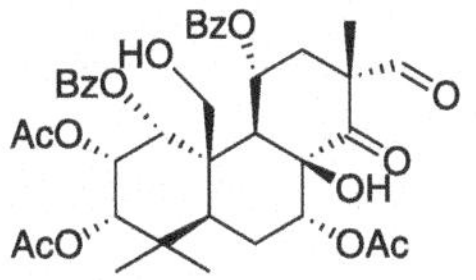

321 (siphonol E)

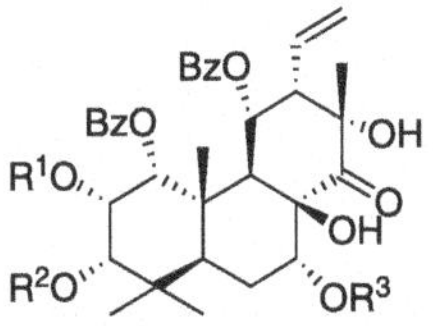

324 (neoorthosiphol A) R^1 = R^3 = Ac, R^2 = H
325 (neoorthosiphols B) R^1 = H, R^2 = R^3 = Ac
326 (staminol A) R^1 = R^2 = R^3 = Ac
327 (staminol B) R^1 = R^2 = Ac, R^3 = H

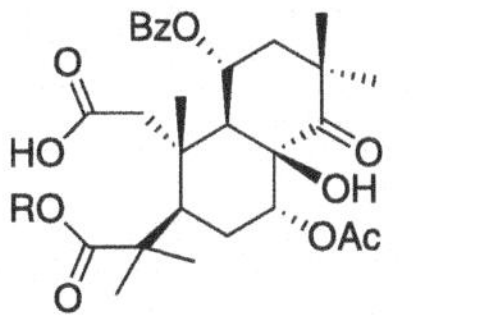

333 (*seco*-orthosiphol A) R = H
334 (*seco*-orthosiphol B) R = CH_3

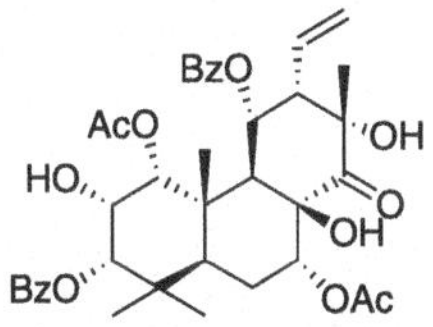

328 (staminol C)

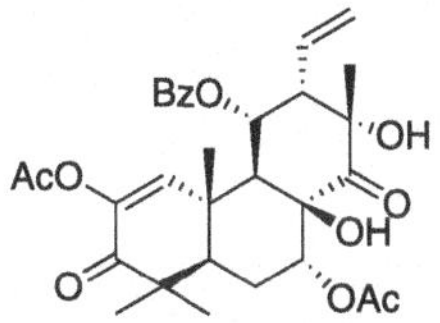

329 (staminol D)

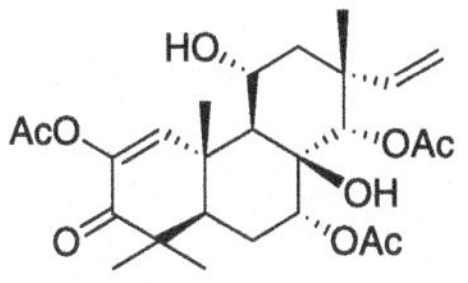

330 (14-deoxo-14-*O*-acetylorthosiphol Y)

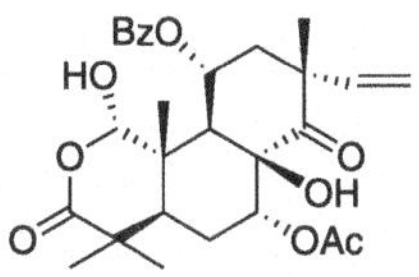

331 (*nor*-orthosiphonolide A)

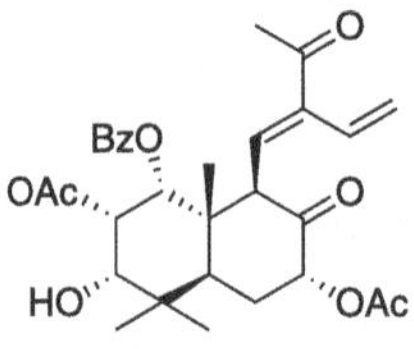

332 (norstaminone A)

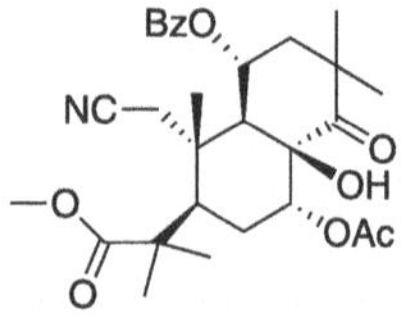

335 (*seco*-orthosiphol C)

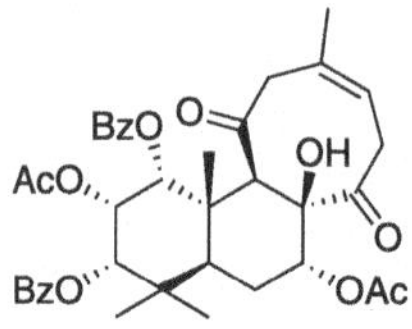

336 (neoorthosiphonone A)

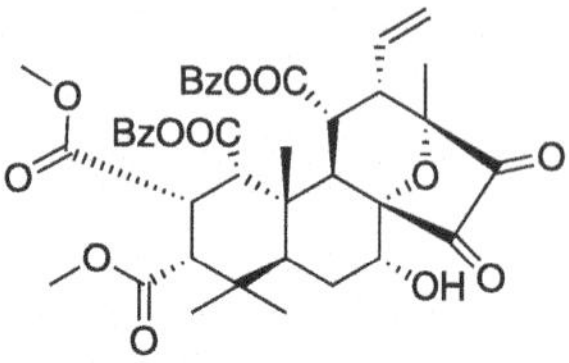

337 (stamilactatone A)

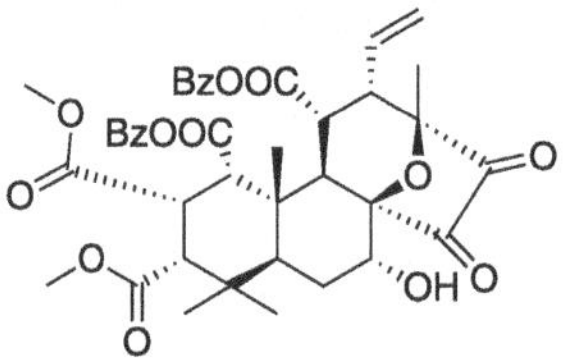

338 (stamilactone B)

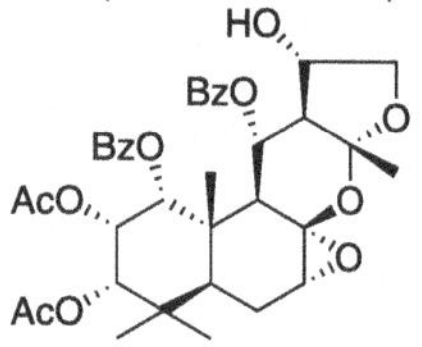

339 (norstaminol A)

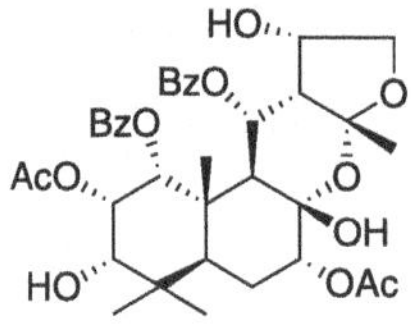

340 (norstaminol B)

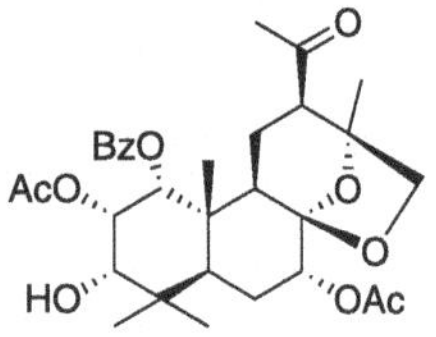

341 (norstaminol C)

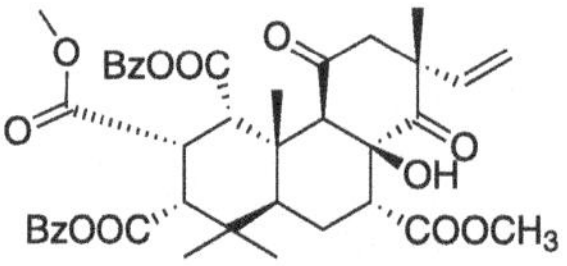

342 (orthosiphonone A)

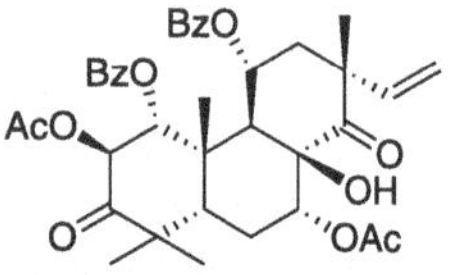

343 (orthosiphonone B)

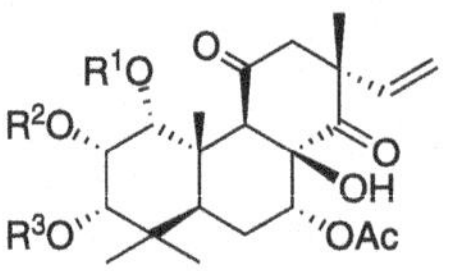

344 (orthosiphonone C) $R^1 = R^3 = H$, $R^2 = Bz$
345 (orthosiphonone D) $R^1 = R^3 = Bz$, $R^2 = H$

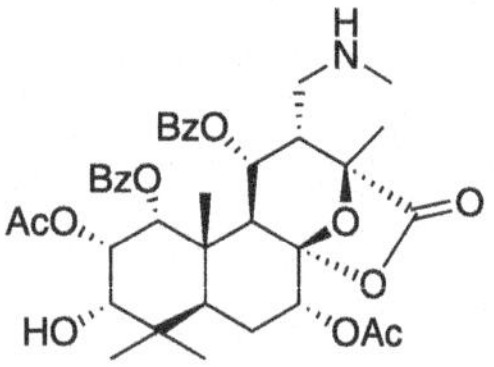

346 (norstaminolactone A)

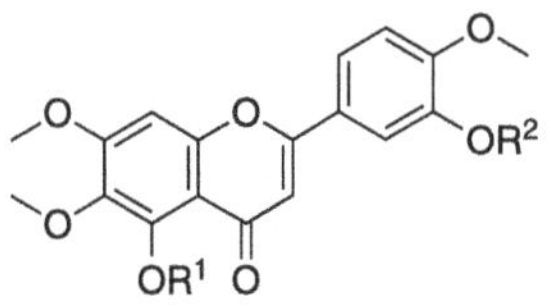

347 (eupatorin) $R^1 = R^2 = H$
348 (sinensetin) $R^1 = R^2 = CH_3$
349 (5-hydroxy-6,7,3',4'-tetramethoxyflavone) $R^1 = H$, $R^2 = CH_3$

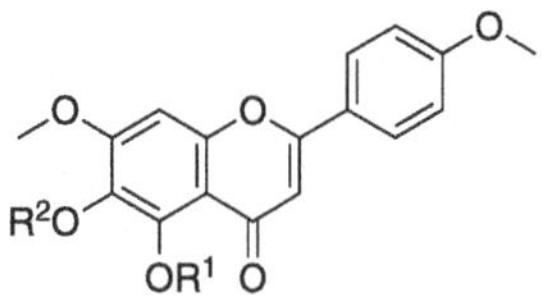

350 (salvigenin) $R^1 = H$, $R^2 = CH_3$
351 (6-hydroxy-5,7,3'-trimethoxyflavone) $R^1 = CH_3$, $R^2 = H$
352 (tetramethylscutellarein) $R^1 = R^2 = CH_3$
353 (ladanein) $R^1 = R^2 = H$

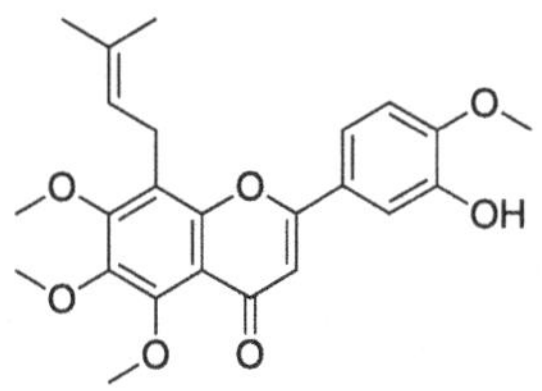

354 (5,6,7,3'-tetramethoxy-4'-hydroxy-8-*C*-prenylflavone)

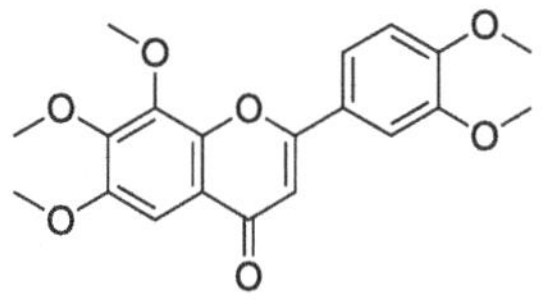

355 (3',4',6,7,8-pentamethoxyflavone)

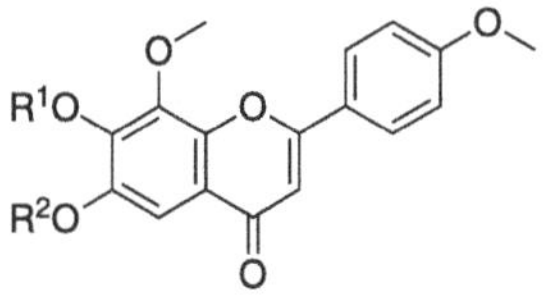

356 (5,6-dihydroxy-4',7-dimethoxyflavone) $R^1 = R^2 = H$
357 (scutellarein tetramethyl ether) $R^1 = R^2 = CH_3$

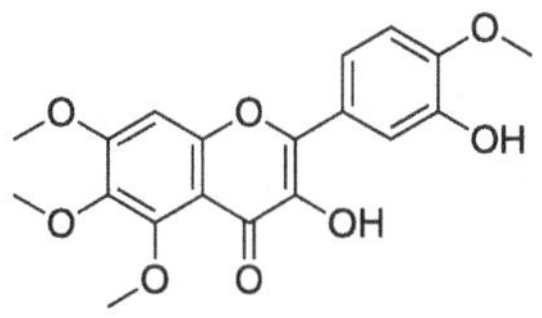

358 (eupatoretin)

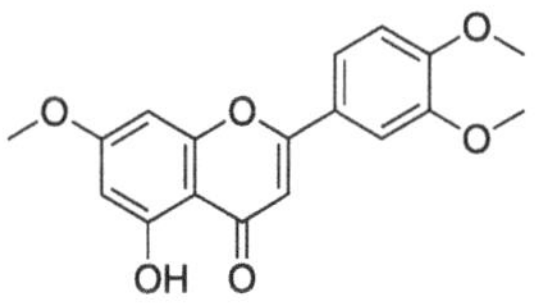

359 (7,3',4'-tri-*O*-methylluteolin)

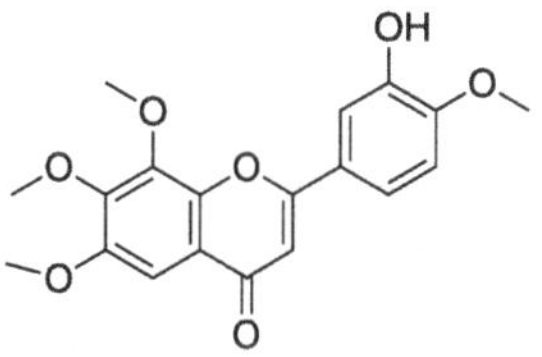

360 (3'-hydroxy-5,6,7,4'-tetramethoxyflavone)

2.6.3 Biological Activities

The studies conducted to date on *Orthosiphon stamineus* are supportive of its therapeutic potential in managing disorders such as diabetes, obesity, and hypertension. In diabetic rodent models, *O. stamineus* leaf extracts showed substantial antihyperglycemic activity, reducing blood glucose levels in streptozotocin/alloxan-induced diabetic rats and mice, comparable to the positive standards used (glibenclamide/metformin) [342–345]. This major effect has been attributed to compounds present, such as the flavonoids eupatorin (**347**) and sinensetin (**348**) as well as phenolic acids. The hypoglycemic activity seems to be due to an insulin-mimetic effect [346] by enhancing glucose uptake in cells, likely through increased GLUT4 translocation to the cell membrane [347]. Additionally, *O. stamineus* was also reported to inhibit the α-glucosidase and α-amylase enzymes, consequently reducing postprandial blood glucose levels by slowing carbohydrate breakdown [348].

In obesity-related studies, *O. stamineus* has been shown to lower total cholesterol, triglycerides, and low-density lipoprotein, while normalizing high-density lipoprotein and antioxidant enzyme levels like superoxide dismutase. *Orthosiphon stamineus* demonstrated antiobesity effects by inhibiting pancreatic lipase, reducing fat absorption, lowering body weight and visceral fat mass, and enhancing lipid metabolism in high-fat diet-induced obese mice [349]. The extracts also significantly reduced triglycerides and low-density lipoprotein cholesterol levels, showing lipid-lowering properties [350]. These effects may be attributed to the presence of flavonoids and other bioactive compounds, making *O. stamineus* a promising natural alternative for weight management [351].

The plant also exhibited potent antihypertensive effects, reducing blood pressure in both animal models and in human patients. In two clinical trials, supplementation with nutraceutical formulations containing *O. stamineus* reduced systolic and diastolic blood pressure in mild and moderate hypertensive patients [352, 353]. This antihypertensive effect may be linked to its influence on vascular relaxation. *Orthosiphon stamineus* was found to modify the activity of α_1-adrenergic and AT_1 receptors, enhancing vasorelaxation through sustained release of nitric oxide and prostacyclin (PGI_2) [354]. Additionally, the hypotensive effect could be attributed to anti-inflammatory effects as shown by a reduction in high-sensitivity C-reactive protein [355].

Overall, *O. stamineus* shows strong potential as a multi-functional therapeutic agent for managing chronic conditions such as diabetes, obesity, and hypertension, with several bioactive compounds contributing to its pharmacological effects. Table 9 summarizes the key findings on the biological activities of *O. stamineus*.

Table 9 Biological activities of *Orthosiphon stamineus*

Extraction method/compound	Results	References
Aqueous	Treatment with the extract reduced plasma glucose by up to 24% in streptozotocin-induced diabetic rats, and improved lipid profiles (increased HDL, reduced triglycerides). In the perfused rat pancreas, 100 μg/cm^3 extract potentiated glucose-induced insulin secretion	[342]
Methanol	Treatment with 800 mg/kg of the extract reduced hyperglycemia in streptozotocin-induced diabetic rats comparable to the effects of glibenclamide, but the level of insulin secretion did not change. The treatment also reduced G6P and G6PD levels in diabetic rats	[343]
Chloroform	The extract given at 1 g/kg reduced the blood glucose levels of rats treated subcutaneously with 150 mg/kg (b.w.) glucose. Fraction *cf*2-B displayed a blood glucose-lowering effect in fasted normal rats after glucose-loading of 150 mg/kg, similar to metformin. Phytochemical analysis revealed the presence of sinensetin (**348**) in the fraction	[345]
Water	Alloxan-induced diabetic rats treated with the water extract showed a significant reduction in blood glucose levels, from an initial 14.6 mmol/dm^3 down to 6.3 mmol/dm^3 after four weeks, comparable to the effect of metformin-treated rats. The treated rats had an average weight gain of 23%, increased insulin levels by 25%, and reduced glucagon levels by 18%, and also decreased excessive food and water intake	[344]
50% Ethanol	The ethanol leaf extract inhibited α-glucosidase (IC_{50}: 4.63 mg/cm^3) and α-amylase (IC_{50}: 36.70 mg/cm^3). Sinensetin (**348**) was more effective (IC_{50}: 0.66 and 1.13 mg/cm^3, respectively) than the extract	[348]
Chloroform subfraction of methanol extract	Several purified pimarane diterpenes including siphonol B (**318**) exhibited potential as natural insulin mimetics by glucose uptake stimulation and inhibition of protein tyrosine phosphatase 1B (PTP1B)	[346]
Methanol and water	The methanol extract enhanced GLUT4 translocation in L6 muscle cells, increasing it by 279% without insulin and 351% with insulin. In a diabetic mice model, the blood glucose levels were reduced from 359 mg/dL in the untreated group, to 1640 and 1740 mg/dm^3 with the water and methanol extracts, respectively. The methanol extract helped diabetic mice maintain body weight and reduce weight loss, when compared to untreated mice	[347]

(continued)

Table 9 (continued)

Extraction method/compound	Results	References
Dry extract combined in a nutraceutical (Diuripres®)	Diuripres® supplementation in Type I hypertension patients over 8 weeks led to reductions in systolic blood pressure by 3.4 mmHg and diastolic blood pressure by 3.1 mmHg, when compared to a placebo. The aortic blood pressure decreased by 5.2 mmHg, and high-sensitivity C-reactive protein (hs-CRP), a marker of inflammation, dropped by 0.03 mg/cm^3	[355]
Nutraceutical tablet with additional components	Treatment with the nutraceutical tablet reduced systolic (-9.50 ± 6.66 mmHg) and diastolic blood pressure (-5.55 ± 4.14 mmHg), when compared to baseline, although the reduction was less when compared to the standard, hydrochlorothiazide	[352]
Phytopharmaceutical preparation (combination with *Apium graviolens*)	Treatment for two weeks in mild and moderate hypertensive patients reduced systolic blood pressure by 24.72 mmHg and diastolic blood pressure by 10.0 mmHg; this was equivalent to amlodipine with no significant side effects	[353]
Sinensetin (**348**)	Sinensetin (**348**) exhibited a vasorelaxant effect involving the NO/sGC/cGMP and indomethacin pathways, calcium and potassium channels, and muscarinic and β-adrenergic receptors	[351]
Water and ethanol	An *O. stamineus* leaf extract reduced vasoconstriction responses by modifying the activity of the α_1-adrenergic and AT_1 receptors. The involvement of the release of endothelial-derived nitric oxide (EDNO) was observed. Additionally, the enhancement of vasorelaxation may be attributed to the sustained release of prostacyclin (PGI_2)	[354]
Ethanol and water	The ethanol leaf extract inhibited pancreatic lipase activity in vitro, however, the activity was much less compared to orlistat. High-fat diet rats were treated with this ethanolic leaf extract, which resulted in weight loss, organ weight loss, and visceral fat loss, comparable to the effects of orlistat	[349]
Ethanol	The ethanol extract inhibited pancreatic lipase activity in a dose-dependent manner (IC_{50}: 5.7 mg/cm^3), comparable to the effects of orlistat. High-fat diet obese mice showed a reduction in adipose tissues comparable to orlistat. The treatment also normalized the total cholesterol, high-density cholesterol level, the triglyceride content, and the low-density lipid level. The superoxidase dismutase and malondialdehyde values were also normalized	[350]

2.7 *Cynometra cauliflora*

Fruit-bearing trees that are indigenous to tropical regions are often cultivated in the relevant climatic zones of Asia, Africa, America, and Australia. Malaysia, located in Southeast Asia, is well-known for its tropical rainforests and the quantity of fruit-bearing trees available. *Cynometra cauliflora* L., sometimes known as "Mountain Ebony", is a medium-sized tree endemic to Southeast Asia. It may reach 30 m and thrives in tropical rainforests and other humid environments [356]. *Cynometra cauliflora* has a densely branched stem and relatively tiny blooms (1.2 cm wide) that cluster directly on the stem. The flattened, crescent-shaped pods resemble the Indonesian pastry "Namu-namu" or "ñam-ñam" in the Maldives. The name *Cynometra* comes from the Greek words "κύων" (kyon, meaning dog) and "μήτρα"

Plate 7 Fruits and flowers of *Cynometra cauliflora* (family Fabaceae). Photograph courtesy W.A. Djatmiko (https://commons.wikimedia.org/wiki/File:Cynom_caulif_Fr_071126-1806_tdp.jpg)

(metra, meaning womb), which relate to the appearance of the fruit pods. The term "*cauliflora*" refers to the fact that the plant blooms on its main stems [357].

The tree, native to Malaysia, is commonly known as "Nam-nam" or "Katak puru". It may grow up to five meters tall and produces fruit on its trunk. Belonging to the family Fabaceae (Leguminosae), the tree prefers full sun with moderate water. *Cynometra cauliflora* leaves are made up of two elliptical to oval-oblong leaflets that are normally 5.5–16.5 cm long and 1.6–5.6 cm wide. The leaflets have a smooth, glabrous surface, while the leaf petiole, which connects the leaf to the stem, measures 2–8 mm in length and provides structural support and flexibility to the leaf structure. Furthermore, the petiolules (the stalks that link the leaflets to the midrib) are relatively short, which contributes to the compact and tidy arrangement of the leaflets. This leaf form represents adaptations for efficient photosynthesis and environmental requirements, which are common in the tropical rainforest habitat of *C. cauliflora* [357]. The fruits of the plant are kidney-shaped fleshy pods containing one brown flattened seed, while the outside skin is rough and wrinkled (Plate 7).

2.7.1 Ethnobotany

The use of *C. cauliflora* in ethnobotany by indigenous groups in various tropical regions of the world shows an intriguing integration of traditional knowledge and ethnomedical practices. Although a majority of the plant parts can be utilized medicinally, some local populations have been observed only to employ the fruits [358, 359]. For example, the Osing tribe in East Java, Indonesia, uses *C. cauliflora* fruits, from which they make a beverage called "Celang Tea", which also includes several other plants, and has been used to alleviate the symptoms of COVID-19 [358]. Local populations of Baturraden, Jawa Tengah, have used the fruits of *C. cauliflora* taken orally to treat coughs [359].

In Malaysia, the fruits of *C. cauliflora* may be consumed raw, incorporated into salads, or occasionally made into a juice, sweetened with sugar, or fried in batter. The tree is cultivated in villages for its nutritional and medicinal benefits. The fruits are also used traditionally to treat appetite loss [356]. Also, the oil derived from *C. cauliflora* seeds is used to treat leprosy and skin ailments [357].

2.7.2 Phytochemistry

A variety of different secondary metabolite constituents of *C. cauliflora* have been reported, including certain coumarins, flavonoids, quinones, saponins, steroids, and terpenoids. Among the individual phytochemicals isolated from the plant are the coumarins 8-methoxypsoralen (**361**) and fraxetin (**362**), the flavonoids naringenin (**363**), malvidin (**364**), cyanidin (**365**), nobiletin (**367**), isorhamnetin (**368**), epigallocatechin (**91**) and apigenin (**87**), and other compounds including amorphigenin (**366**), oenin (**369**), gallate (**370**), β-carotene (**100**), and ascorbic acid (**98**) [360–364]. The structures of the constituents isolated from *C. cauliflora* are shown in their formula drawings.

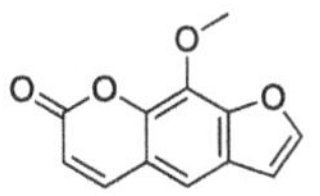

361 (8-methoxypsoralen)

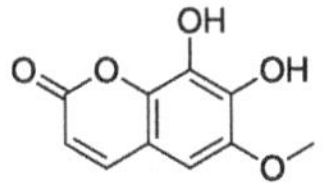

362 (fraxetin)

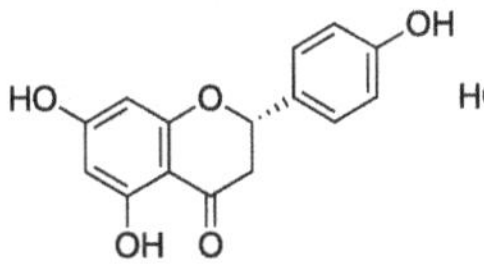

363 (naringenin)

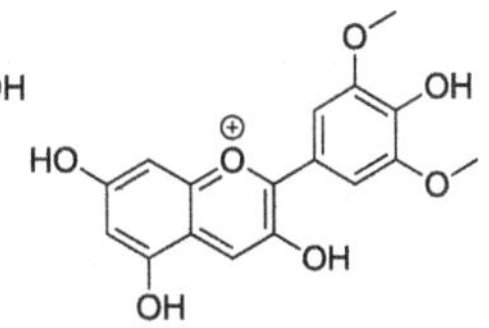

364 (malvidin)

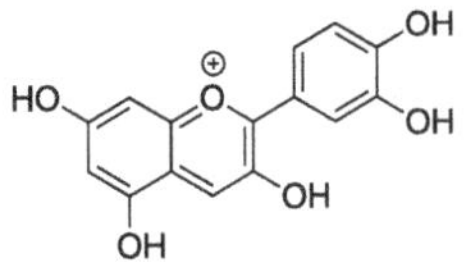

365 (cyanidin)

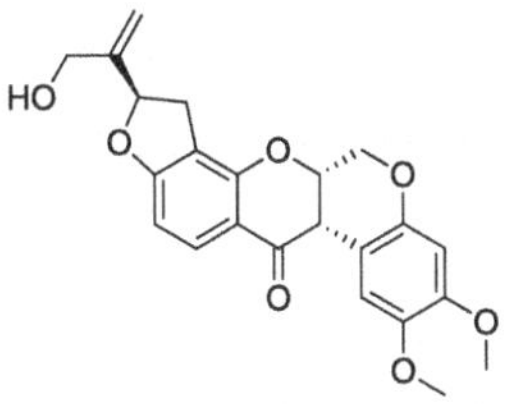

366 (amorphigenin)

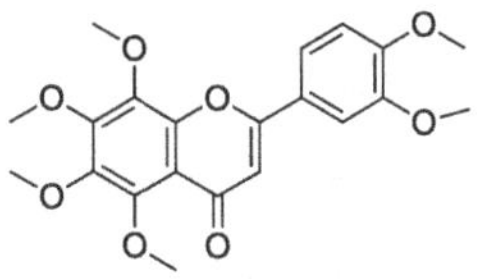

367 (nobiletin)

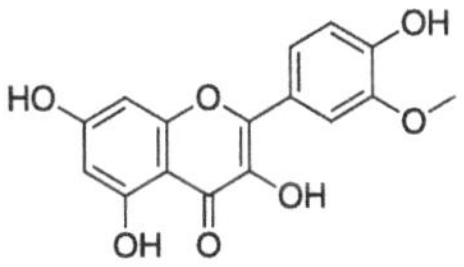

368 (isorhamnetin)

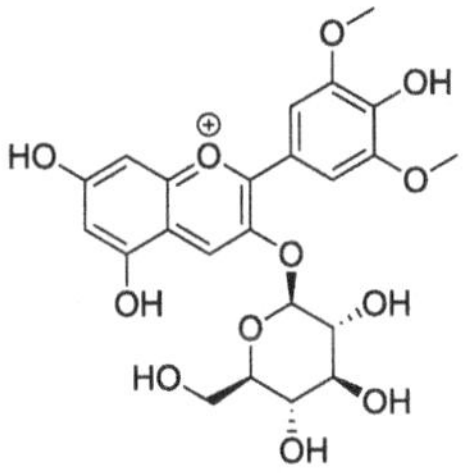

369 (oenin)

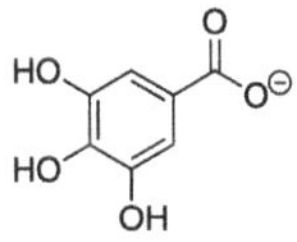

370 (gallate)

2.7.3 Biological Activities

Progress has been made in understanding the biological activities of *C. cauliflora*, including its antimicrobial and antioxidant effects, and its potential involvement in disease prevention. Most of the antimicrobial evaluations have utilized the leaves of the plant [356, 365–368]. However, other parts such as fruits [365, 364], flowers [369], stems [367], twigs [356], and roots [367] also have been investigated. It has been found that *C. cauliflora* extracts showed growth inhibitory activities towards pathogenic bacteria such as methicillin-resistant *Staphylococcus aureus*, *Staphylococcus aureus*, *Escherichia coli*, *Bacillus cereus*, *Proteus vulgaris*, *Micrococcus luteus*, *Pseudomonas aeruginosa*, *Bacillus subtilis*, *Proteus mirabilis*, *Klebsiella pneumoniae* and also for yeasts inclusive of *Candida utilis* and *Candida albicans* [356, 364–370].

So far as antioxidant testing is concerned, *C. cauliflora* has been studied for its ability to scavenge free radicals, which are linked to a variety of chronic illnesses, including cardiovascular problems, neurological ailments, and cancer. Most laboratory work on *C. cauliflora* has been performed on the leaves, with the fruits, twigs and flowers also investigated in this manner. Application of the in vitro DPPH (2,2-diphenyl-1-picrylhydrazyl) assay has shown discernible antioxidant capability in many of the studies conducted ($IC_{50} < 50\ \mu g/cm^3$), with other research work showing weaker antioxidant capability for the leaf and fruit oil and for healthy and ripe fruits [356, 360, 369, 371–375].

Other research has focused on the antidiabetic propensity of *C. cauliflora*, inclusive of its capability to regulate blood sugar levels in comparison with the clinically used drug, metformin. A leaf extract tested in a Sprague–Dawley rat model showed effective reductions in blood glucose levels by 23–34%, decreased plasma triglyceride levels by 13–30%, and increased liver glycogen levels by 68–96%, when compared to a control group [376]. In turn, when this extract was tested on male white rats (*Rattus norvegicus*) pretreated with alloxan, a dose of 2500 mg per rat per day exhibited similar results to metformin after 14 days administration [377]. Also, it was found that an oral treatment of *C. cauliflora* at 200 and 400 mg/kg significantly decreased body weight, adipose tissue, and liver weight, as well as lipid buildup in the liver, when compared to a control group.

A *C. cauliflora* leaf extract was shown to be very weakly cytotoxic with an IC_{50} value of 125.9 $\mu g/cm^3$ using HeLa human cervical cancer cells (ATCC CCL-2) [378]. A leaf extract exhibited antiviral activity with a CC_{50} value of 36 mg/cm^3 when evaluated against a clinical isolate of herpes simplex virus type-1 (HSV-1) [361]. Also, a methanol extract from the plant leaves showed some toxicity against *Artemia salina* (brine shrimp) larvae, with an LC_{50} of 314.6 $\mu g/cm^3$ [360]. Another study with an in vitro MTT assay on MCF-7 human breast cancer cells showed an antiproliferative effect using the twig oil extracted from *C. cauliflora* [356].

Additionally, there have been other studies on *C. cauliflora* regarding its plant growth [367] and on its effects on birth anomalies [379]. A study on the endophytic bacteria from the leaves, stems, and roots of *C. cauliflora* indicated that a sample coded R1S4 produced the most notable effects on indole-3 acetic acid at 20.6 $\mu g/cm^3$, out of 15 isolates tested. Their genetic sequences revealed that samples RIR2, TKS2/R1L3, and R1S4/R1S5 belonged to the bacteria *Methylobacterium radiotolerans*, *Mycobacteroides abscessus*, and *Sphingomonas* sp., respectively [367]. This indicates that *C. cauliflora* contains microorganisms (endophytes) that can help these plants grow more effectively. Research on the impact of a *C. cauliflora* fruit extract on birth abnormalities showed that at a dosage of 291.7 mg/kg, the extract caused more than 50% mortality in chick ED-3 day 3 embryos. Furthermore, the fruit extract caused malformations in ED-3 and ED-6 embryos, such as abnormalities in anterior neuropore openings, heart tubes, and problems with liver and kidney tissues [379]. These findings indicate that consuming *C. cauliflora* fruits is not to be recommended for newborn children or pregnant women due to possible negative effects on fetal development. This underlines the significance of exercising caution while using and consuming *C. cauliflora*, especially during pregnancy and in infant care settings.

The phenolic constituents of *C. cauliflora* have been investigated for their biological activities. Phenols are recognized for their antioxidant qualities and ability to protect cells from oxidative stress. High concentrations of phenolic compounds occur mainly in the plant leaves [371, 380] and fruits [364, 372–374]. The total phenolic compounds were determined as at 996.0 mg/dm^3 [364] and 27.8

± 0.1 mg GAE/g (gallic acid equivalents) fresh weight [373], 428.5 ± 1.3 mg GAE/g fresh weight [374], when using fresh fruits, with a lower concentration at 5.7 mg GAE/g fresh weight found when using dried fruits [372]. This indicates that a significant reduction of the phenolic content in the fruits occurred as a result of drying. In turn, the total phenolic content in the leaves was high at 113.2 ± 0.1 mg [380] and 267.7 mg [371] of GAE/g (fresh weight).

The flavonoid content in the leaves of *C. cauliflora* ranged from 12.3 mg QE/g (quercetin equivalents) fresh weight [371] to 55.0 ± 2.6 mg rutin equivalents/g [380]. The fruit extract showed a concentration of 26.1 ± 1.0 mg QE/100 g fresh weight [374], with the the ripe fruit exhibiting a greater content of 421.1 mg/dm^3 [364]

Two studies have been conducted to investigate the vitamin C (**98**) content of *C. cauliflora* plants. One of these indicated that the fruit contains a total vitamin C (**98**) value of 37.9 ± 1.8 mg/100 g of fresh weight [374]. Another investigation found that the leaves contain a total vitamin C (**98**) content of 203.5 mg/g [371].

Up to the present, research on the content of saponins in *C. cauliflora* trees has been limited. One study found that the concentration of saponin in the leaves is 629.1 ± 2.6 mg diosgenin equivalents/g [380]. A summary of the biological activities found in this species is provided in Table 10.

Table 10 Summary of biological studies on *Cynometra cauliflora*

Activity	Plant part	Findings	Target test	References
Antimicrobial	Leaves	The gelling agent HPMC showed the more potent growth inhibitor activity when compared to carbopol. Increasing the amounts of HPMC and carbopol enhanced viscosity while decreasing spreadability and the inhibitory effect on the test organism	Methicillin-resistant *S. aureus* (MRSA)	[365, 366]
	Fruits	At 100% concentration, *S. aureus* and *E. coli* gave 6.0 and 7.3 mm zones of inhibition, respectively	*S. aureus*, *E. coli*	[364]
	Endophytic bacteria in leaves, stems, and roots	Seven isolates revealed antibacterial activity, with the R1L3 and TKL2 extractives both showing inhibitory effects against *B. cereus*, *E. coli*, and *P. vulgaris*	*B. cereus*, *E. coli*, and *P. vulgaris*	[367]
	Flowers	An extract of *C. cauliflora* flowers displayed antibacterial activities against the three test organisms. The minimal inhibitory and bactericidal concentrations were 18.9 and 300 mg/cm^3, respectively	*S. aureus*, *M. luteus* and MRSA	[369]

(continued)

Table 10 (continued)

Activity	Plant part	Findings	Target test	References
	Leaves	*MIC* values ranged from 6.25 to 100 mg/cm^3	*S. aureus, M. luteus* and MRSA	[368]
	Leaves	A sample prepared from a 20% leaf ethanol extract gave a zone of inhibition of 13.9 mm	*S. aureus* ATCC 25923	[365]
	Leaves	Zones of inhibition were determined against the two test bacteria for both a *C. cauliflora* leaf extract and a gel preparation made from the leaves	*S. aureus, P. aeruginosa*	[370]
	Leaves, fruits and twigs	The *C. cauliflora* twig essential oil inhibited the growth of the microorganisms tested with inhibition zones ranging from 10.3 ± 0.4 to 29.7 ± 0.4 mm The oil displayed low *MIC* and *MBC* values against *S. aureus* (*MIC* 125 mg/cm^3; MBC 250 mg/cm^{3}L) and MRSA (*MIC* 125 mg/cm^3; *MBC* 250 mg/cm^3)	Four Gram-positive bacteria (*B. cereus, B. subtilis, S. aureus,* MRSA), three Gram-negative bacteria (*P. mirabilis, E. coli, K. pneumoniae*) and two yeasts (*C. utilis, C. albicans*)	[356]
Antioxidant	Healthy and ripe fruits	Of six Sri Lankan species tested, *C. cauliflora* exhibited the least potent radical scavenging and total antioxidant activities	DPPH assay	[371, 372]
	Leaves	A leaf extract showed an IC_{50} value of 26.9 µg/cm^3	DPPH assay	[360]
	Leaves, twigs and fruits	The twig essential oil (IC_{50} 37.1 ± 2.8 mg/cm^3) showed a higher antioxidant capacity than those from the leaves (IC_{50} 207.2 ± 3.0 mg/cm^3) and fruits (IC_{50} 461.9 ± 12.6 mg/cm^3) Ascorbic acid, as a positive control, gave an IC_{50} of 22.4 ± 0.8 mg/cm^3	DPPH assay	[356]
	Fruits	The IC_{50} of a methanol extract was 41.74 ± 0.24%	DPPH assay	[373]
	Fruits	An extract gave an IC_{50} value of 8.7 ± 0.3 mg/cm^3	DPPH assay	[374, 375]
Antidiabetic	Leaves	A methanol extract of *C. cauliflora* leaves substantially lowered blood glucose levels by 23–34%, plasma triglyceride levels by 13–30%, and liver glycogen levels by 68–96%, when compared to a control	Sprague Dawley rats aged 3.5–4 months, weighing 200–250 g each	[376]

(continued)

Table 10 (continued)

Activity	Plant part	Findings	Target test	References
	Leaves	Oral dosing of a *C. cauliflora* extract (400 or 200 mg/kg) reduced body weight, adipose tissue, and liver weight, as well as lipid buildup in the liver, when compared to a control high-fat diet group	Male C57BL/6 mice	[380]
	Leaves	A dose of the extract given 2500 mg per rat per day produced results comparable to metformin after 14 days	Male white rats (*R. norvegicus*) treated with alloxan	[377]
Toxicity/ cytotoxicity	Leaves	The extract tested gave a LC_{50} value of 314.6 μg/cm^3	*Artemia salina* larvae	[360]
	Leaves, twigs and fruits	In an vitro MTT experiment, the twig essential oil showed a growth inhibitory effect	Human breast cancer MCF-7 cells	[356]
	Leaves	The cytotoxic concentration (CC_{50}) of *C. cauliflora* extract was 36 mg/cm^3	Clinical isolate of herpes simplex virus type-1 (HSV-1)	[361]
	Leaves	The IC_{50} value was determined as 125.9 μg/cm^3 by 57.5%	HeLa cancer cells (ATCC CCL-2)	[378]
	Leaves	The toxicity of a leaf extract of *C. cauliflora* (LC_{50} = 196.1 ppm) was enhanced after combining with trigona honey (LC_{50} value = 168.2 ppm)	*Artemia salina* larvae	[363]
Plant growth-promoting factors	Endophytic bacteria in leaves, stems, and roots	Fifteen bacterial isolates from *C. cauliflora* were found to produce the plant hormone indole 3-acetic acid (IAA) Among them, R1S4 produced the most IAA (20.6 μg/cm^3). Genetic analysis determined that R1S4 was identified as *Sphingomonas* sp.	Plant growth promoting factor production	[367]
Birth anomalies	Fruits	At a dose of 291.7 mg/kg, a *C. cauliflora* fruit extract increased mortality at the embryonic day-3 stage by more than 50%. Furthermore, the fruit extract caused abnormalities such as in the anterior neuropore apertures and heart tubes, and also liver and kidney tissue malformations	Developing chick embryos	[379]

2.8 *Momordica charantia*

Momordica charantia L. (Cucurbitaceae), commonly known as bitter melon, is cultivated extensively in a wide region spanning parts of Asia, East Africa, and South America [381]. The plant is a slender, climbing species that belongs to the family Cucurbitaceae, and has garnered recognition for its biological and nutritional attributes [382]. This monoecious species has separate male and female flowers, both yellow, which grow from the leaf axils. The plant features long-stalked leaves that are simple and alternate, measuring 4–10 cm, with three to six deeply separated lobes. Its tendrils can either be unbranched or have two branches. The fruits of *M. charantia* are ovoid (Plate 8), resembling a small cucumber, and can be ellipsoidal or spindle-shaped. Each individual fruit is filled with flat seeds encased in pulp, typically with a ridged surface. The fruit can either split irregularly as a fleshy capsule or remain indehiscent. When young, the fruit is green but it turns orange upon ripening. As the fruit ripens, it splits into three irregular valves, curling backward to release numerous reddish brown seeds and as well as the pith, which turns red as the fruit matures [383].

Plate 8 Ripe bitter melon (*Momordica charantia*) (family Cucurbitaceae). Photograph courtesy Balaram Mahalder. (https://upload.wikimedia.org/wikipedia/commons/b/b5/Ripe_bitter_melon_%28Momordica_charantia%29.jpg)

The flowers of the *M. charantia* plant are solitary and feature a shallow hypanthium with five lobes on the calyx as well as five yellow petals, of which some have incurved scales at the base. Typically, three stamens are inserted towards the base of the hypanthium, with broad filaments. Two of these stamens are dithecal, while the other is monothecal, with curved or flexuous cells. The pistillate (female) flowers are usually solitary on a bracteate scape with a hypanthium that is ovoid to spindle-shaped. The perianth of the pistillate flowers is generally smaller than that of the staminate (male) flowers. The pistillate flowers contain numerous horizontal ovules and three stigmas, each with two lobes. Staminodes are either absent or present in a set of three. The seeds are few to numerous, ovate, and typically sculptured [383].

2.8.1 Ethnobotany

In Malaysia, the juice of *Momordica charantia* fruits is used for blood purification, as a gargle for abdominal disorders, and as a tonic for high blood glucose, chronic colitis, and dysentery [227, 265]. The cooked fruits are consumed to treat runny noses and coughs [227]. The leaves are ingested to improve appetite, relieve headaches, and stimulate lactation [17, 384]. A decoction of the leaves (or fruit juice) is used for constipation, fever, and as a vermifuge [17, 227, 265]. The crushed leaves, either as a poultice or ointment, are applied externally to treat skin diseases, burns, scalds, and stomachache in children [17, 265]. The leaves, when used as a decoction, also serve as a skin lotion and aid in wound healing, with the macerated fruit in oil being similarly applied for wound care [17, 265].

Infusions of the flowers are taken to treat asthma [265]. The roots are used as an astringent, abortifacient, and for the external treatment of hemorrhoids [265, 384]. The plant is also used to treat diarrhea in children, as well as flatulence and delirium [265, 385]. Apart from its traditional medicinal uses, the fruits of *M. charantia* are consumed widely in various culinary forms, including being used raw in salads, or juiced, stir-fried with vegetables or meats, or added to soups. The fruits of *M. chariantia* are also commonly pickled or deep-fried to mitigate their natural bitterness.

2.8.2 Phytochemistry

To date, more than 220 different compounds have been identified from different parts of *Momordica charantia* [383]. An aqueous extract of the fruits was reported to contain amino acids, carbohydrates, cardiac glycosides, flavonoids, proteins, terpenoids, saponins, and other constituents. Phytochemical screening on the fruit juice of *M. charantia* has identified various classes of phytochemicals including carbohydrates, flavonoids, phenols, saponins, sterols, and terpenoids [386]. In addition, *M. charantia* accumulates a diverse array of cucurbitacins, a class of secondary metabolites found typically in plants of the family Cucurbitaceae. These compounds share a structural similarity to lanostane derivatives, and are distinguished primarily by the presence of a C-19 methyl group at the C-9 position, as opposed to C-10 in lanostane. Accordingly, cucurbitacins possess a 19-(10$\alpha \rightarrow 9\beta$)-*abeo*-10$\alpha$-lanost-5-ene type ring system [381]. Bitter gourd, like many other species within the Cucurbitaceae family, contains a plethora of triterpenoids, based on the cucurbitane skeleton (**373**), with approximately 200 distinct structures identified to date [387]. These triterpenoids commonly are referred to as momordicosides, although other names such as karavilosides, kuguacins, and goyaglycosides also have been applied [388, 389]. Some important triterpenoids that have been isolated from the plant include charantin (**374**), karaviloside IX (**375**), and momordiscosides A (**380**) and B (**381**) and Q–T (**376–379**). Other types of compounds that have been reported from *M. charantia* include the saponins momordin (**371**), momorcharin (**372**), the nitrogen-containing compounds vicine (**387**), ascorbigen (**388**), pipecolic acid (**390**), 5-hydroxytyptamine (**391**), and the flavonoid luteolin (**389**). Several miscellaneous amino acids (**382–386**) have also been reported from the plant. The structures of the above-mentioned constituents of *M. charantia* are shown in the formula drawings **371–391**.

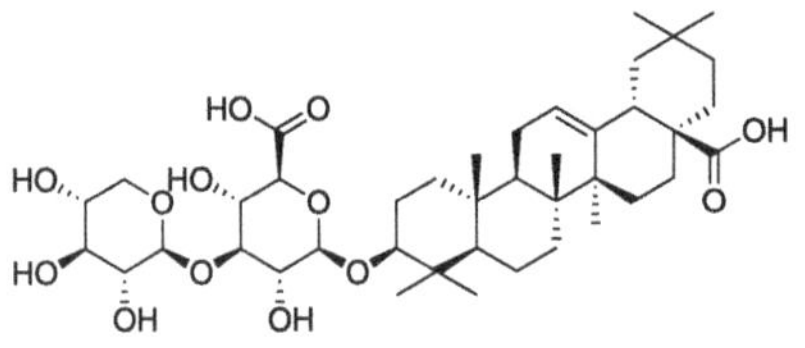

371 (momordin)

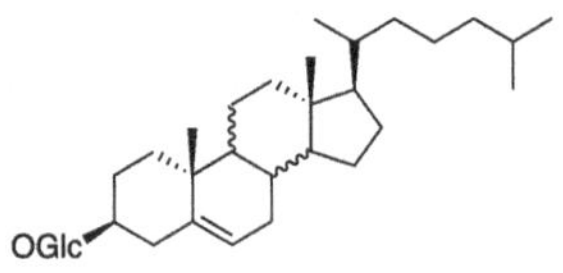

372 (momorcharin)

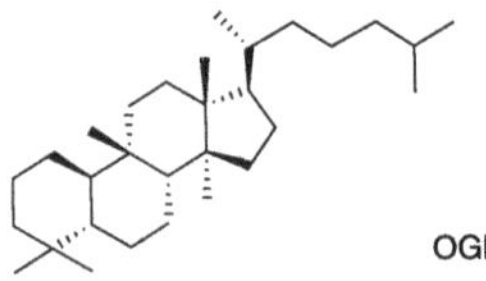

373 (cucurbitane)

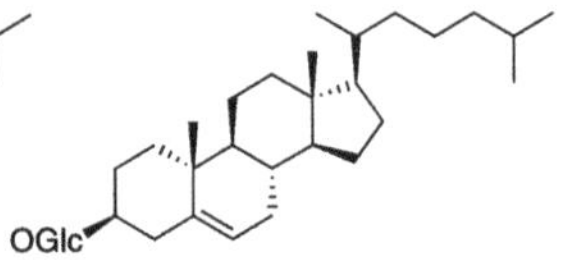

374 (charantin)

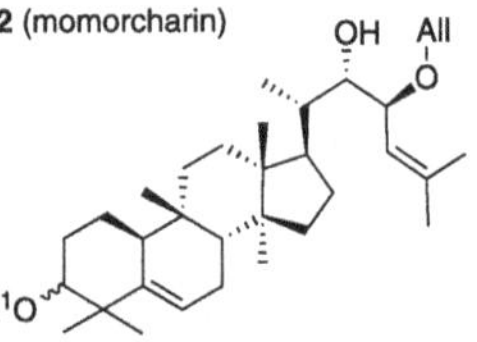

375 (karaviloside IX) R = β-D-glucopyranosyl, All = β-D-allopyranosyl

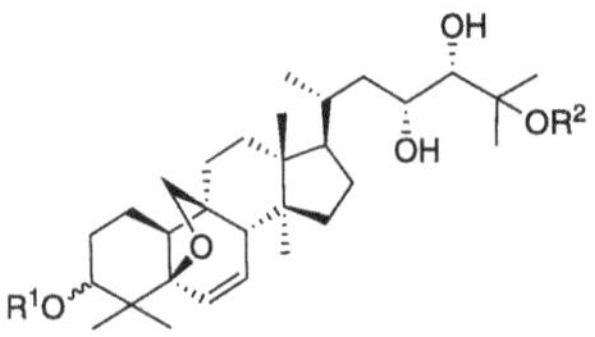

376 (momordicoside Q) R^1 = Glc', R^2 = H
377 (momordicoside R) R^1 = All', R^2 = Glc"
378 (momordicoside S) R^1 = Glc(1→6)-Glc, R^2 = Glc"'
379 (momordicoside T) R^1 = Xyl""'(1→4)-[Glc"(1→6)-Glc, R^2 = Glc""

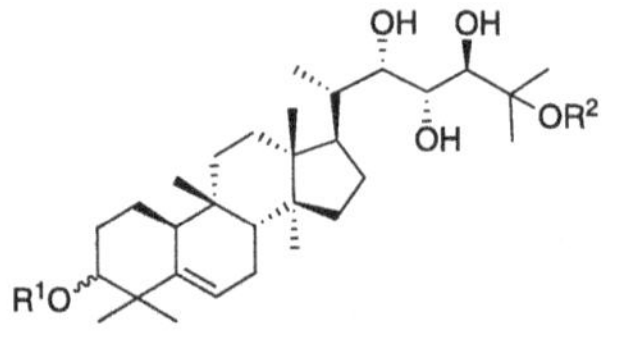

380 (momordicoside A) R^1 = Glc"(1→6)-Glc', R^2 = H
381 (momordicoside B) R^1=Xyl""'(1→4)-[Glc"(1→6)-Glc, R^2 = H
Glc = β-D-glucopyranosyl, Xyl = β-D-xylopyranosyl

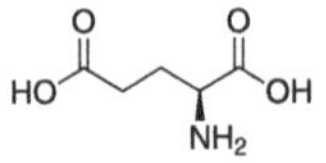

382 (glutamic acid)

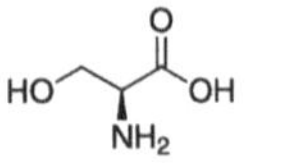

383 (serine)

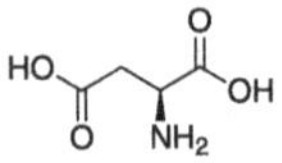

384 (aspartic acid)

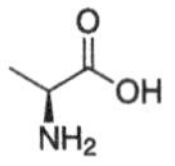

385 (alanine)

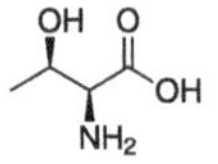

386 (threonine)

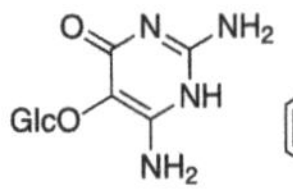

387 (vicine)

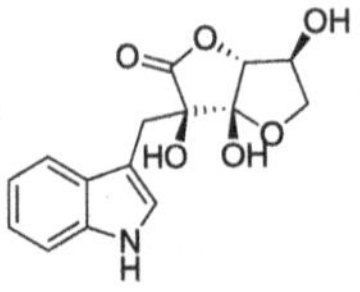

388 (ascorbigen)

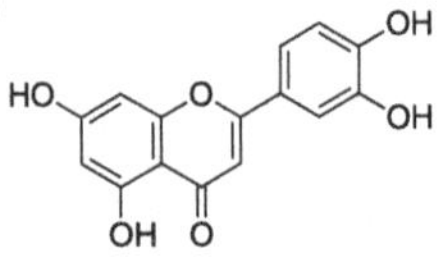

389 (luteolin)

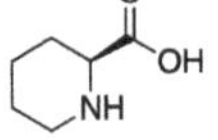

390 (pipecolic acid)

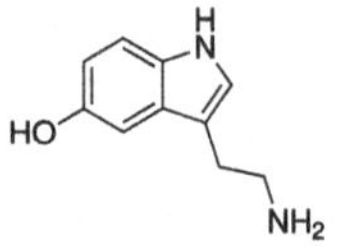

391 (5-hydroxytryptamine)

2.8.3 Biological Activities

Momordica charantia polysaccharides have been investigated for their immunomodulatory activity. One key indicator of immune activation is the proliferation of splenocytes and thymocytes. Hence, when *Momordica charantia* polysaccharides were administered in a test system at a dose of 200 μg/cm^3, the proliferation index of splenocytes was significantly higher compared to that of a control group. Additionally, *Momordica charantia* polysaccharides at a dose of 25 μg/cm^3 exhibited an optimum potency in promoting thymocyte proliferation. These findings suggest a dose-dependent effect of *Momordica charantia* polysaccharides on immune cell proliferation [390]. Further research has confirmed the efficacy of *M. charantia* polysaccharides, specifically MCP1 and MCP2, in promoting the proliferation of spleen lymphocytes. In vitro experiments demonstrated that both normal spleen lymphocytes and those induced by concanavalin A can proliferate effectively in the presence of *Momordica charantia* polysaccharides [389]. Also, silver nanoparticles using an aqueous extract of *M. charantia* L. leaf demonstrate a powerful inhibitory effect on the growth of breast cancer MCF-7 cells, and may have potential applications in the treatment of metastatic breast cancer [397].

Several studies have probed the use of various *M. charantia* extracts for their antidiabetes activity [381, 387, 389, 398]. Thus, *M. charantia* has been accorded with potent antidiabetic properties, as evidenced by studies using cell-based assays, animal models, and human clinical trials [391–393]. The plant parts investigated in this regard include bitter gourd fruits, leaves, peel, juice, powder, and extracts obtained from this plant using various solvents. Several studies have found that different parts or dosages of *M. charantia* in type 1 and type 2 diabetes induced by streptozotocin (STZ), and type 2 diabetes induced by a high-fat diet in rats, have a hypoglycemic effect, which caused a significant reduction of serum glucose, total cholesterol, triglyceride levels, and insulin resistance index [386, 390]. Studies have indicated that *M. charantia* may exert its hypoglycemic action through diverse pathways, including protecting islet β cells, promoting insulin secretion, improving insulin sensitivity, and inhibiting α-amylase and α-glucosidase. Among the known constituents of this plant, polysaccharides have been found to possess effective antidiabetic activities. Hence, a polysaccharide extracted from *M. charantia* (MCPIIa) that was orally administered (100, 200, or 300 mg/kg body weight) to STZ-induced diabetic mice for four weeks significantly reduced glycemia and increased serum insulin [393, 394].

Different extracts from *Momordica charantia*, as well as the phenolic constituents of the fruits and seeds [395], particularly those rich in phenolic acids, have been characterized for their antioxidant activities, showing DPPH radical scavenging activity and effectiveness in various models for scavenging free radicals [382]. The phenolic compounds in *M. charantia* are crucial to its antioxidant properties, underscoring its potential as a natural antioxidant source for therapeutic purposes in managing oxidative stress-related conditions. *Momordica charantia* extracts activate peroxisome proliferator-activated receptors, leading to the upregulation of genes associated with antioxidant activity in hepatoma cells and mouse models [395–398]. The triterpenoids in *Momordica charantia* provide cytoprotective activity against induced hepatotoxicity in cell models [395].

Momordica charantia also exhibits antimicrobial activity. Thus, an ethanol extract of *M. charantia* leaves inhibited the growth of *Bacillus cereus* and *Streptomyces aureus* [381, 386]. An in vivo experiment confirmed that methanol and ether extracts of *M. charantia* leaves effectively prevented a *Salmonella typhi* infection, and improved antibody titer in a mouse model, enhanced leukocyte mobilization, and reduced colonization and infection rates in the blood [394, 395]. An ethanol extract of *M. charantia* leaves and stems inhibited the HSV-1 and SINV viruses. Proteins and sterols from *M. chariantia* have shown antiviral effects, and stimulate the production of MAP30, a protein that suppresses HIV activity, and selectively kill HIV-infected lymphocytes and macrophages [387]. The bioactive components present also exhibit activity against *Helicobacter pylori* [391, 393], Sindbis virus, *Herpes simplex* virus type 1, and anthelmintic activity against *Caenorhabditis elegans* [387, 388]. The antimicrobial activity of *M. charantia* has been attributed to its antimicrobial proteins, seed oil, tannins, triterpenoids, alkaloids, cardiac glycosides, and sterols.

In conclusion, *M. charantia* produces a variety of of biologically active metabolites of the types indicated above from its various plant parts.

2.9 Hibiscus sabdariffa

The blooming herb or subshrub known as *Hibiscus sabdariffa* L. (Plate 9), or roselle, is a perennial plant belonging to the family Malvaceae. Roselle is thought to have originated in West Africa, and was brought to Asia and the West Indies in the sixteenth and early seventeenth centuries, where it subsequently became naturalized. It is referred to as "sorrel" in many English-speaking Caribbean regions, "Florida Cranberry" or "Jamaica sorrel" in the United States, "Saril" or "Flor de Jamaica" in Central America, "Yakuwa" by the Hausa people of northern Nigeria, "Bissap in Wolof" in Senegal, "Wegda" in the Mossi area of Burkina Faso, and "Karkadeh" in Arabic. The top two producers of *H. sabdariffa* worldwide are China and Thailand, with Egypt, Jamaica, Mali, Mexico, Nigeria, Senegal, Sudan, and Tanzania also being significant suppliers. In 2020, the global roselle market was valued at $122.8 million, and it is expected to reach $252.6 million by 2030 [399].

Hibiscus sabdariffa is an herbaceous plant that can be either annual or perennial, with a simple or branched habit. The stems are covered with sparse, simple, bulbous, and spiny hairs. The leaves are ovate in shape at the base, becoming 3–7 lobed in the upper part of the plant, with the lobes being elliptic-lanceolate, coarsely serrated, and nearly glabrous on both sides. The stipules are filiform, measuring 6–8 mm long, and the petiole is 4–15 cm in length, often with spiny protuberances near the top. The flowers are axillary and solitary, and typically subsessile. The epicalyx consists of seven to eight linear segments, each 6–10 mm long, and sparsely covered in prickles. The calyx is fused at the base, measuring 1–2.5 cm in length, woolly, and setose or prickly. The calyx lobes are long, acuminate-aristate, and feature a swollen, linear

Plate 9 Fruits and leaves of *Hibiscus sabdariffa* (family Malvaceae). Photograph courtesy Mokkie (https://upload.wikimedia.org/wikipedia/commons/7/7c/Roselle_%28Hibiscus_sabdariffa%29.jpg)

gland along the central nerve at the base. The corolla is yellow with a crimson center, 3–5 cm in diameter, and the petals are obovate, measuring 4–6 cm long and 2–4 cm wide. The staminal column is prominently inserted. The fruit is a conical capsule, 1.5–2 cm long and about 1 cm wide, beaked and covered with appressed setose hairs. The seeds are numerous, 2–3 mm in length, and brown in color [400, 401].

2.9.1 Ethnobotany

Hibiscus sabdariffa calyces, either fresh or dried, are used to prepare a variety of drinks, including wine, other fermented preparations, hot and cold beverages, and herbal teas. In Malaysia, the whole plant, including the calyces, is used traditionally to prepare decoctions for their diuretic, tonic, and antiscorbutic properties. On the other hand, the leaves are known for their emollient effects, and a poultice made from these is applied to abscesses and ulcers [402]. The leaves are also cooked as a vegetable added to traditional Malaysian dishes, such as "asam pedas" (a spicy and sour fish stew) and "singgang ikan" (a tangy fish soup or stew), wherein they impart a distinctive sour flavor. In addition, the leaves of *H. sabdariffa* are utilized in the production of cakes, puddings, ice cream, chocolates, jams, and jellies [403].

Hibiscus sabdariffa has a wide range of ethnomedical uses within the different regions where it occurs. In Africa, India, and Latin America, infusions of the leaf or calyx are valued for their diuretic, choleretic, febrifugal, and hypotensive properties, and for promoting gastrointestinal motility and reducing blood viscosity. In Senegal, such infusions are used specifically as a hypotensive, while in Egypt, formulations prepared from the calyx of *H. sabdariffa* are utilized to treat neurological and heart disorders and to support urine output. In Egypt and Sudan, such infusions have use to lower the body temperature. In Guatemala, the plant is used to treat intoxication, while in North Africa, the leaf pulp is employed to heal external wounds and abscesses. Calyx preparations are also used for coughs, sore throats, and genital issues. In India, seed decoctions have use to treat dyspepsia and urinary discomfort, while in Brazil, the roots are considered as emollient and stomach-soothing. In traditional Chinese medicine, *H. sabdariffa* is employed for liver conditions and high blood pressure, while in Iran, sour hibiscus tea is used for hypertension. In Nigeria, seed decoctions of *H. sabdariffa* are believed to enhance lactation [403].

2.9.2 Phytochemistry

Hibuscus sabdariffa produces common organic acids, including citric acid, ascorbic acid, maleic acid, hibiscic acid, oxalic acid, and tartaric acid, which contribute to its distinct sour taste [399]. The calyces also produce amino acids, anthocyanins, minerals, and organic acids [404, 405], while the seeds contain quantities of lipids and proteins [405–407]. With a 17–20% oil content, *H. sabdariffa* seeds represent an affordable edible oil source that includes a high concentration of unsaturated fatty acids.

A previous study investigated how crushing, heating, and fermentation affected the nutritional and phytochemical content of *H. sabdariffa* seeds. The results showed that heating significantly degraded key phytochemicals in the seeds, including flavonoids, phlobatannins, and other phenolic compounds [405, 407]. The flavonol, gossypetin (**393**), and the anthocyanin hibiscin (**395**) are the flavonoids that are most prevalent in roselle petals. This plant has diuretic properties that lower blood pressure, stimulate bowel movements, and decrease blood viscosity [404]. Some significant phytochemicals (**392–396**) isolated from *H. sabdariffa* are shown in the formula drawings.

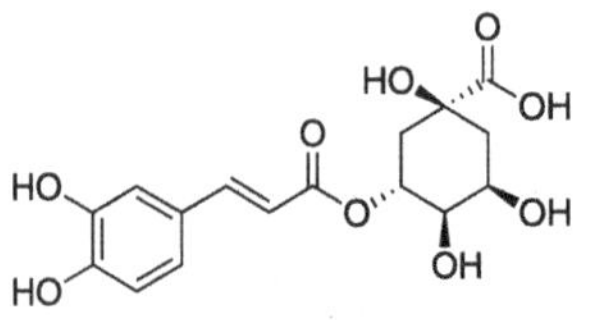

392 (chlorogenic acid)

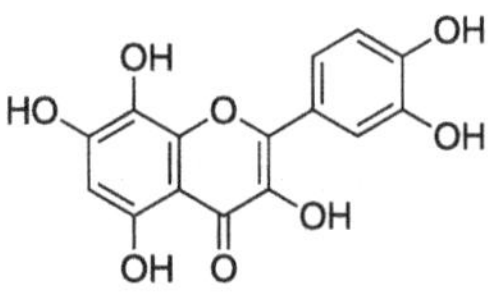

393 (gossypetin)

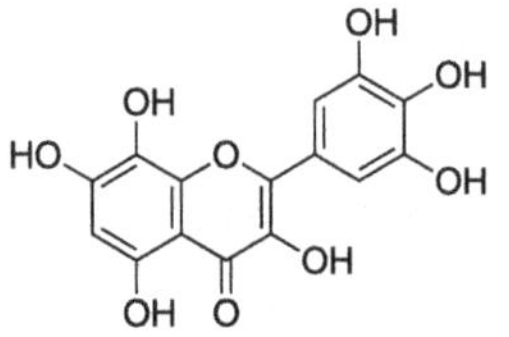

394 (hibiscetin)

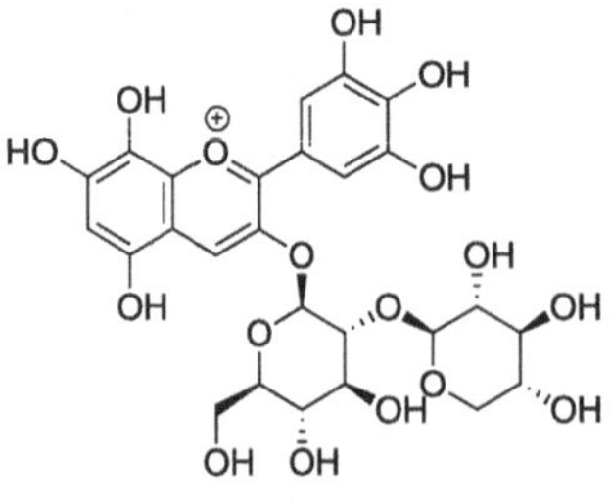

395 (hibiscin)

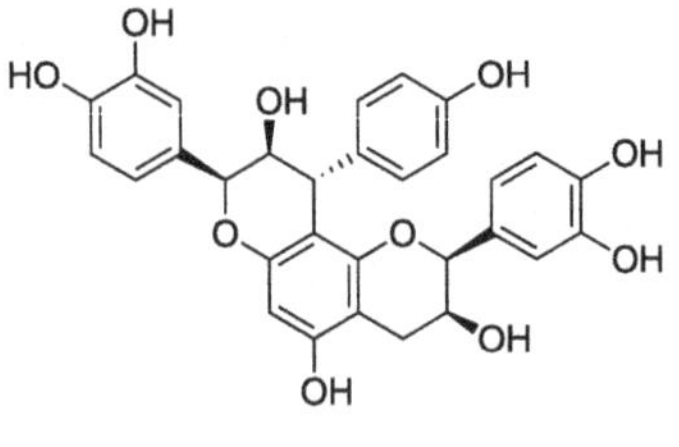

396 (phlobatannin)

2.9.3 Biological Activities

Among the different types of biological activity investigated thus far, the calyx samples of *H. sabdariffa* have been studied for their antihypertensive activity [404, 408, 409]. According to the research reported, the consumption of hibiscus tea for six weeks reduced average systolic and diastolic readings by 3.2 and 3.1 mm/Hg, respectively [404, 410]. Its diuretic qualities, which aid in reducing blood volume, and its capacity to suppress the activity of the angiotensin-converting enzyme, which helps regulate blood pressure, are two of the parameters that explain such efficacy in decreasing blood pressure. Furthermore, the anthocyanin and other flavonoid constituents of *H. sabdariffa* enhance vasodilation, which lowers blood pressure [411].

The effectiveness of a 70% (*v*/*v*) methanol extract of *H. sabdariffa* leaves in a lipopolysaccharide-treated murine macrophage RAW 264.7 cell for an anti-inflammatory model. Nitric oxide synthase was shown to be dose-dependently inhibited in the cell type used, with inhibition levels ranging from 9 to 58%. The main components found in this roselle leaf extract, such as quercetin (**85**), kaempferol (**88**), and chlorogenic acid (**392**), are supportive of its promising anti-inflammatory action [399, 405]. These compounds may aid in preventing the synthesis of pro-inflammatory mediators, which are important for the inflammatory response and include cytokines and enzymes like COX-2.

Recent investigations of the cancer-related potential of *H. sabdariffa* by cytotoxic and apoptotic induction assays were undertaken [412, 413], with such activities being related to the flavonoids and terpenoids present [408]. Also, the leaves, stems, and roots of *H. sabdariffa* are abundant in flavonoids, which could explain their anti-inflammatory, anti-fever, antioxidant, and spasm-inhibiting properties. The flowers are also well-known for their calming qualities, which enable them to effectively treat inflammation, reduce spasms, ease menstrual cramps, and lessen overall cramping [414].

Accordingly, studies have demonstrated that extracts of *H. sabdariffa*, particularly from its calyces, may have use in the management and/or prevention of various ailments. The antioxidant properties demonstrated to date are consistent with an ability to help combat oxidative stress, which is a common underlying factor in many chronic diseases. The antihypertensive effects are due mainly to the diuretic action of the plant, and its ability to inhibit the angiotensin-converting enzyme [414].

2.10 *Zingiber zerumbet*

Zingiber zerumbet (L.) Roscoe ex Sm. (Plate 10), commonly known as "bitter ginger" or "shampoo ginger", is a well-recognized member of the family Zingiberaceae [415], which is renowned for its aromatic and medicinally valuable species, including such examples as *Zingiber officinale* (common ginger) and *Curcuma longa* (turmeric). However, *Z. zerumbet* has distinctive properties and uses, and hence has attracted

Plate 10 Leaves and inflorescences of *Zingiber zerumbet* (family Zingiberaceae). Photograph (https://commons.wikimedia.org/wiki/File:ZingiberCrucero.jpg)

significant attention ethnomedically. As a perennial herbaceous plant, *Z. zerumbet* is native to Southeast Asia and is characterized by its preference for humid, tropical climates. This is due largely to the plant's evolutionary adaptation to moist environments, where rainfall is abundant and temperatures remain consistently warm, facilitating the plant's growth and survival in dense, forested areas [416].

The distribution of *Z. zerumbet*, known folklorically as "Asian ginger" or" 'bitter ginger", spans across a wide geographic range, since it has been now introduced to several in tropical and subtropical regions around the world. The plant is distributed mainly in India, Bangladesh, Malaysia, Nepal, and Sri Lanka, and grows naturally in damp, shaded parts of lowland or hill slopes [417]. This wide distribution range highlights its resilience and adaptability to humid environments characterized by high temperatures and abundant rainfall. Over time, the plant has extended beyond its native range through naturalization and cultivation, largely due to its valuable medicinal and ornamental uses. Notably, *Z. zerumbet* has become established in other tropical and subtropical regions around the world, including the Pacific Islands, Hawaii, the Amazon basin in Brazil, and parts of Africa [418, 419].

In the Amazon region of Brazil, *Z. zerumbet* also has been incorporated into traditional medicine practice by indigenous communities. The plant's adaptability to the Amazonian climate, which mirrors the tropical conditions of Southeast Asia, has allowed it to flourish in this region. In Brazil, *Z. zerumbet* is used to treat a range of ailments, from digestive issues to inflammatory diseases. The plant's utilization in Brazil underscores its importance as a global medicinal resource, bridging traditional knowledge across continents and cultures [418].

The morphology of *Z. zerumbet* (Plate 10) is defined by its upright, clumping growth habit, which can reach heights of up to two meters. The plant's narrow, lance-shaped leaves typically are arranged in an alternate pattern along the stems [420]. However, it is the rhizomes that hold the most interest for the medicinal and scientific communities. When cut, the rhizomes emit a strong aroma, a characteristic feature attributed to the presence of essential oils within its tissues. The distinctive scent is often considered as an indicator of the biological potency of the rhizomes [416].

Zingiber zerumbet is distinguished by a pulvinus located between the base of its petiole and ligule. This variegated wild ginger species features upright, rounded stems that grow to a height of 1–2 m. These stems are annual and encased in smooth leaf sheaths. Its leaves and inflorescence emerge from a thick, knobby rhizome and a subterranean stem situated just beneath the soil surface. The leaves, often tinged with purple underneath young shoots, are slender, measuring about 25–35 cm in length, with prominent midribs on the underside. The petioles are around 6 mm long, while the ligules are broad, delicate, and measure 1.5–2.5 cm in length. The leaflets are arranged alternately along an arching pseudostem of 1–2 m. The rhizomes are the plant's most notable feature, and are thick, scaly, aromatic, and pale-yellow inside, and serve as the sole means of propagation for this asexual reproducing species [415].

2.10.1 Ethnobotany

The practical applications of *Z. zerumbet* extend beyond traditional medicine, as the plant is being utilized increasingly in foods, beverages, and even cosmetics. The liquid found within its cone-shaped bracts has earned the plant the nickname "shampoo ginger" for the milky substance present, and has a long-established use as a natural hair cleanser [421]. The diverse range of health-related applications of *Z. zerumbet* are described in Ref. [416].

The rhizomes of *Z. zerumbet* are macerated or boiled to produce medicinal decoctions. In Southeast Asia, particularly in Malaysia, Indonesia, and Thailand, *Z. zerumbet* has been used for centuries to treat a wide array of ailments, from digestive disorders to inflammatory conditions. Traditional practitioners often employ the rhizomes in various forms, including decoctions, poultices, and infusions. The variety of preparation methods reflects the plant's adaptability to different cultural contexts and to its perceived effectiveness in treating both internal and external health maladies [415, 416].

In many cultures where *Z. zerumbet* is indigenous or has been introduced, the plant plays a multifaceted role beyond having only medicinal applications. The rhizomes, in particular, represent a foundational element in traditional healing practices in Southeast Asia, where they are valued not only for their therapeutic properties but also for their functional benefits in daily life. Its diverse uses in regions of Indonesia,

Malaysia, Thailand, and beyond highlight a blend of medicinal, cultural, and ritualistic applications that underscore the plant's broad significance. Traditional healers commonly use the rhizomes of *Z. zerumbet* in several forms, with each form serving specific medicinal purposes, tailored to the needs of the person or the condition being treated. For example, a rhizome paste is often used topically to relieve pain and inflammation in conditions such as arthritis or injuries, while decoctions made by boiling the rhizomes in water are ingested to alleviate digestive disorders, infections, and fevers [415, 418].

In Malaysia, *Z. zerumbet* is locally known as "Lempoyang", where it holds a prominent place in traditional medicine. The rhizomes of *Z. zerumbet* are utilized there for treating digestive ailments, and for pain relief, and to reduce inflammatory conditions. Traditional applications include using the plant's essential oil for its antinociceptive effects, in order to alleviate pain such as toothache as well as muscle sprains [422]. Additionally, the plant's bioactive compounds show antioxidant, antimicrobial, and anti-inflammatory effects, which align with its role in traditional Malay healthcare for the treatment of infections and overall health maintenance [423]. The rhizomes are used as appetizer and food flavoring in various types of Malaysian cuisine. Furthermore, an extract of the rhizomes is used in Malay traditional medicine to treat inflammatory- and pain-mediated diseases, worm infestations, and diarrhea [424].

Another culturally significant use of *Z. zerumbet* is evident in mountains of Hawaii, where it is known as "Awapuhi" or "Shampoo ginger". Native Hawaiians have used traditionally the plant's liquid-rich flower bracts as a natural hair dressing and body cleanser. The viscous liquid is squeezed out from the bracts and applied directly to the hair or skin, where it serves as a moisturizer and cleanser [424]. In India, *Z. zerumbet* is known as "Yaiimu", and the rhizome powder is mixed with ripe *Morinda citrifolia* for the treatment of severe pain. The cooked and softened rhizomes of *Z. zerumbet* are used to treat toothache, coughs, asthma, worm infestations, leprosy, and other skin diseases, and the ground and strained rhizomes are mixed with water and drank to treat stomachache [424].

In Thai traditional medicine, *Z. zerumbet* (known as "Hiao dam") is also an important ingredient in herbal compresses and balms, as used in traditional Thai massage. The anti-inflammatory and analgesic properties of *Z. zerumbet* make it an effective natural treatment for muscle pain and joint stiffness. Additionally, the rhizome extract is applied in poultices for skin infections, underscoring the plant's antimicrobial capabilities, which are supported by modern research [415].

2.10.2 Phytochemistry

The phytochemical profile of *Z. zerumbet* is diverse, containing many types of bioactive compounds, such as terpenoids, phenols, and alkaloids.

Dev isolated in 1962 initially zerumbone (**397**) from the essential oil of *Z. zerumbet* and elucidated its structure [425]. In a subsequent study, Nigam and Levi analyzed in 1963 eight essential oil fractions and identified several key terpenes, including borneol (**398**), α-pinene (**399**), camphor (**400**), linalool (**401**), zerumbone (**397**), limonene (**402**), α-humulene (**403**), and β-caryophyllene (**404**) [425, 426]. Further studies by Damodaran and Dev in 1968 characterized additional sesquiterpenes, such as humulene epoxides I (**405**), II (**406**), and III (**407**), and humulenols I (**408**) and II (**409**) as well as caryophyllene oxide (**410**) [427].

More recent studies have expanded on these earlier findings. In 2013, Dai et al. identified 46 compounds in the essential oil of *Z. zerumbet* rhizomes, with the more abundant components including zerumbone (**397**) (11.0%), sabinene (**412**) (33.0%), and β-myrcene (**421**) (13.3%) [428]. Their analysis revealed that the essential oil sample studied consisted predominantly of monoterpenes (76.1%), including camphene (**411**) (16.3%), sabinene (**412**) (14.6%), citral (**413**) (26.1%), zingiberene (**414**) (7.2%), and lavandulyl acetate (**415**) (6.7%). In another study, the essential oil demonstrated a lower zerumbone content (1.2%). Zerumbone (**397**) is a monocyclic sesquiterpene with three double bonds, of which two are conjugated with a carbonyl group [429–431].

In 2009, Bhuiyan et al. confirmed through GC–MS analysis the presence of terpenes, such as α-pinene (**399**), camphor (**400**), linalool (**401**), zerumbone (**397**), limonene (**402**), camphene (**411**), β-caryophyllene (**404**), 3-carene (**416**), 4-terpineol (**417**), and eucalyptol (**418**) [432]. In a similar study to quantify the constituents present, it was found that sesquiterpenes were the most abundant, with zerumbone (**397**) (36.1%) being the predominant compound, followed by humulene (10.0%) [433]. Gas chromatography-MS analysis also was used to identify and quantify monoterpenes such as borneol (**398**) (4.8%), zingiberene (**414**) (3.7%), camphor (**400**) (4.2%), linalool (**401**) (1.1%), camphene (**411**) (14.39%), eucalyptol (**418**) (3.9%), α-terpinene (**419**) (2.0%), and α-phellandrene (**420**) (1.6%). Yu et al. also confirmed in 2008 that sesquiterpenes constitute 85.8% of the essential oil, with zerumbone (**397**) (48.1%) and α-humulene (**403**) (17.2%) being the most prominent compounds, alongside other sesquiterpenoids in lower concentrations [421].

Chien et al. isolated in 2016 various flavonoids from the rhizome of *Z. zerumbet*, including kaempferol-3-*O*-(2,4-di-*O*-acetyl-*α*-L-rhamnopyranoside) (**422**) and kaempferol-3-*O*-(3,4-di-*O*-acetyl-*α*-L-rhamnopyranoside) (**423**) [435]. These phenolic compounds were extracted using ethyl acetate and identified via HPLC–DAD [434]. The composition of *Z. zerumbet* extract was also analyzed, showing the presence of several other flavonoids, including kaempferol-3-*O*-rhamnoside (**424**), kaempferol-3-*O*-(2-*O*-acetyl)rhamnoside, kaempferol-3-*O*-(3-*O*-acetyl)rhamnoside and kaempferol-3-*O*-(4-acetyl)rhamnoside (**425**) [423]. Jang et al. also investigated the aromatic compounds and flavonoids from *Z. zerumbet*, isolating kaempferol-3,4-*O*-dimethylether (**426**), kaempferol-3-*O*-methylether (**427**), kaempferol-3,4,7-*O*-trimethylether (**428**), 4-*O*-acetylafzelin (**429**), 2,4-*O*-diacetylafzelin (**430**), and 3,4-*O*-diacetylafzelin (**431**) [435]. They noted that the composition of *Z. zerumbet* can vary depending on its cultivation region, leading to different concentrations of these compounds [432]. The structures of the compounds isolated from *Z. zerumbet* are shown in their formula drawings.

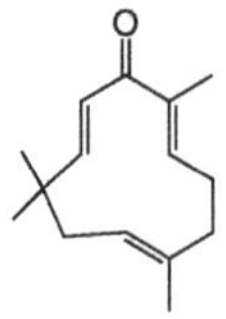
397 (zerumbone)

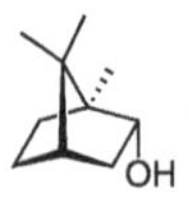
398 (borneol)

399 (α-pinene)

400 (camphor)

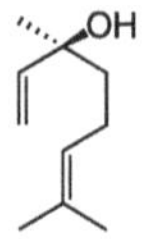
401 (linalool)

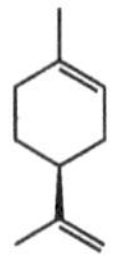
402 (limonene)

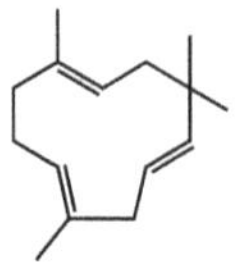
403 (α-humulene)

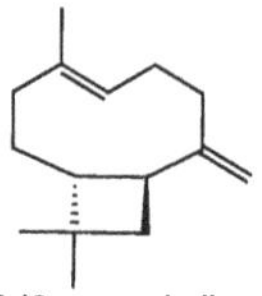
404 (β-caryophyllene)

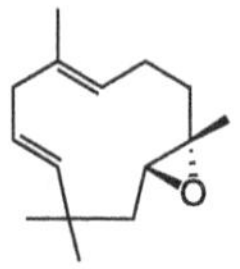
405 (humulene epoxide I)

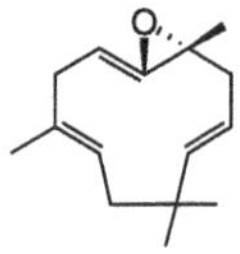
406 (humulene epoxide II)

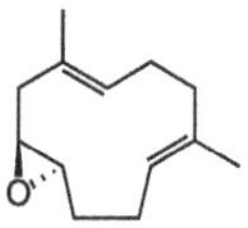
407 (humulene epoxide III)

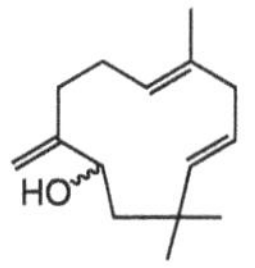
408 (humulenol I)

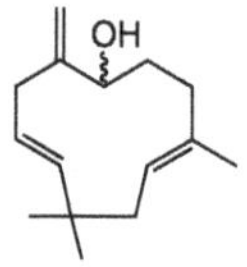
409 (humulenol II)

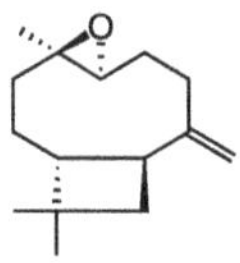
410 (caryophyllene oxide)

411 (camphene)

412 (sabinene)

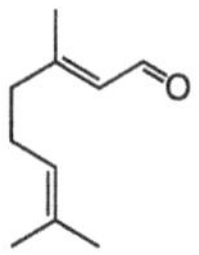
413 (citral)

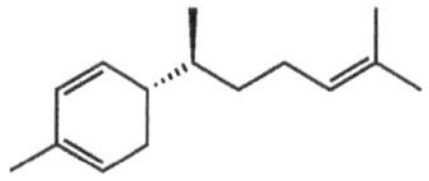
414 (zingiberene)

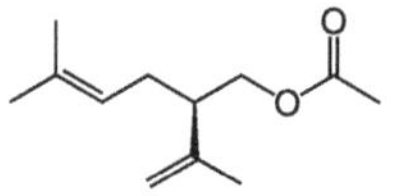
415 (lavandulyl acetate)

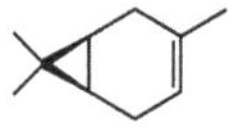
416 (3-carene)

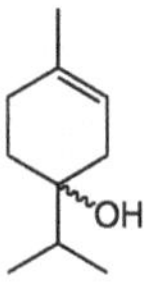
417 (4-terpineol)

418 (eucalyptol)

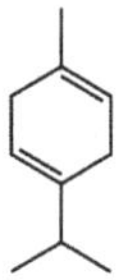
419 (α-terpinene)

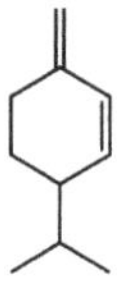
420 (α-phellandrene)

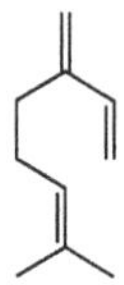
421 (β-myrcene)

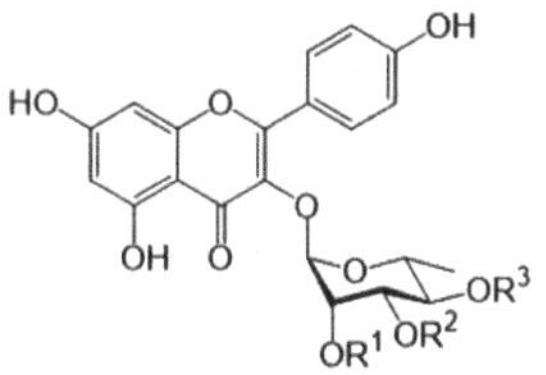

422 (kaempferol-3-*O*-(2,4-di-*O*-acetyl-α-L-rhamnopyranoside)) $R^1 = R^3 = Ac$, $R^2 = H$
423 (kaempferol-3-*O*-(3,4-di-*O*-acetyl-α-L-rhamnopyranoside)) $R^1 = H$, $R^2 = R^3 = Ac$
424 (kaempferol-3-*O*-rhamnoside) $R^1 = R^2 = R^3 = H$
425 (kaempferol-3-*O*-(4-acetyl)rhamnoside) $R^1 = R^2 = H$; $R^3 = Ac$

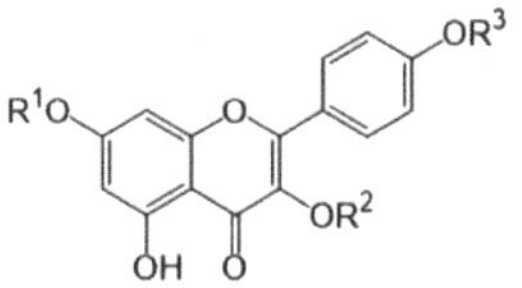

426 (kaempferol-3,4-*O*-dimethylether) $R^1 = H$, $R^2 = R^3 = CH_3$
427 (kaempferol-3-*O*-methylether) $R^1 = R^3 = H$, $R^2 = CH_3$
428 (kaempferol-3,4,7-*O*-trimethylether) $R^1 = R^2 = R^3 = CH_3$

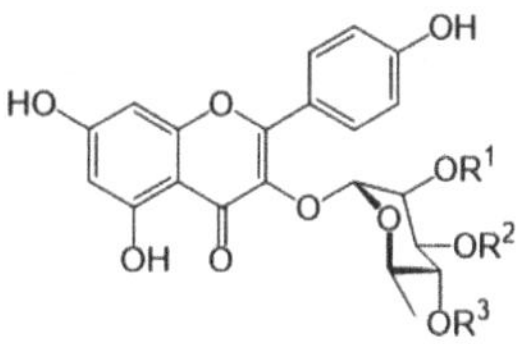

429 (4-*O*-acetylafzelin) $R^1 = R^2 = H$, $R^3 = Ac$
430 (2,4-*O*-diacetylafzelin) $R^1 = R^3 = Ac$, $R^2 = H$
431 (3,4-*O*-diacetylafzelin) $R^1 = H$, $R^2 = R^3 = Ac$

2.10.3 Biological Activities

The main bioactive compound found in the ethanol extract and essential oil is zerumbone (**397**) (2,6,9,9-tetramethyl-(2*E*,6*E*,10*E*)-cycloundeca-2,6,10-trien-1-one), which is a crystalline compound [436]. Zerumbone (**397**) is recognized for its multi-targeted therapeutic potential, including having anti-inflammatory, antioxidant, and cancer-related effects [437]. This sesquiterpene interacts with cellular targets involved in inflammation [438] cancer progression [439], antibacterial activity [440], and oxidative stress [441]. Another major constituent in the *Z. zerumbet* rhizome essential oils is the sesquiterpene α-humulene (**403**) [417]. This exhibited an anti-asthmatic activity by reducing eosinophil recruitment into the airways of BALB/c mice induced with allergic inflammation [442].

In addition to sesquiterpenes, the rhizomes of *Z. zerumbet* are a source of monoterpenes such as camphene (**411**), which exhibits anti-inflammatory, antioxidant, and antimicrobial effects. Camphene (**411**) is able to to interact with respiratory biological targets, consistent with the traditional use of *Z. zerumbet* in treating coughs and colds in Southeast Asian and Pacific cultures. Monoterpenes like **411** enhance the overall efficacy of *Z. zerumbet* extracts by providing a broad pharmacological profile [418].

Phenolic compounds in *Z. zerumbet*, including kaempferol (**88**) and other flavonoids, are known to possess in vitro and in vivo activities via their antioxidant [438], antimicrobial [439], antimutagenic [440] and cytotoxic effects [416, 441].

Thus, the pharmacological properties of *Z. zerumbet* and its main bioactive compound, zerumbone (**397**), have been studied widely, focusing primarily on their analgesic, anti-inflammatory, antidiabetic, antihyperlipidemic, and cancer-related effects [443–447]. Additional studies have explored their antioxidant [448], antipyretic, hepatoprotective, nephroprotective, and gastroprotective properties, as well as their potential to reduce locomotor activity. Among these, the analgesic and anti-inflammatory activities are the most extensively researched [419]. The biological activities of *Z. zerumbet* are summarized in Table 11.

Table 11 Biological activities of *Zingiber zerumbet*

Activity	Assay(s)	References
Hyperlipidemia	Zerumbone (**397**) decreased hepatic mRNA levels of fatty acid synthase, sterol-regulatory element binding protein, and 3-hydroxy-3-methyl-glutaryl-CoA reductase This compound upregulated hepatic mRNA expression of peroxisome proliferator-activated receptor-α and its target gene	[443]
Antibiofilm	Zerumbone (**397**) targeted key components of the extracellular matrix and enhanced antifungal delivery to reduce resistance in fluconazole-resistant *Candida albicans* biofilms.	[444]
Anticarcinogenic/cytotoxic	Dietary administration of zerumbone (**397**) suppressed both mouse colon and lung carcinogenesis by enhancing apoptosis and mediating the expression of NF-κB and heme oxidase-1	[445]
	Zerumbone (**397**) showed a potential preventive effect in a rat lung cancer model, and protected against the combined effects of 2-diethylnitrosamine and 2-acetylaminofluorene	[446]
	The in vitro-sensitizing effects of zerumbone (**397**) were shown in a clonigenic survival assay and by flow cytometry, with altered cross-linking of heat shock protein 27 (Hsp27) evident	[447]
Antiallergic/immune regulation	Oral administration of zerumbone (**397**) reduced the Th1 response and influenced the ovalbumin-induced Th2 response and airway inflammation in mice	[449]

(continued)

Table 11 (continued)

Activity	Assay(s)	References
	α-Pinene (**399**) demonstrated activity in an ovalbumin-sensitized allergic rhinitis mouse model and increased the levels of nasal immunoglobulin E	[449]
	Demonstrated protective effects against a dinitrophenylated human serum albumin-induced passive cutaneous anaphylaxis reaction in ICR mice	[450]
Osteoarthritis	Zerumbone (**397**) suppressed nitric oxide, prostaglandin E_2, and matrix metalloproteinase production in lipopolysaccharide (LPS)-stimulated RAW 264.7 cells. Also, this compound reduced paw edema and the pain response in a rat osteoarthritis model	[434]
Neuropathic pain	Zerumbone (**397**) attenuated pain symptoms in a mouse model, and inhibited interleukin 1 beta (IL-1β) and interleukin-6 (IL-6) and affected tumor necrosis factor alpha (TNF-α) production in the blood plasma and spinal cord tissues	[451, 452]
Diabetic retinopathy	An ethanol extract of the rhiizomes of *Z. zerumbet* protected against retinal inflammation-induced diabetes, using a streptozotocin-induced diabetic rat model, showing also a blood-retinal barrier breakdown and increased vascular permeability This extract acted at the p38 mitogen-activated protein kinase enzyme in the retina, and inhibited retinal nuclear factor kappa-light-chain-enhancer of activated B cells activation	[353, 354]
Hyperglycemia	No significant reduction in blood glucose levels in hyperglycemic rats were found for an aqueous extract of *Z. zerumbet* at a dose range of 50–150 mg/kg body weight	[453]

2.11 *Piper sarmentosum*

Piper sarmentosum Roxb. (Plate 11), a prominent member of the Piperaceae family, is a perennial creeping herb widely recognized for its dual culinary and medicinal applications. This species is closely related to economically significant plants in the same genus, such as *Piper nigrum* (black pepper) and *Piper betle* (betel leaf), making it an important subject of ethnobotanical research [455, 456]. Indigenous to Southeast Asia, *P. sarmentosum* has long been incorporated into traditional medicine

Plate 11 Leaves of *Piper sarmentosum* (family Piperaceae). Photograph courtesy Raffi Kojian (https://commons.wikimedia.org/wiki/File:Gardenology.org-IMG_7747_qsbg11mar.jpg)

systems in countries such as Malaysia, Thailand, and Indonesia, where it is commonly employed to treat gastrointestinal ailments, inflammatory diseases, and respiratory conditions. Furthermore, its constituents such as alkaloids and flavonoids, and its essential oil have biological activities, including antioxidant, anti-inflammatory, and antimicrobial properties, which continue to attract scientific interest [456, 457].

Piper sarmentosum is widely distributed in China and Southeast Asia [458, 459] and native to Malaysia, Thailand, Indonesia, and northeastern India [458]. The species thrives in humid, tropical environments and is found commonly in shaded or semi-shaded habitats such as forest margins, rural gardens, and home landscapes [456]. The plant has considerable ecological adaptability, allowing it to grow in a variety of soil types, including acidic tropical soils and moderately fertile loams [460]. Cultivation practices in Southeast Asia often involve integrating *P. sarmentosum* into agroforestry systems, where it serves multiple purposes, including as a dietary supplement, medicinal resource, and ecological contributor. Its ability to propagate through vegetative methods, such as stem cuttings, has ensured its widespread availability across diverse ecological and agricultural landscapes [461].

The cultural significance of *P. sarmentosum* is evident in the various names attributed in its native regions. The plant is known locally as "Chaplu" in Thailand, "Sirih duduk", "Akar buguor" or "Mengkadak" in Indonesia, and "Pokok Kadok" and "Kaduk" in Malaysia [455, 462]. It is recognized by similar vernacular names associated with its traditional medicinal roles. Its common names within different regions highlight an association with ethnomedicinal practices and its role as a health-promoting plant. This widespread recognition indicates its integral place in the dietary and therapeutic practices of various Southeast Asian communities [463].

Piper sarmentosum is a dioecious, upright shrublet that can grow up to 1.5 m in height, often exhibiting stoloniferous (sarmentose) growth. The young parts are finely puberulent. Its stems are swollen at the nodes, with short aerial roots present on the lower nodes. The leaves (Plate 11) have considerable variation in their shape and size, with the lower leaves broadly heart-shaped, measuring up to 15 × 14 cm, having a cordate base, acuminate apex, and long stalks, while the upper leaves are ovate-oblong, with a truncate to cuneate base, acute apex, and short stalks. The leaves feature seven primary, palmate veins, with the apical vein pair typically originating 0.5–2 cm above the base, prominently raised on the underside and slightly raised on the upper surface. The spike is oblongoid, measuring 2–5 cm in length and 0.8–1 cm in thickness, with a peduncle as long as the spike [464].

2.11.1 Ethnobotany

In Malaysia, the raw tender leaves of *P. sarmentosum* are consumed for their therapeutic properties in treating malaria and coughs, while the crushed leaves are applied to alleviate kidney stones [457, 465]. The fruits serve as an expectorant for respiratory disorders [455]. In Malay traditional medicine, *P. sarmentosum* is used to manage diabetes mellitus, joint pain, hypertension, gum diseases, acne, and to reduce a white discharge during the menstrual cycle [466]. Indonesian practices similarly value *P. sarmentosum* for its efficacy in relieving chest pain, rheumatism, and other conditions such as joint and menstrual pain [457, 463].

In Thailand, this species is utilized as a relaxant to reduce muscle pain and to manage hyperglycemia, particularly in the southern regions of the country [466, 467]. The leaves have been used traditionally as a carminative, while the entire plant serves as an expectorant, antispasmodic, throat refresher, appetite enhancer, antiflatulence remedy, and for treating muscle pain, asthma, and coughs [465, 468]. The leaves are also applied in poultices as counter-irritants for bone pain and headaches. A mixture of the crushed leaves and water is used during bathing to help with kidney-related urinary difficulties and kidney stones [469]. Its use extends to Panama, where it has utility for skin infections, snakebites, and uterine bleeding [470].

The roots are used as a stomachic and carminative, while the leaves and fruits act as expectorants [459]. The roots and leaves are considered remedies for headaches, toothache, coughs, asthma, pleurisy, and fungal dermatitis on the feet. A decoction of the roots and leaves is used for muscle weakness and bone pain [466], while the boiled leaves are employed to relieve toothache, bad breath, influenza, rheumatism, and coughs. The fruits and roots are consumed to treat dysentery, while the roots alone are used for lumbago, influenza, enuresis, pleurisy, and coughs [471].

2.11.2 Phytochemistry

The phytochemistry of *P. sarmentosum* is highlighted by its well-characterized essential oil, comprising 89 volatile compounds that are primarily sesquiterpenes, monoterpenes, phenylpropanoids, and alkanes. Among the sesquiterpenes, 40 compounds have been identified, including *trans*-caryophyllene (**464**) and spathulenol (**465**) [472, 473]. Monoterpenes, another significant group, include 15 compounds such as linalool (**401**) and α-pinene (**399**), which contribute to the characteristic aroma of the plant. Phenylpropanoids, a group of simple aromatic compounds, including asaricin, *trans*-asarone (**446**), and myristicin (**466**), have been isolated in high concentrations from specific regions of growth [474]. Thus, the essential oil content of *P. sarmentosum* varies with its geographical origin, with Vietnamese specimens being rich in benzyl benzoate (**646**), benzyl alcohol, and (2*E*)-butenyl benzene [475], while Malaysian varieties contain substantial amounts of sesquiterpene olefins [473].

The alkaloid and amide profile of *P. sarmentosum* is characterized by 26 identified compounds, including unsaturated pyrrolidone amides, piperidines, and aliphatic chain amides. Notable alkaloids include piperine (**471**), sarmentine (**469**), sarmentosine (**472**), and brachyamide B (**473**) [454, 476]. Piperine (**471**), found in the roots, stems, leaves, and fruits, is among the most abundant alkaloids present. Sarmentine (**469**) and sarmentosine (**472**) are unsaturated compounds with a terminal alkyl chain [477]. Amides like pellitorine (**470**) and guineensine (**474**), although present in smaller quantities, add to the chemical diversity. Structurally, many alkaloids of *P. sarmentosum* contain key functional characteristics such as having cinnamyl and pyrrolidone units, which make them distinct within the genus *Piper* [478].

Flavonoids, with 13 identified compounds, are another major chemical group in *P. sarmentosum*, and notable for their structural variety. Myricetin (**89**), quercetin (**85**), and naringenin (**92**) are the primary flavonoids, with rutin (**86**) being a significant flavonoid glycoside identified in an ethanol extract [470]. Sarmentosumins A (**475**), B (**476**), C (**477**), and D (**478**) are *C*-benzylated dihydroflavones recently isolated, adding to the new chemistry of the plant [479]. Flavonoids are built on a C-6–C-3–C-6 skeleton, which forms the basis for variations like hydroxylation, methylation, and glycosylation. The total flavonoid content has been estimated at 120.5 mg/kg, with high proportions of apigenin and kaempferol derivatives present [480]. Beyond flavonoids, other compounds like β-sitosterol (**479**) and stigmasterol (**109**) and the bioactive alkenylphenol sarmentosumol A (**480**) add to the plant's natural product profile [481]. The structures of the isolated additional phytochemicals from *P. sarmentosum* are shown in the formula drawings **432–481**.

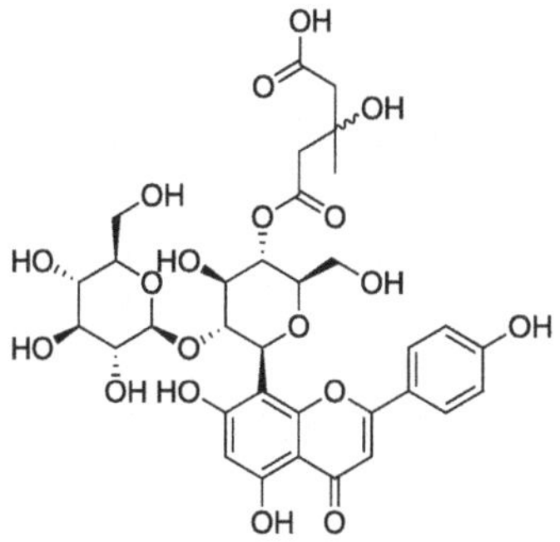

432 (4"-(3-hydroxy-3-methylglutaroyl)-2"-β-D-glucopyranosyl vitexin)

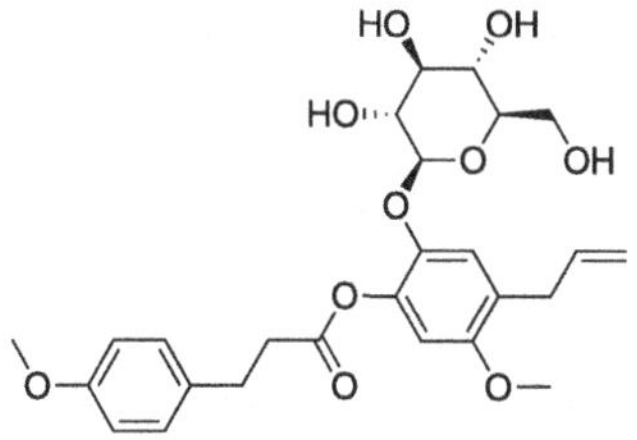

433 (kadukoside)

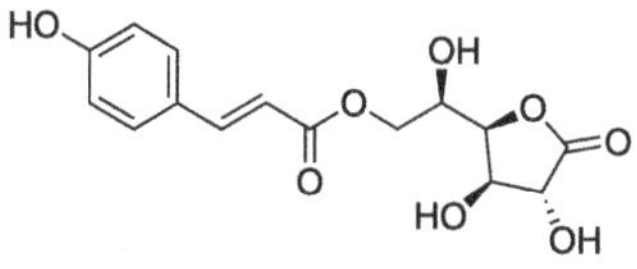

434 (6-*O*-(*E*)-*p*-coumaroyl-D-glucono-1,4-lactone)

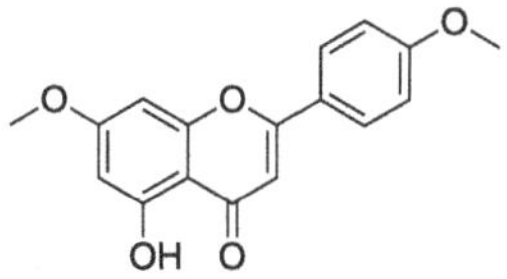

435 (5-hydroxy-7,4'-dimethoxyflavone)

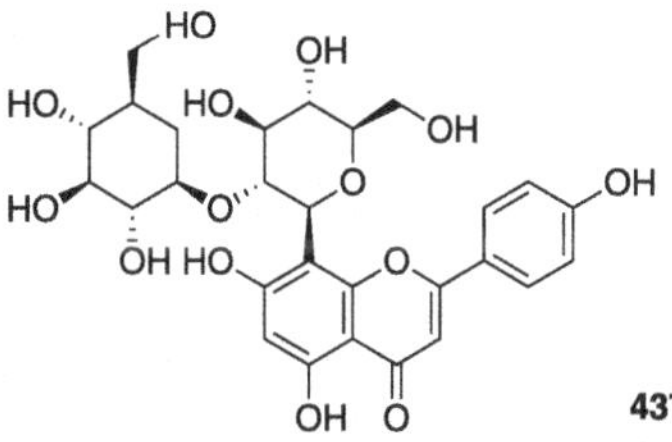

436 (2"-glucosylvitexin)

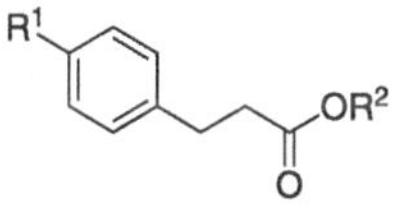

437 (hydrocinnamic acid) $R^1 = R^2 = H$

438 (3-(4-methoxyphenyl)propanoic acid) $R^1 = OCH_3$, $R^2 = H$

439 (methyl 3-(4-methoxyphenyl)propionate) $R^1 = OCH_3$, $R^2 = CH_3$

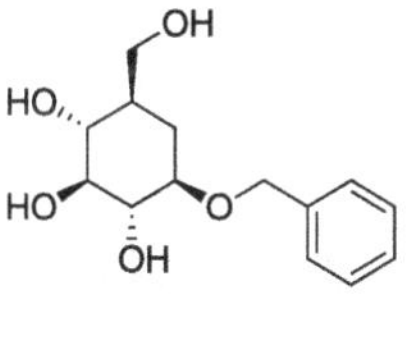

440 (benzyl-β-D-glucopyranoside)

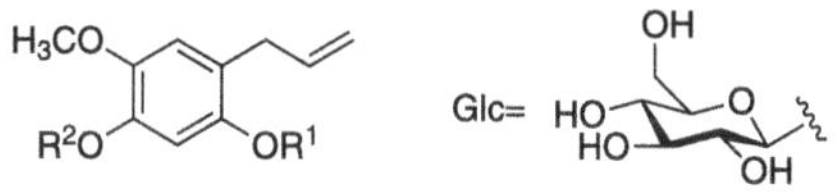

441 (citrusin C) $R^1 = H$, $R^2 = \beta$-Glc

442 (4-allyl-2,5-dimethoxyphenol-1-β-D-glucopyranoside) $R^1 = CH_3$, $R^2 = \beta$-Glc

443 (isoasarone) $R^1 = CH_3$, $R^2 = CH_3$

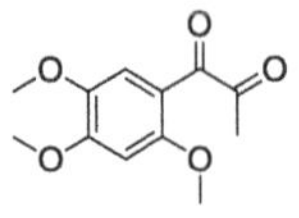

444 (1-(2,4,5-trimethoxyphenyl)-1,2-propanedione)

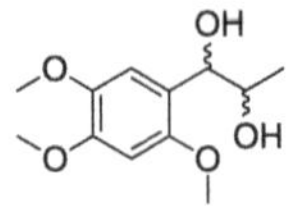

445 (1',2'-dihydroxyasarone)

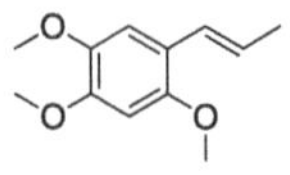

446 ((*E*)-asarone)

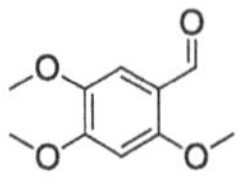

447 (asaraldehyde)

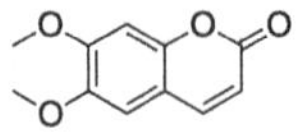

448 (scoparone)

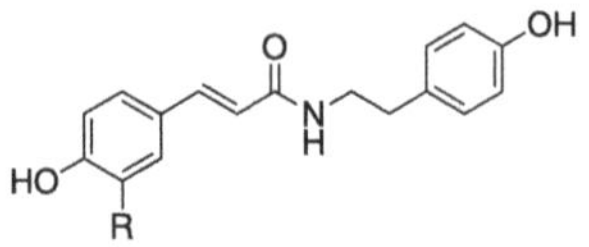

449 ((*E*)-*N*-feruloyltyramine) R = OCH_3
450 (paprazine) R = H

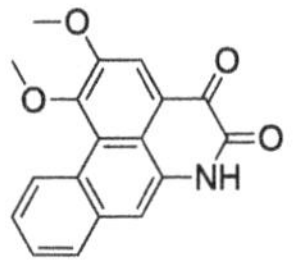

451 (norcepharadione B)

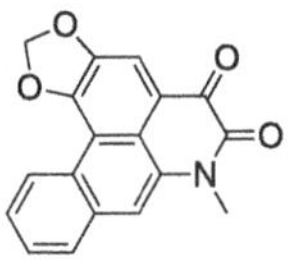

452 (cepharadione A)

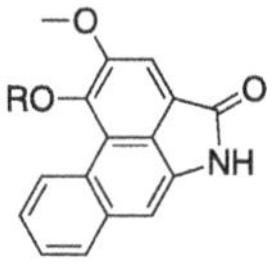

453 (aristolactam B II) R = CH_3
454 (piperolactam A) R = H

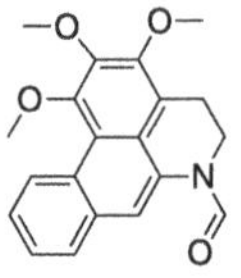

455 (dehydroformouregine)

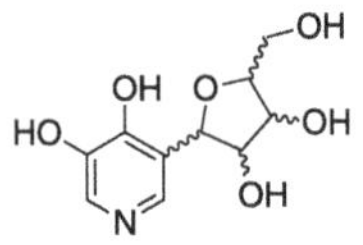

456 (3-[2,3-dihydroxy-4-(hydroxymethyl)-tetrahydrofuran-1-yl]-pyridine-4,5-diol)

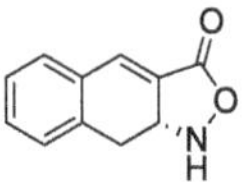

457 (naphthisoxazol A)

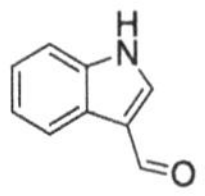

458 (indole-3-carboxaldehyde)

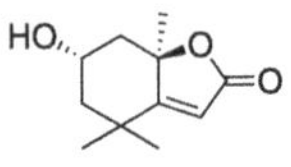

459 (loliolide)

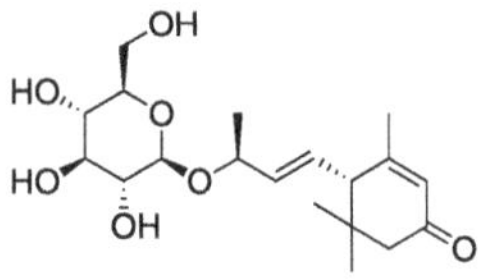

460 (reseoside)

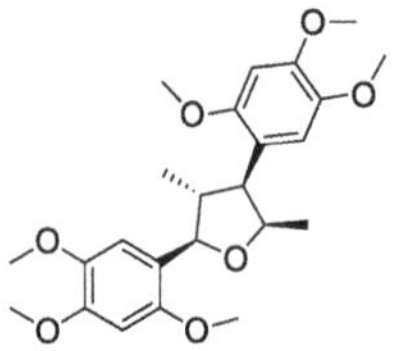

461 (magnosalicin)

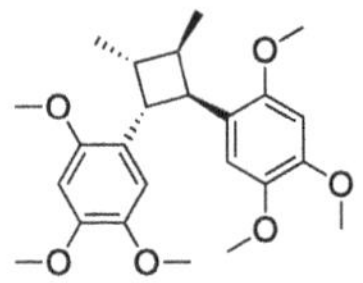

462 (andamanicin)

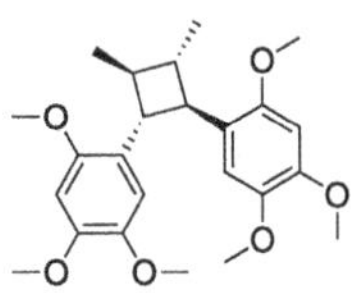

463 (magnosalin)

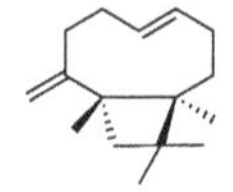

464 (*trans*-caryophyllene)

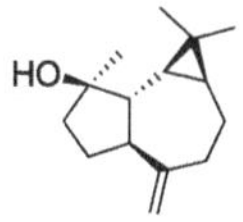

465 (spathulenol)

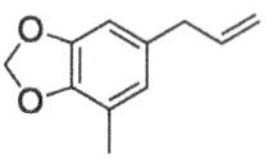

466 (myristicin)

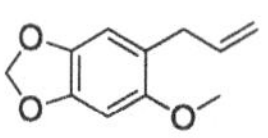

467 (asaricin)

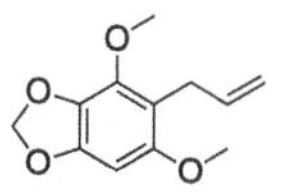

468 (1-allyl-2,6-dimethoxy-3,4-methylenedioxy-benzene)

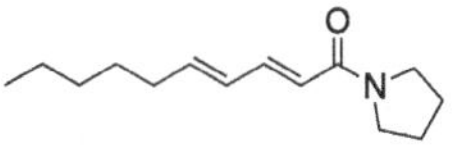

469 (sarmentine)

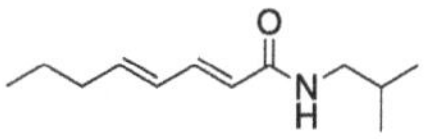

470 (pellitorine)

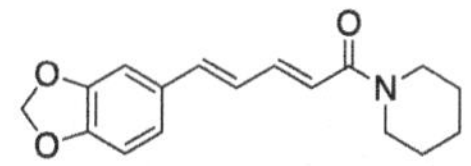

471 (piperine)

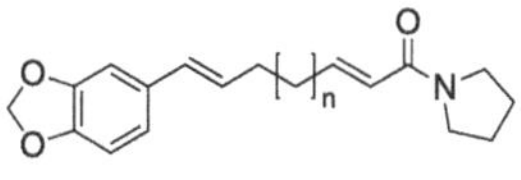

472 (sarmentosine) n = 1
473 (brachyamide B) n = 3

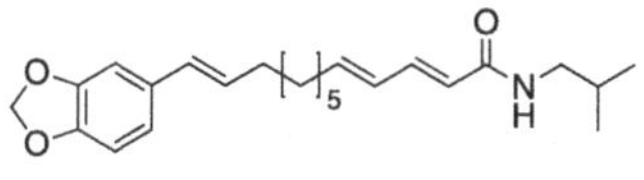

474 (guineensine)

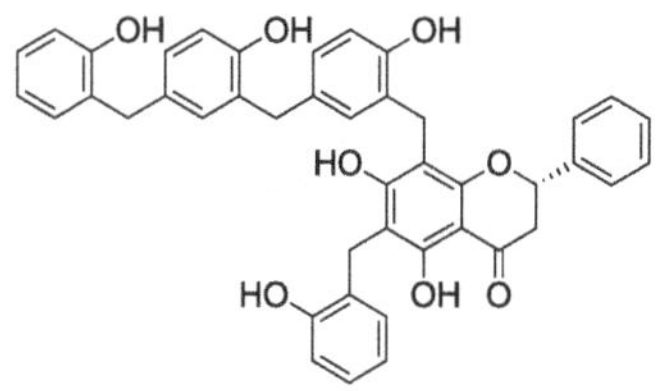

475 (sarmentosumin A)

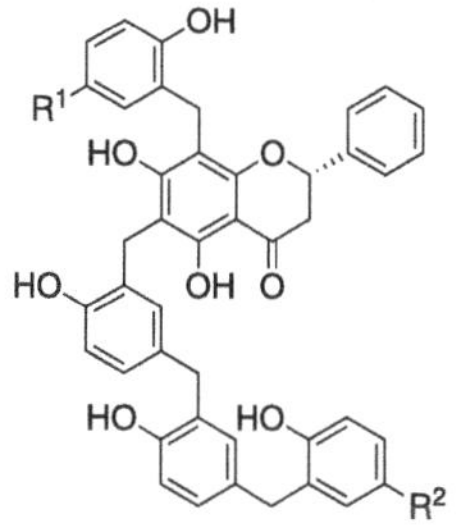

476 (sarmentosumin B) $R^1 = R^2 = H$
477 (sarmentosumin C) $R^1 = H$, $R^2 = C_6H_5OH$
478 (sarmentosumin D) $R^1 = C_6H_5OH$, $R^2 = H$

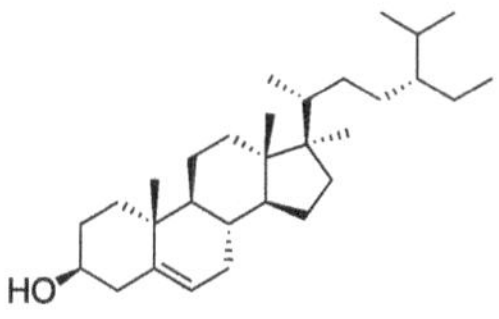

479 (β-sitosterol)

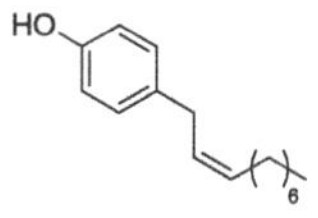

480 (sarmentosumol A)

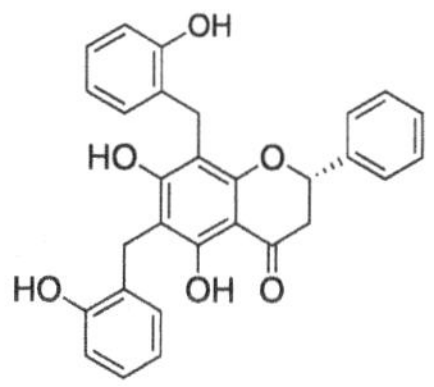

481 (dichamanetin)

2.11.3 Biological Activities

The pharmacological activities of *P. sarmentosum* are quite extensive, ranging from antibacterial and antifungal effects to promoting fracture healing and hypoglycemic activities. The plant's utility in fracture healing has been supported through animal studies showing its capacity to accelerate callus formation, cartilage ossification, and mineralization in osteoporotic models. Specifically, a water extract of *P. sarmentosum* at 125 mg/kg improved bone strength, trabecular structure, and reduced glucocorticoid-induced bone loss in rats [482, 483]. Flavonoids and alkaloids in the extracts are likely to contribute to reducing oxidative stress and enhancing bone density, although further studies are needed to clarify the active components and their precise mechanisms of action [484].

The antibacterial and antifungal activities of *P. sarmentosum* extracts have shown the potential of this species in controlling microbial infections. Its methanol and ethanol extracts exhibit growth inhibitory activities against pathogens such as *Escherichia coli*, *Staphylococcus aureus*, and *Cryptococcus neoformans*. Specific compounds like sarmentosumol A (**480**) inhibit bacterial growth at low concentrations [481, 485]. Additionally, bioformulations derived from *P. sarmentosum* have demonstrated effectiveness in agricultural applications, such as controlling rice pathogens and enhancing seedling vigor [486].

The anti-inflammatory properties of *P. sarmentosum* have been studied extensively using carrageenan-induced inflammation methodology. Water and ethanol extracts reduced paw swelling and inflammation in rats, with effects comparable to aspirin [457, 487]. These anti-inflammatory activities have been linked to the modulation of inflammatory mediators such as COX-2 and the prostaglandins as well as downregulation of pro-inflammatory cytokines like TNF-α and IL-1β [488]. Myristicin (**466**) and piperine (**471**) contribute to these effects, but further work is required to elucidate more precisely the cellular pathways involved and to confirm their efficacy in human models [489].

In terms of neuroprotective and antidepressant activities, an ethyl acetate extract of *P. sarmentosum* was found to regulate the hypothalamic–pituitary–adrenal axis and enhance the expression of brain-derived neurotrophic factor (BDNF) in a rodent model of depression [490]. Additionally, these extracts reduced neuroinflammation by inhibiting P38α MAPK and tau phosphorylation in an Alzheimer's disease model [488]. While acetylcholinesterase inhibition has been observed, specific active compounds contributing to these effects from *P. sarmentosum* remain unidentified thus far.

The activities of *P. sarmentosum* in inhibiting cancer cell growth and inducing apoptosis have been investigated. An ethanol extract demonstrated an antiproliferative effect against the human hepatoma (HepG2) cell line, with an IC_{50} value of 12.5 μg/cm^3 [491]. Sarmentosumins A (**475**), B (**476**), C (**477**), and D (**478**), when purified from a chloroform extract of *P. sarmentosum* aerial parts, were found to induce apoptosis in HT-29 human colon cancer cells by reducing the mitochondrial transmembrane potential (ΔΨm), with IC_{50} values ranging from 2.5 to 6.8 μg/cm^3 [481]. The previously known *C*-benzylated flavanone dichamanetin (**481**), when obtained from *P. sarmentosum*, inhibited the growth of three human cancer cell lines by a mechanism involving oxidative stress and NF-κB [492]. Additionally, *trans*-asarone (**446**) and isoasarone (**443**) were found to be cytotoxic to the MDA-MB-231 human cancer cell line by inducing apoptosis and by increasing reactive oxygen species (ROS) and affecting mitochondrial-dependent pathways [474].

Insecticidal activities of *P. sarmentosum* have been attributed to its essential oil and myristicin (**466**) and *trans*-caryophyllene (**464**). These compounds exhibit larvicidal and ovicidal effects against insect pests such as *Aedes aegypti* and *Brontispa* *longissima*. Myristicin (**466**) showed dose-dependent contact and fumigation toxicity, achieving 100% mortality in mosquito larvae and preventing egg hatching [472]. The acetylcholinesterase inhibitory activity of asaricin (**467**) and isoasarone (**443**) has been determined as a key mechanism for their insecticidal effects [474].

The hypoglycemic activity of *P. sarmentosum* is another traditional use that has been investigated pharmacologically. In streptozotocin-induced diabetic rats, a water extract significantly reduced blood glucose levels, comparable to the standard drug glibenclamide [493]. Mechanistic studies suggested that this plant extract inhibits α-glucosidase [494]. Furthermore, an extract of the plant was shown to protect against diabetes-associated complications, including oxidative stress-induced liver damage, by maintaining carbohydrate metabolism and reducing lipid peroxidation [495, 498]. Some reported pharmacological activities of *P. sarmentosum* are summarized in Table 12.

Table 12 Biological activities of *Piper sarmentosum*

Activity	Assay(s)	References
Antibacterial	A methanol extract of the fruits and leaves of *P. sarmentosum* inhibited the plant pathogens *Xanthomonas oryzae* and *Pseudomonas fuscovaginae*	[486]
	An extract inhibited the honey bee pathogens *Paenibacillus larvae* and *Ascosphaera apis*, and was not toxic to adult bees	[496]
	Sarmentosumin A (**475**) inhibited the growth of *Staphylococcus aureus*	[481]
Antifungal	The *P. sarmentosum* alkaloids brachyamide B (**473**), piperine (**471**), inhibited the growth of *Cryptococcus neoformans*	[485, 497]
	A methanol extract displayed antimicrobial activity against *Ascosphaera apis*	[497]
Anti-inflammatory	An aqueous plant extract reduced carrageenan-induced paw edema in rats in a dose-dependent manner	[457]
Antioxidant	High total flavonoid and phenolic content in extracts showed strong radical-scavenging activity, reducing oxidative stress markers	[470, 498]
Cytotoxic	A hexane extract of *P. sarmentosum* was cytotoxic for the HeLa and MCF-7 cell lines	[475]
	An ethanol extract of *P. sarmentosum* inhibited HepG2 cell proliferation	[491]

(continued)

Table 12 (continued)

Activity	Assay(s)	References
	Dichamanetin (**481**) induced apoptosis in HT-29 and MDA-MB-231 cancer cells by decreasing the mitochondrial membrane potential and inhibiting NF-κB and PARP-1	[492]
	Asaricin (**467**) and isoasarone (**443**) induced mitochondrial apoptosis in MDA-MB-231 breast cancer cells in a dose- and time-dependent manner	[499]
Antidepressant/neuroprotective	An ethyl acetate fraction of *P. sarmentosum* upregulated CREB and BDNF expression, and improved HPA axis function in stressed rats. A polar extract reduced P38α MAPK and phosphorylated tau protein, reducing neuroinflammation in microglial cells	[488, 490]
Antiatherosclerotic	An aqueous extract of *P. samentosum* increased eNOS, mRNA and NO production in HUVEC cells, protecting against endothelial dysfunction	[459]
	A methanol extract of *P. sarmentosum* reduced atherosclerotic lesions and foam cell infiltration in cholesterol-fed mice	[499, 500]
Hypoglycemic	An aqueous extract decreased blood glucose levels in diabetic rats	[492, 495]
Insecticidal	Asaricin (**467**) and isoasarone (**443**) caused 100% larval mortality in *Aedes aegypti* and inhibited egg hatching (95%)	[474]
	The essential oil of *P. sarmensosum* leaves showed dose-dependent toxicity to the beetle, *Brontispa longissima*	[472]
Antiosteoporosis	A water extract of *P. sarmentosum* enhanced bone strength and stiffness in ovariectomized rats by increasing the trabecular bone volume	[482]
Antituberculosis	Alkaloids inclusive of pellitorine (**470**) and sarmentine (**469**) from *P. sarmentosum* inhibited *Mycobacterium tuberculosis*	[455]
Antiparasitic	Antiparasitic activity against *Plasmodium falciparum, Leishmania amazonensis* and promastigotes form of *L. amazonensis*	[501]
Larvicidal	A hexane extract of *P. sarmenotosum* showed larvicidal activity against *Aedes aegypti*, with a LC_{50} value of 39.0 μg/cm^3	[502]

2.12 *Moringa oleifera*

Moringa oleifera Lam. (Plate 12), often referred to as the "drumstick tree" or "horseradish tree", is a member of the Moringaceae family and a subject of widespread research due to its broad medicinal, nutritional, and industrial applications. Originating from the sub-Himalayan regions of India, Pakistan, and Bangladesh, it exhibits an adaptability to diverse climates, and now flourishes in tropical and subtropical areas worldwide [503, 504]. Its ability to withstand harsh conditions, including prolonged drought and moderate frost, coupled with its minimal maintenance requirements, underscores its role as a sustainable resource. In degraded soils where conventional crops often fail, *M. oleifera* can thrive, offering multiple benefits, including combating malnutrition, enhancing soil fertility, and providing a raw material for the cosmetic, pharmaceutical, and biofuel industries [505, 506].

Moringa oleifera has been cultivated extensively in Africa, Southeast Asia, the Caribbean region, and Latin America, and was introduced primarily through colonial trade and migration patterns. The plant's resilience to low rainfall in requiring only 250 mm annually, and its capacity to thrive in nutrient-poor soils, have made it a cornerstone crop in arid and semi-arid regions [504, 507]. Countries like Nigeria, Kenya, and Haiti have carried out its cultivation, recognizing its potential to alleviate

Plate 12 Leaves of *Moringa oleifera* (family Moringaceae). Photograph courtesy Krish Dulal (https://commons.wikimedia.org/wiki/File:Moringa_oleifera_NP.JPG)

food insecurity, combat malnutrition, and restore degraded ecosystems. Additionally, the growth of *M. oleifera* has been linked to smallholder agricultural initiatives, which focus on sustainable farming practices to support rural livelihoods [508].

The plant is known by various vernacular names that reflect its significance across cultures. The name "Drumstick tree" originates from its long, slender pods, which resemble drumsticks and are used widely for culinary purposes. Also, "Horseradish tree" highlights the similarity in flavor between its roots and horseradish. In Malaysia, it is referred to as "Kelor", "Kaylor", or "Marango", and in India, it is termed "Munagaa", "Sahjan", "Shevga", and "Murungai", depending on the region and language used. In turn, in Pakistan, it is widely known as "Sohanjna" or "Sonjhna", while in Bangladesh, it is called "Shajina" or "Sohanjana", aligning with its Bengali linguistic roots [425]. Beyond these descriptive names, the term "Miracle tree" is used widely, emphasizing its multifaceted benefits, including its nutritional richness and utility in all of ethnomedicine, agriculture, and environmental conservation [502, 505].

Moringa oleifera is a fast-growing deciduous tree characterized by a slender, upright trunk that can reach a height of six meters [508]. Its branches, adorned with pale green, feathery leaves arranged in a tripinnate pattern, contribute to its esthetic appeal and functional adaptability. The tree produces fragrant, white flowers in dense clusters, which develop into elongated pods. These pods house dark-brown, round seeds encased in a distinctive papery coat [425]. This robust morphology (Plate 12) enables *M. oleifera* to thrive in varied climates and soil conditions, further enhancing its value as an agricultural resource [506, 507].

2.12.1 Ethnobotany

Moringa oleifera has been recognized for centuries in traditional medicine systems due to its extensive therapeutic applications. The plant leaves, seeds, pods, bark, roots, and flowers are all utilized to treat various diseases and conditions. People worldwide have incorporated *M. oleifera* into their diets and health practices, because of its therapeutic value [509, 510]. Various medicinal products derived from the plant are consistent with its ethnomedicinal uses, and alleviate several health conditions, including hypertension [511], anxiety [512], diarrhea [513], and diuresis [514]. The leaves of *M. oleifera*, for instance, are used to produce poultices for inflammatory conditions such as glandular swelling, bronchitis, and headaches [515]. Its pods are employed in treating hepatitis and joint pain [505], while its roots are utilized for kidney stones [516], liver ailments [517], and earache and toothache [518]. The gum has traditional applications in treating fever and also for inducing abortions [519]. The roots have a role in the treatment of helminthiasis [520]. The plant seeds, known for their laxative properties, are used also in treating tumors, prostate issues, and bladder problems [507], while also having use in arthritis treatment by reducing oxidative stress and inflammation [521]. The plant's leaves are particularly valued for improving the health of nursing mothers, aiding malnourished infants, and treating insomnia [522] and wounds [523]. Moreover, *M. oleifera* has also found extensive applications in the cosmetic industry, with its historical use in ancient Egyptian dermal ointments further testifying to its versatility and cultural importance [524].

Building on its widespread use, *M. oleifera* is entrenched in traditional medicinal systems across various cultures as a versatile remedy for different ailments. In Ayurvedic medicine, the leaves are considered effective for managing digestive disorders, respiratory tract infections, and joint pain due to their anti-inflammatory and antimicrobial properties. Similarly, traditional medicine in parts of Africa has incorporated *M. oleifera* leaves into the diet to combat malnutrition and strengthen the immune system, particularly among vulnerable groups such as children and pregnant women. The plant contains high levels of vitamins, minerals, and antioxidants, supporting its use as both a food and medicinal resource [509]. Additionally, *M. oleifera* seed powder has long been used in South Asian and African communities for water purification, leveraging its coagulating properties to remove impurities and enhance public health [525]. These traditional applications have inspired modern scientific research into determining the bioactive natural products of this species, thereby offering some support for its antioxidant, anti-inflammatory, and immune system-boosting effects [509].

2.12.2 Phytochemistry

Moringa oleifera is recognized for its phytochemical diversity, which underpins its broad biological activities. Over 100 compounds have been identified from different parts of the plant, including flavonoids, glucosinolates, phenolic acids, and other derivatives The leaves produce bioactive flavonoids such as quercetin (**85**), kaempferol (**88**), and isoquercetin (**751**), known for their potent antioxidant and anti-inflammatory effects [526]. These flavonoids act as scavengers of reactive oxygen species (ROS), thus protecting cells from oxidative damage. In addition, phenolic acids, such as ferulic (**489**), caffeic (**490**), and gallic (**95**) acids have demonstrated free-radical scavenging properties, adding to their potential role in mitigating chronic conditions like cardiovascular disease and diabetes [527]. The plant's leaves also have a high content of vitamins and carotenoids, which enhance their utility in functional foods [528].

Glucosinolates are another key class of the constituents of *M. oleifera*, and are found predominantly in its seeds and leaves. Glucomoringin (**493**), is a major glucosinolate and precursor to the isothiocyanates, which have been researched extensively for their cancer-related and anti-inflammatory effects [529]. Isothiocyanates induce apoptosis in cancer cells and suppress inflammatory pathways [530]. Moreover, moringin (**507**) can be produced from *M. oleifera*, and has been linked to hepatoprotective, antihypertensive, and antioxidant activities [531]. These compounds not only highlight the therapeutic potential of *M. oleifera* but also reflect on its adaptogenic qualities, which may improve the body's resilience to stress [532].

The seeds of *M. oleifera* are a source of fatty acids such as oleic (**496**) and palmitic acids (**498**), which are antimicrobial and anti-inflammatory compounds [533]. These fatty acids contribute to the plant's ability to regulate lipid profiles, and may have a role in aiding cardiovascular health. Sterols like β-sitosterol (**479**), extracted from both the seeds and leaves, have potential in lowering cholesterol and promoting heart health [531]. *Moringa oleifera* is useful in traditional medicine for a wide range of conditions [534]. The structures of the significant phytochemicals of *M. oleifera* are shown in their formula drawings **482–509**.

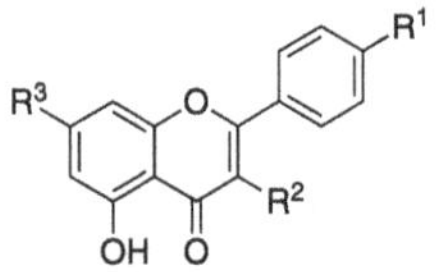

482 (kaempferide) R^1 = OMe , R^2 = R^3 = OH
483 (kaempferitrin) R^1 = OH, R^2 = R^3 = Rha

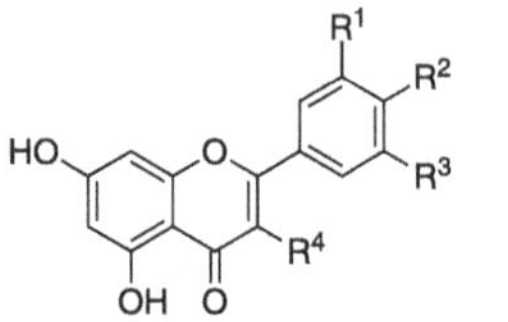

484 (isorhamnetin) R^1 = OMe, R^2 = OH, R^3 = H, R^4 = OH
485 (astragalin) R^1 = R^3 = H, R^2 = OH, R^4 = Glc
486 (quercetin dihydrate) R^1 = H, R^2 = R^4 = R^3 = OH

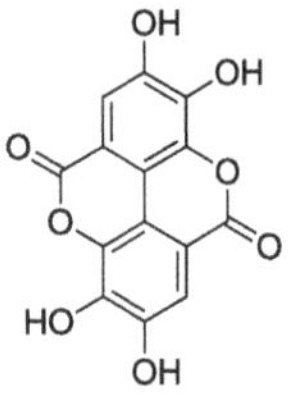

487 (ellagic acid)

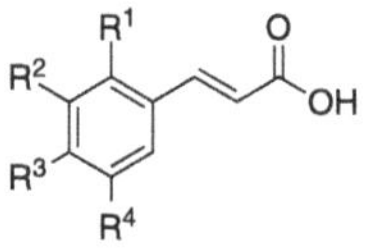

488 (*o*-coumaric acid) R^1 = OH, R^2 = R^3 = R^4 = H
489 (ferulic acid) R^1 = R^4 = H, R^2 = OMe, R^3 = OH
490 (caffeic acid) R^1 = R^4 = H, R^2 = R^3 = OH
491 (sinapic acid) R^1 = H, R^2 = R^4 = OMe, R^3 = OH

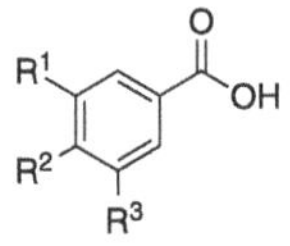

492 (syringic acid) R^1 = R^3 = OMe, R^2 = OH

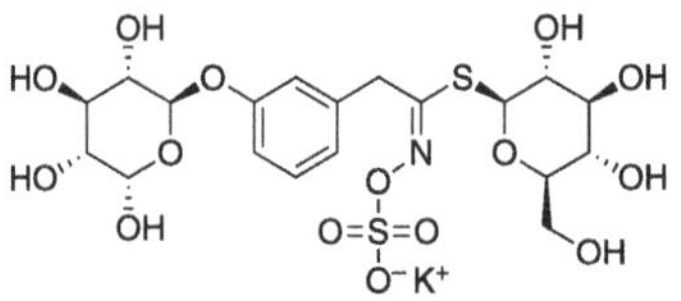

493 (glucomoringin)

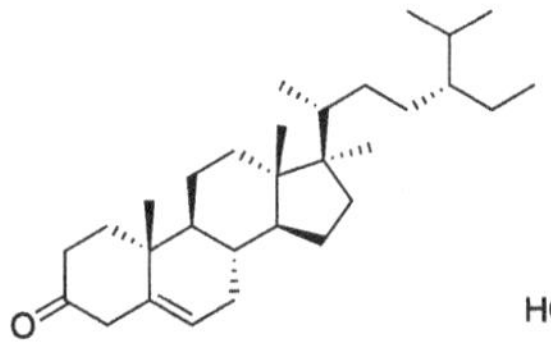

494 (β-sitosterone)

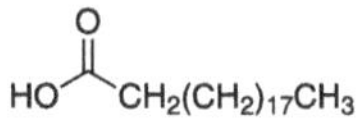

495 (arachidic acid)

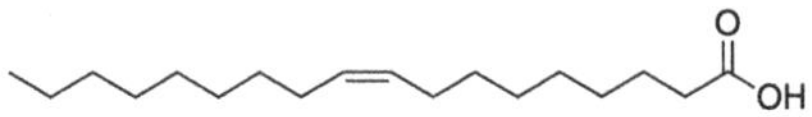

496 (oleic acid)

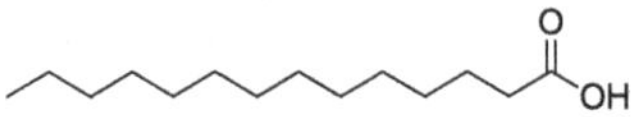

497 (myristic acid)

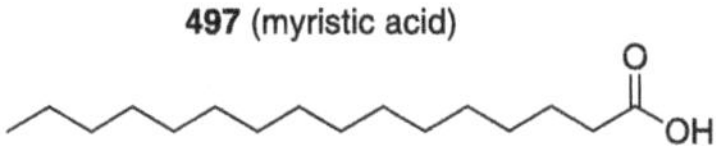

498 (palmitic acid)

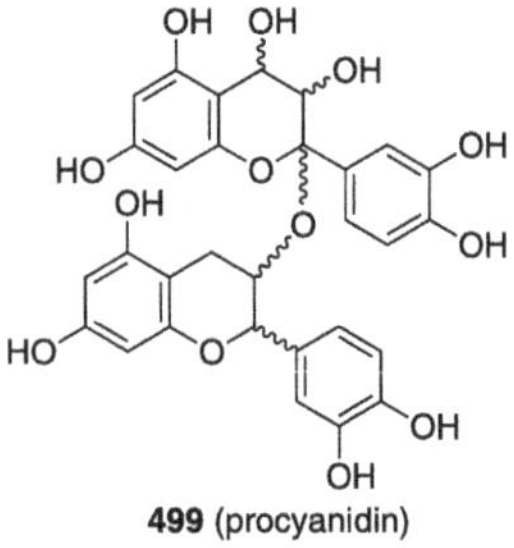

499 (procyanidin)

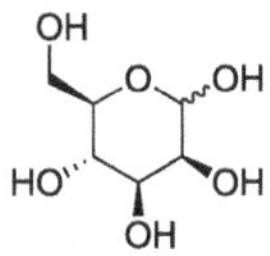

500 (D-mannose)

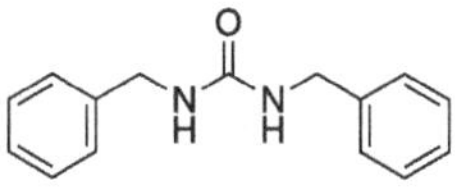

501 (1,3-dibenzylurea)

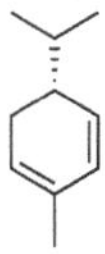

502 (α-phellandrene)

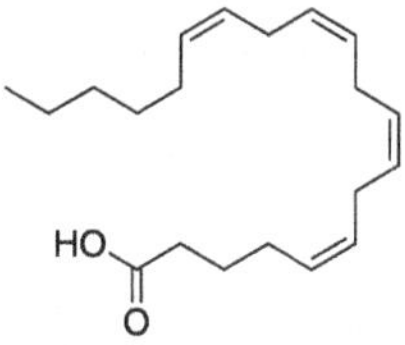

503 (arachidonic acid)

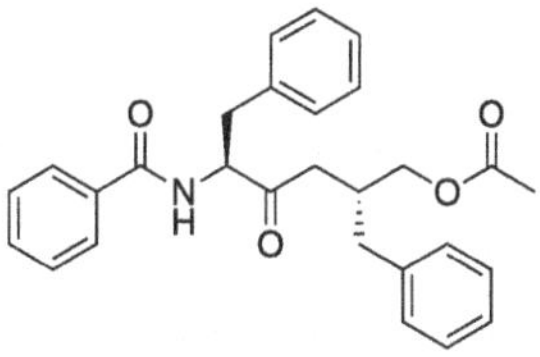

504 (aurantiamide acetate)

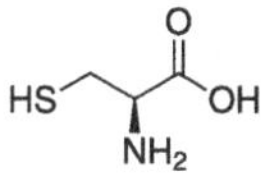

505 (cysteine)

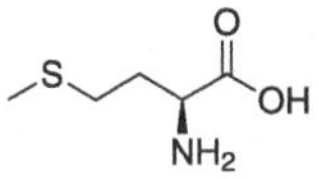

506 (methionine)

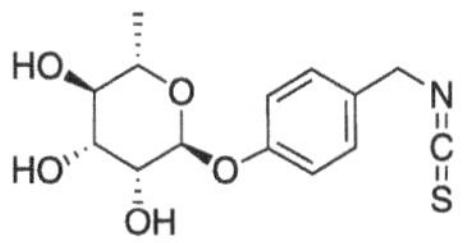

507 (moringin)

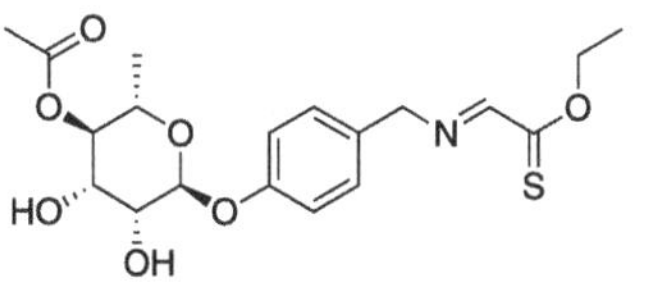

508 (niaziminin)

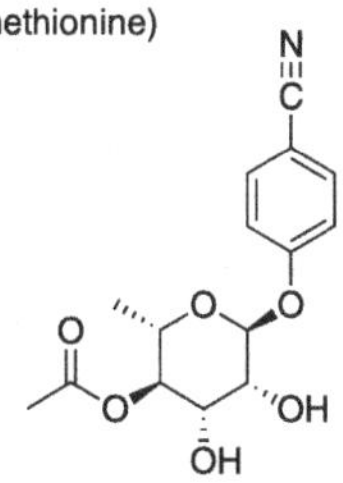

509 (niazirinin)

2.12.3 Biological Activities

The anti-inflammatory potential of *M. oleifera* is well-documented, particularly through the presence of bioactive compounds found in the leaves and pods. For example, the thiocarbamate niaziminin (**508**) inhibits nitric oxide production, reducing inflammation using an in vitro bioassay [535]. Additionally, an extract of the plant blocks the nuclear factor κB pathway, with relevance to treating chronic inflammatory conditions. An additional study has revealed effectiveness in managing inflammation-associated disorders such as arthritis and atopic dermatitis [536].

Research has shown the ability of *M. oleifera* to scavenge free radicals and reduce oxidative stress, through compounds such as quercetin (**85**), kaempferol (**88**), and myricetin (**89**), from its leaves and seeds [537]. A study in an animal model has demonstrated protective effects against oxidative liver damage induced by drugs like methotrexate and diclofenac sodium, suggesting potential applications in hepatoprotection [538]. Furthermore, synergistic combinations of *M. oleifera* with curcumin have shown promise in combating oxidative stress-related diseases [539].

The cytotoxic potential of *M. oleifera* arises from isothiocyanates and thiocarbamates that act as inhibitors of tumor cell growth and apoptosis inducers in cancer cells. An extract of the plant leaves selectively targeted breast cancer cells, while niaziminin (**508**) showed a preventive effect against chemical carcinogenesis [540]. An additional study has identified the binding affinity of the *M. oleifera* constituent rutin (**86**) to BRCA-1, indicating a possible role in breast cancer therapy [523]. The reported biological activities of *M. oleifera* are summarized in Table 13.

Table 13 Biological activities of *Moringa oleifera*

Activity	Assay(s)	References
Anti-inflammation	An extract of *M. oleifera* flowers reduced the activation of mediators of inflammation by inhibiting the NF-κB pathways in a lipopolysaccharide (LPS) RAW 264,7 macrophages In a nanoparticle form, β-sitosterol (**479**) from the stems of *M. oleifera* suppressed the secretion of inflammatory factors from keratosis and macrophages induced by peptidoglycan, TNF-α, and LPS	[533, 535]
Antioxidant	Due to the presence of compounds such as quercetin (**85**), kaempferol (**88**), and ascorbic acid (**98**), *Moringa* scavenged free radicals, reduced oxidative stress, and protected against lipid peroxidation in a mouse speeen model, and protected against methotrexate-induced oxidative stress	[541]
Cytotoxic	A fraction of a *M. oleifera* leaf methanol extract showed selective cytotoxic activity for the MCF-7 breast cancer cell line, and this effect was attributed to apoptosis induction involving the upregulation of the expression of Bax, p53, and caspase 8	[540]
Hepatoprotective	A *M. oleifera* leaf ethanol extract protected against liver toxicity in a rat model that was induced by exposure to a high dose of diclofenac sodium	[538]
Cardioprotective	It has been proposed that extracts of *M. oleifera* can prevent myocardial damage and vascular dysfuncton due to the presence of flavonoids such as quercetin (**85**) and quercetrin (**578**)	[542]

(continued)

Table 13 (continued)

Activity	Assay(s)	References
Antidiabetic	It was found that a dose of 200 mg/kg of a *M. oleifera* leaf aqueous extract decreased the blood glucose levels of normal test animals in both a fasting blood glucose model and in an oral glucose tolerance test, by over 25% in each case	[543]
Antiulcer/ gastroprotective	A leaf and fruit extract of *M. oleifera* showed a healing effect in a rat acetic acid-induced chronic gastic ulcer model. In addition, extracts from the leaves produced a gastric antisecretory effect in pylorus-ligated rats	[519]
Neuroprotective	An ethanol extract of the plant enhanced learning and memory, reduced anxiety, and exhibited anticonvulsant properties in mice. It modulated neurotransmitters like GABA, and appears to offer a potential benefit for treating epilepsy	[544]
Antimicrobial/ antifungal	A plant extract was effective against bacterial strains (e.g., *Staphylococcus aureus*), fungi, and viruses. This extract contained bioactive agents like flavonoids and other phenolic compounds that can disrupt microbial membranes	[545]
Diuretic	An extract of *M. oleifera* root wood promoted increased urine output in rats and prevented calcium oxalate crystal deposition in the kidneys, thus aiding in the treatment of kidney stones	[516]
Anti-hypertensive	Several thiocarbamate derivatives from *M. oleifera* leaves were found to target angiotensin-converting enzyme (ACE) activity, which is a major regulatory enzyme of the renin-aniotensin system	[546]
Analgesic/ antipyretic	Leaf extracts of different polarities displayed antinoceptive activity in rats using a formalin test, when taken orally in the dose range 30–100 mg/kg	[547]
Anti-obesity	*Moringa oleifera* is known to promote weight loss by modulating lipid metabolism, reducing adipogenesis, and improving insulin sensitivity	[548]
Anti-venom	The leaves of *M. oleifera* neutralized *Naja ngricollis* snake venom toxin by inhibiting protease and phospholipase enzymes, and reducing systemic damage	[549]
Immunomodulatory	*Moringa oleifera* has been found to enhance immune cell activity and reduce pro-inflammatory markers, improving resilience against infections	[550]
Wound healing	A *M. oleifera* leaf extract in a composite file accelerated wound closure, reduced inflammation, and stimulated collagen production for tissue regeneration in humans	[551]

2.13 *Cassia alata*

Cassia alata L., locally known as "pokok gelenggang", is a medicinal plant in the Fabaceae family [552]. Common names include "Candle Bush", "Acapulo", "Emperor's Candlestick", "Christmas Candle", "Ringworm Bush", and "Calabra Bush", and the plant is indigenous to tropical and subtropical regions of the Americas, including Central and South America and the Caribbean islands [553]. It grows rapidly throughout the tropics [554], and is found commonly in open areas, roadsides, and disturbed habitats, being well-adapted to various soil types. This fast-growing shrub or small tree can reach heights up to four to nine meters, and is distinguished by its large bright green leaves, which are pinnately arranged, about 30–60 cm in length, with five to seven pairs of leaflets having a single terminal leaflet. The plant produces large striking yellow flowers arranged in cylindrical spikes (Plate 13), which attract bees and butterflies. The seeds are green in color when they are unripe and turn black when they ripen [555].

Plate 13 Leaves and flowers of *Cassia alata* (family Leguminoseae). Photograph courtesy Len Worthington (https://commons.wikimedia.org/wiki/File:Cassia_alata_(6347438828).jpg)

2.13.1 Ethnobotany

Cassia species are used frequently for ornamental and nutritional purposes. In traditional medicine, they are utilized to treat anemia, dermatitis, constipation, fever, menstrual problems, wounds, and leprosy. Traditionally, the Malays and Chinese have applied a paste made from the pounded leaves mixed with lime to the skin as a remedy for ringworm and other skin conditions. Additionally, a decoction of the leaves is sometimes consumed as a purgative [555].

The leaves of *Cassia alata* have been employed traditionally for various medicinal purposes in different parts of the world. For example, a paste made by mixing the leaves with lime juice is applied to the infected skin for its antimicrobial effects and the ground leaves are mixed with coconut oil to create a paste used for its antifungal properties. Additionally, a herbal tea from the leaves can be gargled two to three times daily to soothe aphthous ulcers or boiled for 10 to 15 min and consumed three times daily for the relief of asthma [554]. In Nigeria, decoctions from the leaves, stems, and roots are used to treat infections, and as a wound treatment, and also for burns and constipation [555]. Also, leaf decoctions are used for alleviating abdominal pain, stress, and toothache. In Egypt, these decoctions aid bowel movement, which promotes peristaltic shrinkage and prevents constipation. Furthermore, in India, China, and the Philippines, the leaf decoctions are utilized in treating syphilis, hemorrhoids, inguinal hernia, diabetes, and intestinal parasitosis [554]. Additionally, the juice from the young leaves has use for managing eye infections [552].

2.13.2 Phytochemistry

Phytochemical screening of the roots of *C. alata* has confirmed the presence of flavonoids, anthraquinones, and other constituents [556]. Two flavonoid glycosides were isolated from the seeds, chrysoeriol-7-*O*-(2″-*O*-β-D-mannopyranosyl)-β-D-allopyranoside (**510**) and rhamnetin-3-*O*-(2″-*O*-β-D-mannopyranosyl)-β-D-allopyranoside (**511**) [557]. The anthraquinones emodin (1,6,8-trihydroxy-3-methyl-anthraquinone) (**532**) and chrysophanol (chrysophanic acid) (**529**) were purified from the roots using HPLC [558]. Physcion (**512**), betulinic acid (**513**), stigmasterol (**109**), 1-hydroxy-7-methoxy-3-methylanthraquinone (**514**), 8-*O*-methylchrysophanol (**515**), and aloe-emodin (**516**) were also identified [559]. Additionally, sequential HPLC methods yielded 1,3,8-trihydroxy-2-methyl-anthraquinone (**517**), rhein (**518**), 1,5-dihydroxy-8-methoxy-2-methyl-anthraquinone-3-*O*-β-D-(+)-glucopyranoside (**519**), and emodin (1,6,8-trihydroxy-3-methyl-anthraquinone) (**520**). 1,5,7-Trihydroxy-3-methyl-anthraquinone (**521**) was purified from the stems while lunatin (**528**), 7,4′-dihydroxy-5-methoxyflavone (**522**), luteolin (**389**) and *trans*-dihydrokaempferol (**523**) were obtained from the twigs using column chromatography [557].

Additionally, flavonoids such as kaempferol (**88**), kaempferol-3-*O*-gentiobioside (**524**), and kaempferol-3-*O*-β-D-glucopyranoside (**525**) [555] and the flavones 3,5,7,4′-tetrahydroxyflavone (**526**) and 2,5,7,4′-tetrahydroxyisoflavone (**527**) [560] were also iolated. Rhein (**518**), along with hydroxymethyl anthraquinones and chrysophanic acid (**529**), were also purified from an ethanol extract [560]. Several constituents have been isolated from an ethyl acetate extract of the leaves, such as methyl 2,4,6-trihydroxybenzoate (**530**), and kaempferol-3-*O*-glucoside (**531**). The chemical constituents of *Cassia alata* are displayed in the formula drawings **510–533**.

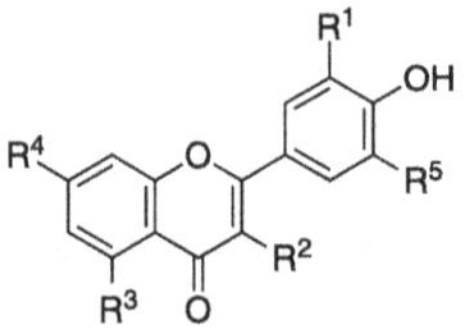

510 (chrysoeriol-7-*O*-(2''-*O*-β-D-mannopyranosyl)-β-D-allopyranoside) R^1 = OMe, R^2 = R^3 = OH, R^4 = 2''-*O*-β-D-mannopyranosyl)-β-D-allopyranoside, R^5 = H
511 (rhamnetin-3-*O*-(2''-*O*-β-D-mannopyranosyl)-β-D-allopyranoside) R^1 = H, R^3 = R^5 = OH, R^2 = ''-*O*-β-D-mannopyranosyl)-β-D-allopyranoside, R^4 = OMe
522 (7,4' dihydroxy-5-methoxyflavone) R^1 = R^2 = R^4 = R^5 = H, R^3 = OMe
524 (kaempferol-3-*O*-gentibioside) R^1 = R^5 = H, R^3 = R^4 = OH, R^2 = *O*-gentibioside
525 (kaempferol-3-*O*-β-D-glucopyranoside) R^1 = R^4 = R^5 = H, R^3 = OH, R^2 = *O*-β-D-glucopyranoside
526 (3,5,7,4'-tetrahydroxy-flavone) R^1 = R^4 = R^5 = H, R^2 = R^3 = OH
531 (kaempferol-3-*O*-glucoside) R^1 = R^4 = R^5 = H, R^3 = OH, R^2 = *O*-glucoside

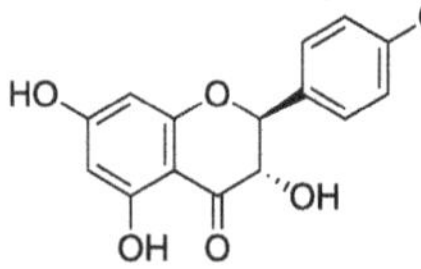

523 (*trans*-dihydrokaempferol)

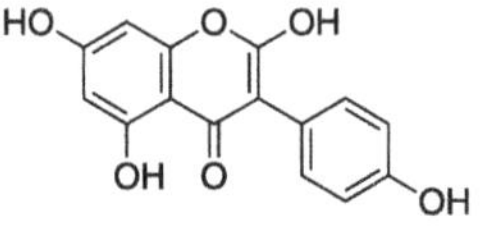

527 (2,5,7,4'-tetrahydroxy isoflavone)

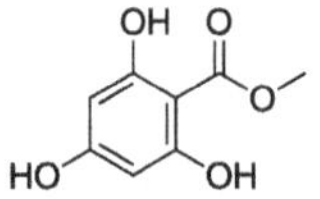

530 (methyl 2,4,6-trihydroxy benzoate)

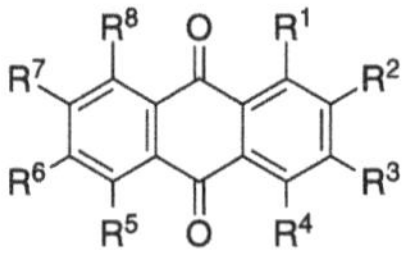

512 (physcion) R^1 = R^3 = R^6 = R^8 = H, R^4 = R^5 = OH, R^2 = OMe, R^7 = CH_3
514 (1-hydroxy-7-methoxy-3-methylanthraquinone) R^1 = R^2 = R^4 = R^6 = R^8 = H, R^5 = OH, R^3 = OMe, R^7 = CH_3
515 (8-*O*-methylchrysophanol) R^1 = R^2 = R^3 = R^6 = R^8 = H, R^4 = OMe, R^5 = OH, R^7 = CH_3
516 (aloe-emodin) R^2 = R^4 = R^5 = R^7 = R^8 = H, R^1 = R^6 = OH, R^3 = CH_2OH
517 (1,3,8-trihydroxy-2-methyl-anthraquinone) R^1 = R^3 = R^6 = R^8 = H, R^4 = R^5 = OH, R^2 = OMe, R^7 = CH_3
518 (rhein) R^1 = R^2 = R^3 = R^6 = R^8 = H, R^4 = R^5 = OH, R^7 = COOH
519 (1,5-dihydroxy-8-methoxy-2-methyl-anthraquinone-3-*O*-β-D-glucopyranoside) R^2 = R^3 = R^8 = H, R^1 = R^5 = OH, R^4 = OMe, R^6 = CH_3, R^7 = *O*-β-glucopyranoside
520 (1,6,8-trihydroxy-3-methyl-anthraquinone) R^2 = R^4 = R^5 = R^7 = H, R^1 = R^3 = R^8= OH, R^6 = CH_3
521 (1,5,7-trihydroxy-3-methyl-anthraquinone) R^2 = R^4 = R^6 = R^8 = H, R^1 = R^3 = R^5 = OH, R^7 = CH_3
528 (lunatin) R^2 = R^4 = R^5 = R^7 = H, R^1 = R^6 = R^8 = OH, R^3 = OMe
529 (chrysophanic acid) R^1 = R^2 = R^3 = R^6 = R^8 = H, R^4 = R^5 = OH, R^7 = CH_3
532 (emodin) R^2 = R^4 = R^5 = R^7 = H, R^1 = R^6 = R^8 = OH, R^3 = CH_3

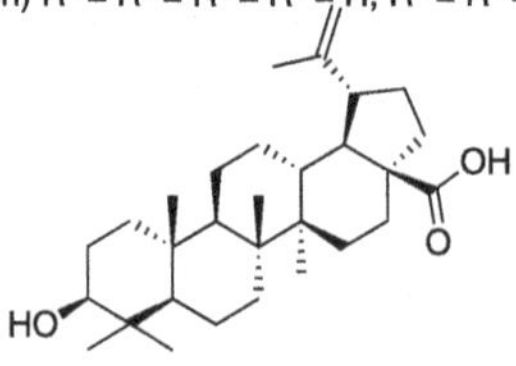

513 (betulinic acid)

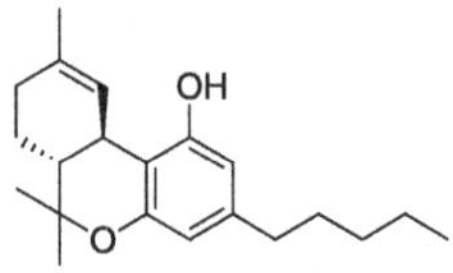

533 (dronabinol)

2.13.3 Biological Activities

Every plant part of *Cassia alata* [562] demonstrated a considerable range of biological activities, including antibacterial, antioxidant, antifungal, dermatophytic, antilipogenic, antidiabetic, antimalarial, antihyperlipidemic, anthelmintic, anti-inflammatory, and wound-healing properties [561]. Among the reported bioactive compounds, the flavonoids kaempferol (**88**) and luteolin (**389**), along with aloe-emodin (**516**) and other anthraquinone [560] derivatives, have exhibited antibacterial activity, while 2,5,7,4′-tetrahydroxyisoflavone (**527**) and 3,5,7,4'-tetrahydroxyflavone (**526**) showed antifungal properties. Some of the bioactivities of *C. alata* constituents are summarized in Table 14.

Table 14 Biological activities of *Cassia alata*

Activity	Assay(s)	References
Antibacterial	The antibacterial potential of an ethyl acetate extract of the roors and stems of *C. alata* showed efficacy against *Staphylococcus aureus*, a causative organism against cellulitis	[553]
	A diethyl ether extract of both the leaves and roots of *C. alata* showed inhibition against *Neisseria gonorrhoea*	[554]
Antioxidant	An aqueous extract of *C. alata* showed potent DPPH radical scavenging activity, in addition to reducing power and lipid peroxidation effects. These properties have been attributed to the presence in the plant of flavonoids and anthraquinones	[561, 562]
Antifungal	Methanol, ethanol, *n*-hexane, and chloroform extracts of *C. alata* were found to act by damaging fungal cell membranes and causing cell death, with their effectiveness shown against *Epidermophyton floccosum* and *Trichophyton rubrum*	[563, 564]
	The compounds 2,5,7,4′-tetrahydroxy isoflavone (**525**) and 3,5,7,4′-tetrahydroxy flavone (**524**) inhibited the growth of fungi including *Candida albicans* and *Aspergillus niger*. The anthraquinones aloe-emodin (**514**) and emodin (**530**) inhibited the growth of *Microsporeum gypseum*. *T. rubrum*, and *T. mentagrophytes*. Additionally, a cannabinoid (**531**) extracted from *C. alata* seeds using ethanol, reduced the growth of *C. albicans* and *A. niger*	
Antimalarial	A crude extract of *C. alata* leaves showed antimalarial activity against *P. falciparum* strain 3D7, with an IC_{50} value of 17.3 µg/cm^3 and a maximum inhibition of 65.6%	[565]
Antihelmintic	An acetone extract of *C. alata* leaves showed both ovicidal and larvicidal activities against the parasite *Haemonchus contortus*	[565]
Anti-inflammatory	The dimeric indole alkaloid cassiaindoline from *C. alata* leaves showed a 57.1% anti-inflammatory activity at a dosage of 75 mg/kg mouse	[561, 566–568]

2.14 *Gynura procumbens*

Gynura procumbens (Lour.) Merr. is a medicinal plant found commonly in tropical Asian countries such as Malaysia, Indonesia, Thailand, Vietnam, and parts of China. It is known as "Longevity Spinach", and "Sambung Nyawa" in Malay, and "Daun Dewa" in Indonesian. This is a herbaceous plant belonging to the family Asteraceae. It can grow to a height of one to three meters, and has a fleshy stem with a purple hue. The leaves are ovate-elliptic, measuring 3.5–8 cm in length and 0.8–3.5 cm in width. The flowering heads are yellow, narrow, and range from 1 to 1.5 cm long, arranged in panicles. This plant grows readily from stem cuttings when the seeds are not available. A semi-shade is preferred by this plant although it can be slowly adapted to grow in full sun, provided the plant does not dry out at the roots [569]. Plate 14 shows the leaves of *G. procumbens.*

Plate 14 Leaves of *Gynura procumbens* (family Asteraceae). Photograph courtesy Shahidul Hasan Roman (https://commons.wikimedia.org/wiki/File:Gynura_procumbens_Diabetes_plant)_(4).jpg)

2.14.1 Ethnobotany

In the food industry, *G. procumbens* is incorporated into coffee powder, tea, kimchi, chewing gum, chocolate, and candy. This plant is not only valued for its medicinal properties, but also serves as an attractive ornamental plant, featuring long trailing stems and bright green leaves [569], making it a popular choice for hanging baskets or ground cover. Additionally, it is employed in the preparation of Chinese traditional medicine to treat various illness, including chronic skin ulcers, uterine cancer, and cervical spondyloses. For culinary purposes, the leaves can be eaten raw in salads or cooked as a leafy green vegetable that possesses a slightly bitter taste. Also, *G. procumbens* use in personal care and cosmetic products includes handwash solutions and sanitizers, oral sprays, facial masks, and skincare creams. Furthermore, *G. procumbens* is also included in specialized diets for patients with heart and liver conditions [569]. It has also been utilized to treat various ailments, including rashes, hypertension, constipation, migraine, urinary infections, diabetes mellitus and cancer and, in Indonesia and Vietnam, to alleviate kidney discomfort and for fevers [570].

2.14.2 Phytochemistry

Phytochemical studies on *Gynura* species have identified a wide range of compounds, including flavonoids, phenolic compounds (quinic acid derivatives), and sterols. Interestingly, many *Gynura* species are used in both cooking and traditional medicine, with various conditions such as inflammation, migraine, fever, rashes, rheumatism, herpes, hypertension, constipation, hemostasis, kidney cancer, and diabetes mellitus being treated [570, 571].

Four types of solvents, ethyl acetate, ethanol, chloroform, and *n*-butanol, were utilized for the extraction of *G. procumbens*. It was found that quercetin (**85**), rutin (**86**), and kaempferol (**88**) were present in all extracts, whereas apigenin (**87**) and myricetin (**89**) were absent from the chloroform and *n*-butanol extracts. Five phenolic acids, namely, vanillic (**534**), gallic (**95**), protocatechuic (**535**), hydrobenzoic (**536**), and syringic (**492**) acids were identified from the ethanol and ethyl acetate extracts [572]. Hydroxycinnamic acids were also reported, including caffeic (**537**), ferulic (**487**), sinapic (**489**) and (*E*)-coumaric (**538**) acids, which were detected in all extracts with the exception of chloroform [573]. Several additional compounds were also identified using high-performance liquid chromatography, including oxypeucedanin (**539**), isoimperatorin (**540**), indole-3-carboxylic acid (**541**), kynurenic acid (**542**), (*Z*)-coumaric acid (**543**), lutein (**544**), and luteolin (**389**) [572].

Another study identified a number of further constituents of *G. procumbens* as 15,16-dihydroxy-(9Z),3,5-*O*-dicaffeoylquinic acid (**545**), (12*Z*)-octadecadienoic acid (**546**), 3,4-dicaffeoylquinic acid (**547**), 3-*O*-methyl gallic acid (**548**), 3-carene (**416**), 5-*O*-(*E*)-caffeoyl-galactaric acid (**549**), 4-*O*-methyl-gallic acid sulfate (**550**), chlorogenic acid (**392**), choline (**551**), and cynarine (**552**). Other constituents reported for *G. procumbens* have been eriocitrin (**553**), ferulic acid (**489**), feruloylquinic acid (**554**), isobioquercetin (**555**), a genkwanin isomer, kaempferol-3-*O*-rhamnosyl-(1→6)-glucoside (**556**), limonene (**402**), malic acid (**557**), neochlorogenic acid (**558**), oxooctadecanoic acid (**559**), *p*-coumaroylquinic acid (**560**), *p*-hydroxybenzoic acid

(**561**), phenylalanine (**562**), protocatechuic acid (**535**), quercetin 3-*O*-rhamnosyl-(1→2)-galactoside (**563**), quercetin (1→6)-glucoside (**564**), and quercetin 3-*O*-rutinoside (**565**). Also, trimethyl gallic acid glucuronide (**566**), and α-pinene (**399**) were reported [574].

The leaves of *G. procumbens* accumulate compounds such as quercetin (**85**), rutin (**86**), kaempferol (**88**), chlorogenic acid (**392**), astragalin (**485**), kaempferol-3-*O*-*β*-D-glucopyranoside (**525**), kaempferol-3-*O*-rutinoside (**567**), and 3,5-dicaffeoylquinate methyl ester (**568**) [573]. Three compounds were isolated from an ethyl acetate-soluble fraction of a methanol crude extract of this plant through purification on silica gel, namely, stigmasterol (**109**), phytol (**569**), and lupeol (**620**) [575]. The structures of constituents **534**–**567** of *G. procumbens* are displayed in the formula drawings below.

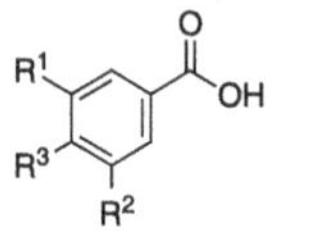

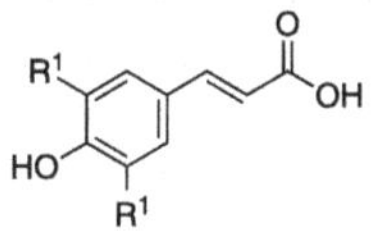

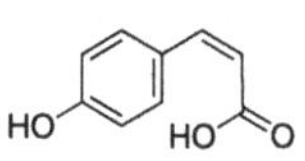

534 (vanillic acid) R[1] = OMe, R[2] = H, R[3] = OH
535 (protocatechuic acid) R[1] = R[3] = OH, R[2] = H
536 (hydrobenzoic acid) R[1] = R[2] = H, R[3] = OH

537 (caffeic acid) R[1] = OH, R[2] = H
538 ((*E*)-coumaric acid) R[1] = R[2] = H

543 ((*Z*)-coumaric acid)

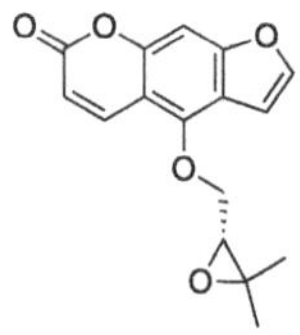
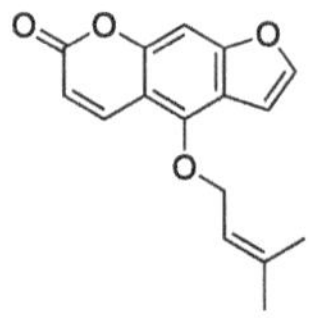
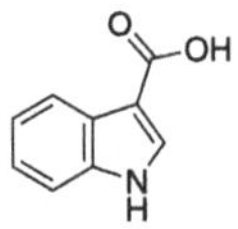

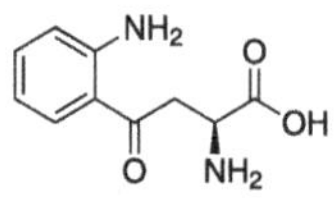

539 (oxypeucadin) **540** (isoimperatorin) **541** (indole-3-carboxylic acid) **542** (kynurenic acid)

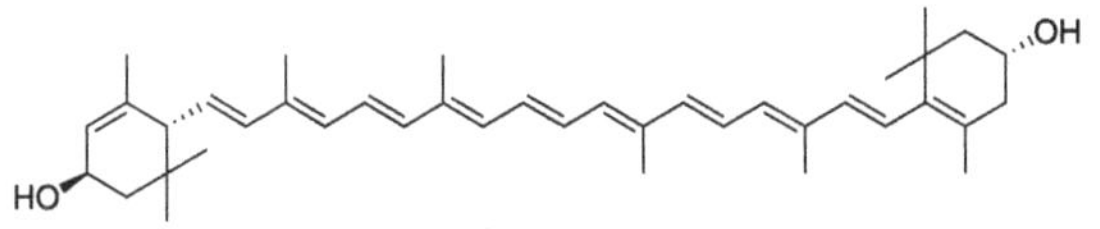

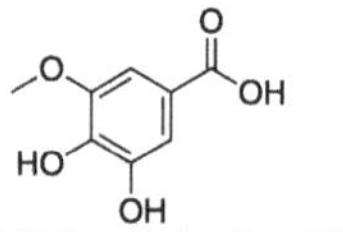

544 (lutein)

548 (3-*O*-methyl gallic acid)

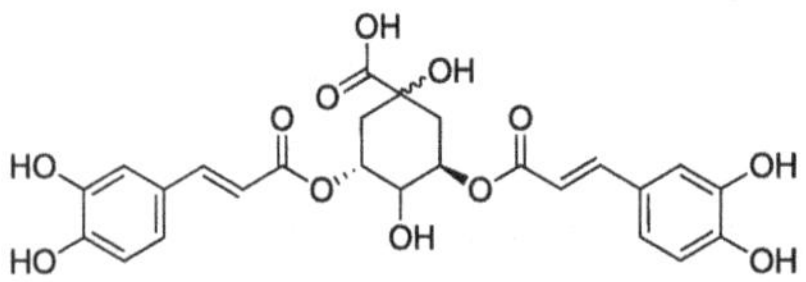

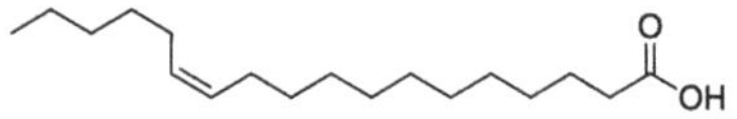

545 (15,16 dihydroxy-(9*Z*),3,5-*O*-dicaffeoylquinic acid

546 ((12*Z*)-octadecanoic acid)

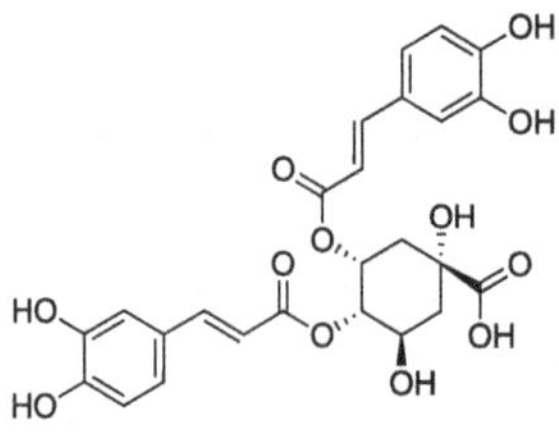

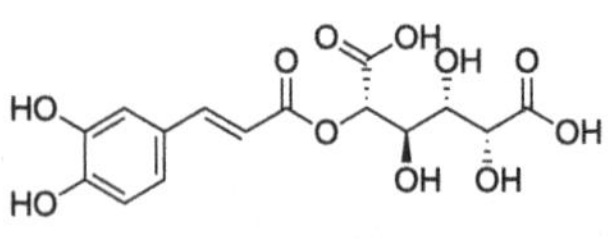

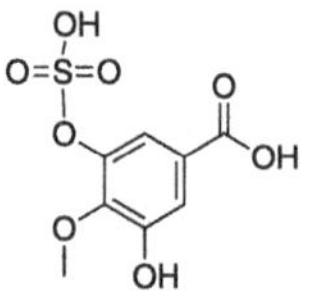

547 (3,4-dicaffeoylquinic acid) **549** (5-*O*-(*E*)-caffeoyl-galactaric acid) **550** (4-*O*-methyl-gallic acid sulfate)

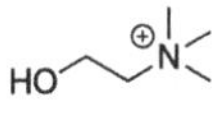

551 (choline)

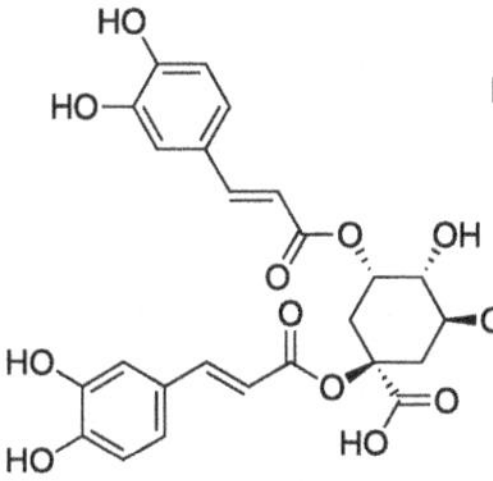

552 (cynarine)

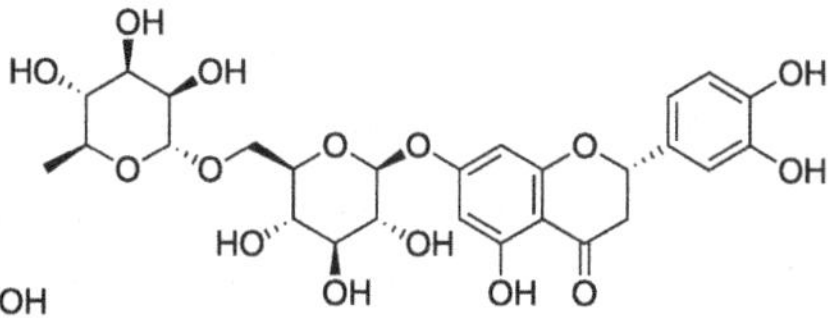

553 (eriocitrin)

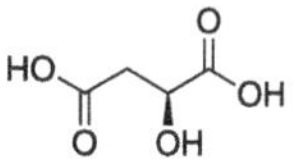

557 (malic acid)

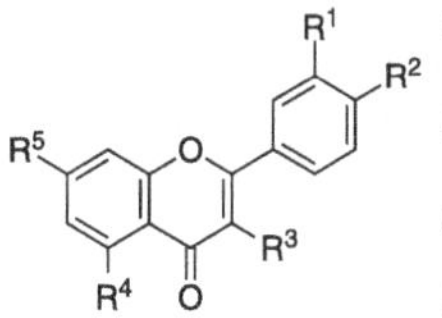

555 (isobioquercetin) R¹ = R² = R⁴ = R⁵ = OH, R³ = OH
556 (kaempferol-3-*O*-rhamnosyl-(1→6)-glucoside)
R¹ = H, R = R⁴ = R⁵ = OH, R³ = Glc-Rha-Glc
563 (quercetin 3-*O*-rhamnosyl-(1→2)-galactoside)
R¹ = R² = R⁴ = R⁵= OH, R³ = Glc-Rha-Rha
564 (quercetin (1→6)-glucoside)
R¹ = R² = R⁴ = R⁵ = OH, R³ = Glc-Rha
565 (quercetin-3-*O*-rutinoside)
R¹ = R² = R⁴ = R⁵ = OH, R³ = Ru
567 (kaempferol-3-*O*-rutinoside)
R¹= H, R² = R⁴ = R⁵ = OH, R³ = Glc-Rha

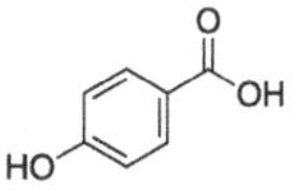

561 (*p*-hydroxybenzoic acid)

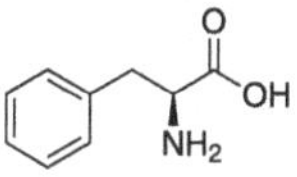

562 (phenylalanine)

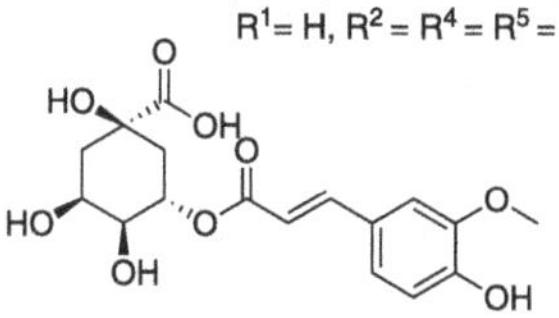

554 (feruloylquinic acid)

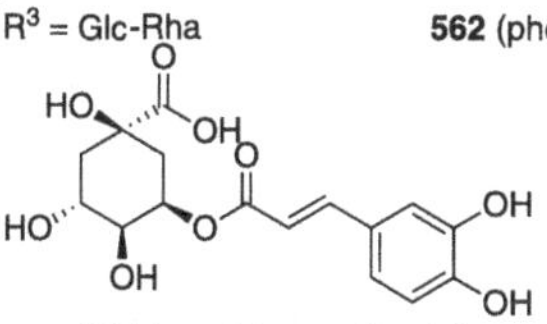

558 (neochlorogenic acid)

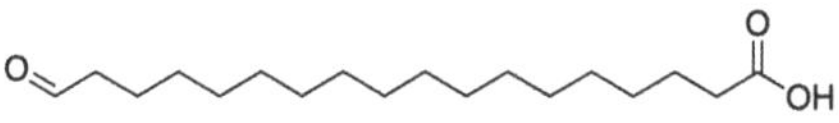

559 (oxooctadecanoic acid)

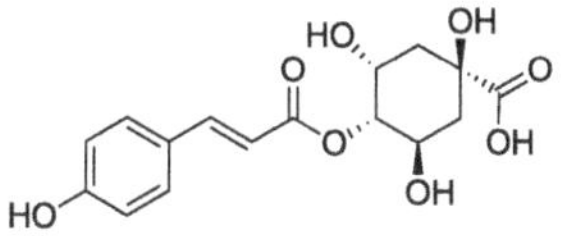

560 (*p*-coumaroylquinic acid)

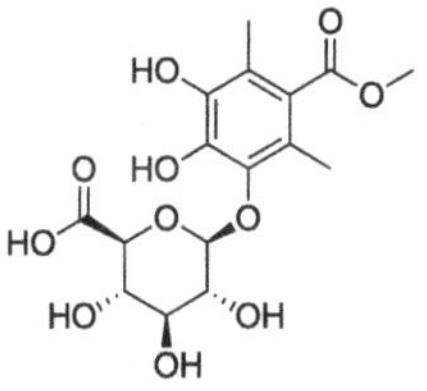

566 (trimethyl gallic acid glucuronide)

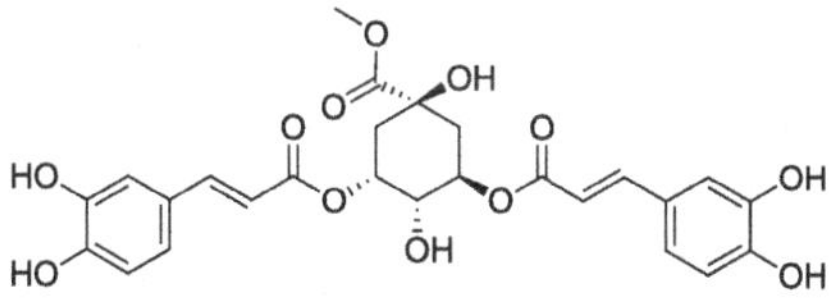

568 (3,5-dicaffeoylquinate methyl ester)

2.14.3 Biological Activities

Gynura procumbens has been reported to exhibit antihypertensive, cardioprotective, antihyperglycemic, fertility enhancement, cytotoxic, antimicrobial, antioxidant [594], organ protective, and antiinflammatory activities (Table 15), indicating its potential for the development of novel therapeutic agents [597–599].

Table 15 Biological activities of *Gynura procumbens*

Activity	Assay(s)	References
Antihypertensive/cardioprotective	A petroleum ether extract of the leaves of *G. procumbens* showed the lowering of systolic blood pressure and of the mean arterial pressure in hypertensive rats	[576]
	Extracts of the plant gave a decrease in the heart rate, and strong negative chronotropic and negative ionotropic effects in the right atrium and left atrium, respectively, using hypertensive rats	[577, 578]
	Inhibition of angiotensin II-induced contraction of rat aortic rings occurred when treated with a fraction of *G. procumbens*	[579, 580]
Antihyperglycemic	Extracts of *G. procumbens* induced hypoglycemic effects in diabetic rats and caused fasting blood glucose levels to be decreased as well as the suppression of glucose elevation during glucose tolerance testing	[581, 582]
	The plant stimulated glucose uptake in 3T3 adipocytes, and an enhancement of activity was found in the presence of insulin	[583]
	Potent hypoglycemic effects were observed in a rat model when *G. procumbens* was combined with *Azadirachta indica* or *Andrographis paniculata*	[584]
Fertility enhancement	An aqueous extract of *G. procumbens* increased sperm count, sperm motility, and reduced the degree of sperm mortality using diabetic rats	[585]
	An aqueous extract increased the sperm count and motility and promoted testicular lactate dehydrogenase activity in diabetic rats	[586]
Anticarcinogenic/cytotoxic	Short-term (10 week) treatment with an ethanolic extract of *G. procumbens* suppressed the progression in rats of nitroquinoline 1-oxide-induced tongue carcinogenesis when given during the initiation phase. Administration for a longer period (26 weeks) led to a greater suppression of oral carcinogenesis	[587]
	An ethanol extract inhibited cell proliferation and suppressed the invasive and migratory abilities of osterosarcoma cells	[588]

(continued)

Table 15 (continued)

Activity	Assay(s)	References
	Effective suppression in the proliferation of breast cancer and mammalian gland epithelial cells occurred when an ethanol extract was co-administered with doxorubicin	[589]
Antiangiogenic	An antiangiogenic activity of the leaves of *G. procumbens* resulted in the inhibition of the expression of vascular endothelial growth factor and the prevention of the formation of new blood vessels in fertilized chicken eggs	[590, 591]
Antimicrobial	An ethanol extract of the plant aerial parts exhibited virucidal and antireplicative activities against the herpes simplex viruses HSV-1 and HSV-2. This was validated in a clinical trial on patients with recurrent herpes labialis where treatment with a *G. procumbens* gel reduced the number of patients infected with HSV	[592]
Antioxidant	An ethanol extract exhibited the highest percentage of DPPH inhibition (52.81%) among the *G. procumbens* plant extracts tested	[593, 594]
	Inhibited lipid peroxidation with the median effective concentration of 2.75 mg/cm^3	
Organ-protective effects	Administration of ethanol extract significantly lessened the extent of ethanol-induced gastric ulcers in rats, with a reduction of submucosal edema and infiltration of leucocytes being observed	[595]
	An aqueous extract of the plant was found to cause inhibition of human kidney mesangial cell proliferation and DNA synthesis	[596]
Anti-inflammatory	Prevented the increase in ear thickness of mice caused by croton oil-induced inflammation	[597]
	The treatment of ethanolic leaf extracts of *G. procumbens* at 0.1 and 1.0 μg/cm^3 caused higher proliferation of CD4+CD25+, CD4+CD62L−, CD4+CD62L+, CD8+CD62L−, and CD8+CD62L+ T cells but lower proliferation of B220+ cells when compared to the higher dosage at 10 μg/cm^3. However, at a dosage of 10 μg/cm^3, it was shown to promote high proliferation of B cells. The concentration used plays an important role in determining whether *G. procumbens* acts as an immunostimulant or immunosuppressant	[598–600]

This plant has been used as a herbal remedy to treat numerous diseases, especially in Southeast Asia. Some of the bioactive compounds reported are flavonoids, such as kaempferol (**88**), myricetin (**89**), astragalin (**483**), rutin (**86**), and quercetin (**85**), which have antihypertensive and cardioprotective properties. The major compounds of this plant are from the flavonoid group, which are known for their appreciable activities in various assays.

2.15 *Melastoma malabathricum*

Melastoma malabathricum L. (Melastomataceae) is a flowering plant found in tropical and subtropical regions. It is known commonly as "Senduduk" in Malaysia, as well as "Indian rhododendron" or "Singapore rhododendron". This species is native to Asia and is found in countries such as Malaysia, Indonesia, the Philippines and India. It is a shrub (Plate 15) that can reach a height of up to three meters, and has simple, ovate leaves, producing attractive purple flowers with five petals. This species has at least three varieties, i.e., having large, medium, and small-sized flowers with dark purple-magenta petals, light pink-magenta petals, and the rare white variety, respectively. Each fruit is a small round berry that is edible and sometimes used in traditional medicine. In general, various parts of this plant including the leaves, roots, and bark are used widely in Indonesia, Malaysia, and India to treat ailments like diarrhea, leukorrhea, dysentery, hemorrhoids, miscellaneous infections, stomachache, toothache, and thrush [601].

2.15.1 Ethnobotany

Melastoma malabathricum is often cultivated as an ornamental plant due to its attractive purple flowers, thereby adding an esthetic value to gardens and landscapes. Its dense growth habit and robust root system make it an effective tool for erosion control, particularly on slopes and banks. The plant also plays an essential role in supporting wildlife, providing food and a habitat for birds and insects. This species has been used for centuries by various communities ethnomedically.

Melastoma malabathricum, has been used for medicinal purposes in Malaysia, India, Thailand, Indonesia, and the Philippines. Each part of the plant is employed in different ways for treating specific ailments. The leaves can be chewed, pounded, or chopped to create a paste or juice to treat cuts, stop bleeding, prevent smallpox scarring, and treat dysentery. The young leaves are eaten to stop diarrhea, while the immature leaves are consumed raw for dysentery [601]. In Malaysia, the leaves are often used in an aqueous extract form and given to postpartum women for healing and womb strengthening. They are also applied as a poultice to aid in healing processes and reduce hemorrhoid discomfort. Similarly, in India, the Phillipines, Thailand, and Indonesia, the plant has been used to reduce smallpox scars, treat dysentery, skin diseases, diarrhea, ulcers, toothache and hemorrhoids.

Plate 15 Flowers of *Melastoma malabathricum* (family Melastomataceae). Photograph courtesy Wiki Farazi (https://commons.wikimedia.org/wiki/File:Melastoma_malabathricum_at_Elmina_West_20231022_105450.jpg)

The roots of the plant also play a significant role in traditional medicine. In Malaysia, a decoction made from the roots is utilized for epilepsy, rheumatoid arthritis, tenderness in the legs, and diarrhea. Also, a paste made from the roots in India is applied to cuts and wounds, and employed for stomach disorders and fevers. In Thailand, a root decoction is commonly used to treat fever, while in Indonesia, the roots are used to treat rheumatoid arthritis, toothache, and epilepsy [603].

The fruits of the plant in Malaysia are made into a juice, which can relieve dry lips, while the shoots are used to address puerperal infections, blood pressure, diabetes, toothache, and leucorrhea. The shoots can also be consumed to treat diabetes. Additionally, the stem bark of the plant is used in India for skin diseases [603] and the juice from the shoots can be used as a mouthwash for toothache relief and to treat leukorrhea [602].

2.15.2 Phytochemistry

In terms of its phytochemical constituents, few studies have been reported on this plant. GC/MS analysis of a methanol crude extract of *M. malabathricum* resulted in the identification of 5-hydroxymethylfurfural (**569**), pyrogallol (**99**), phytol (**570**), palmitic acid (**498**), palmitic acid methyl ester (**571**), 8,11-octadecadienoic acid methyl ester (**572**), stearic acid methyl ester (**573**), *all*-(*E*)-squalene (**574**), and tocopherol (**575**) [603]. Samad et al. identified and characterized phytochemical constituents inclusive of patriscabratine (**576**), α-amyrin (**577**), auranamide (**578**), quercitin (**85**), quercetrin (**579**), and kaempferol-3-*O* (2″,6″-di-*O*-*p*-(*E*)-coumaroyl)-*β*-glucoside (**580**) through purification using repeated chromatographic techniques [604]. The constituents of *M. malabathricum* are displayed in formula drawings **568–588**.

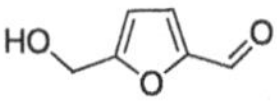

569 (5-hydroxymethylfurfural)

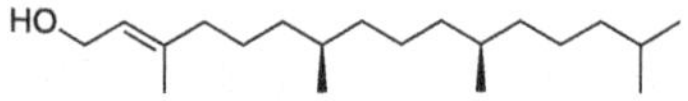

570 (phytol)

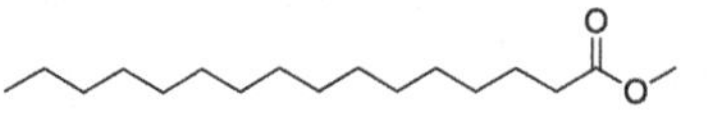

571 (palmitic acid methyl ester)

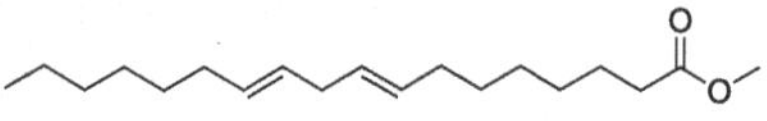

572 (8,11-octadecadienoic acid methyl ester)

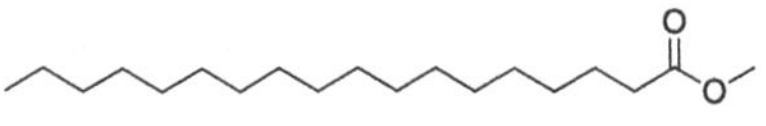

573 (stearic acid methyl ester)

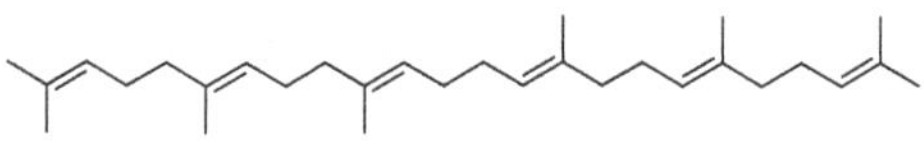

574 (*all*-(*E*)-squalene)

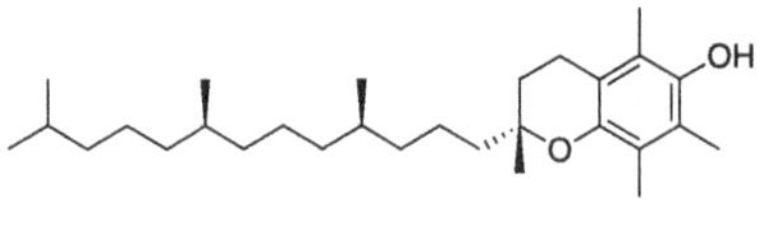

575 (tocopherol)

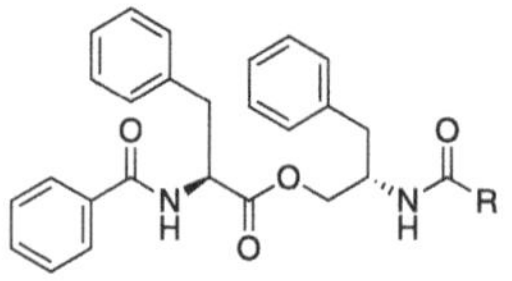

576 (patriscabratine) R = CH_3
578 (auranamide) R = benzene

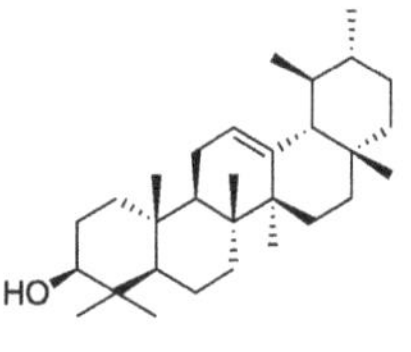

577 (α-amyrin)

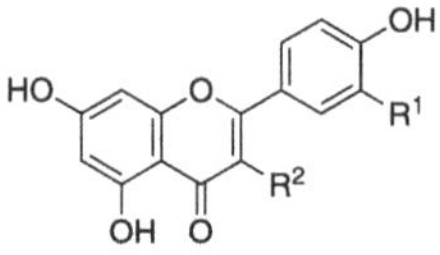

579 (quercitrin) R^1 = OH, R^2 = *O*-Rha
580 (kaempferol-3-*O*-(2″,6″-di-*O*-*p*-(*E*)-coumaroyl)-β-glucoside) R^1 = H, R^2 = *O*-(2″,6″-di-*O*-*p*-(*E*)-coumaroyl)-*β*-glucose

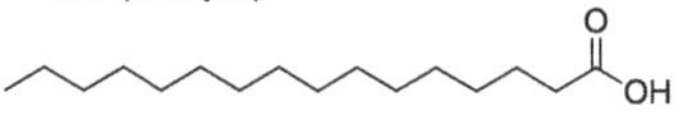

581 (hexadecanoic acid)

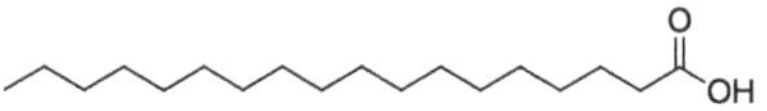

582 (stearic acid)

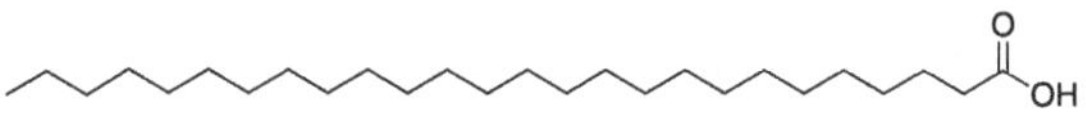

583 (hexacosanoic acid)

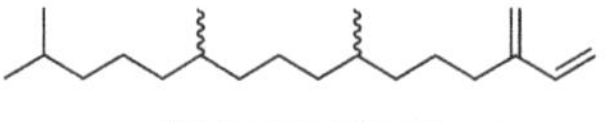

584 (neophytadiene)

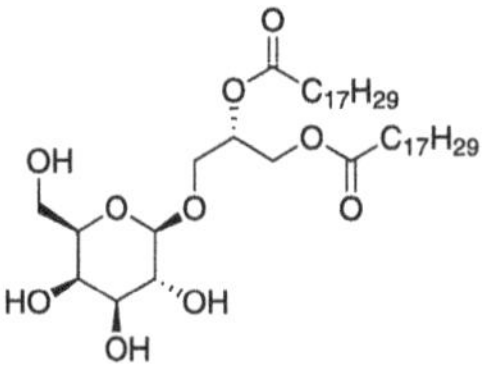

585 (1,2-di-*O*-linaloyl-3-*O*-*β*-D-galactopyranosyl-*sn*-glycerol)

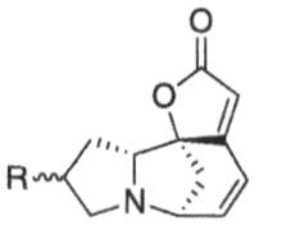

587 (4-methoxy-*nor*-securinine) R = OCH_3
588 (*nor*-securinine) R = H

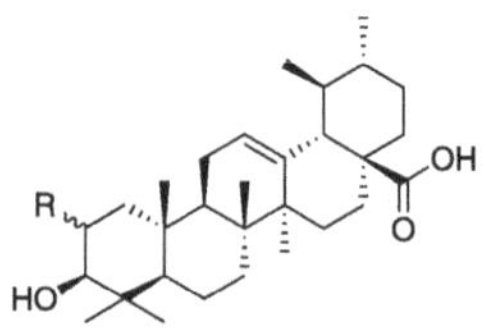

586 (ursolic acid) R = H
589 (2*α*-hydroxyursolic acid) R = OH

2.15.3 Biological Activities

Melastoma malabathricum has been shown to exhibit antimicrobial, anti-inflammatory, antioxidant, and cytotoxic activities, among others (see Table 16). Phytochemical studies identified a variety of bioactive compounds, including, flavonoids, fatty acids, alkaloids and sterols. Examples include α-amyrin (**577**), betulinic acid (**511**), kaempferol-3-*O*-(2″,6″-di-*O*-*p*-(*E*)-coumaroyl)-*β*-glucoside (**580**), and naringenin (**92**), further supporting the plant's potential in the treatment of various diseases, due to their demonstrated wide range of biological activities.

Table 16 Biological activities of *Melastoma malabathricum*

Activity	Assay(s)	References
Antimicrobial	An acetone extract of *M. malabathricum* stem back showed growth inhibitory activity against *Streptococcus mutans*. α-Amyrin (**576**), β-sitosterol (**479**), hexadecenoic acid (**580**), stearic acid (**581**), and hexacosanoic acid (**582**) were present in an active fraction of the plant extract	[601, 602]
	A methanol extract of *M. malabathricum* showed antibacterial activities aganst *Bacillus subtilis*, *Escherichia coli*, *Pseudomonas aeruginosa*, and *Staphylococcus aureus*, while an aqueous extract was not active in this regard	[606]
Antiinflammatory	α-Amyrin (**576**) and betulinic acid (**511**) showed inhibitory effects on platelet-aggregating factor (PAF) in rabbit platelets (IC_{50} values of 20.0 and 22.2 μ*M*, respectively). Additionally, quercetin (**85**) and quercitrin (**578**) showed lower potencies in this in vitro bioassay	[605, 607]
Antioxidant	An ultrasound extract of *M. malabathricum* showed free radical scavenging activity in a DPPH assay, with the activity related to the high flavonoid content present	[605]
Cytotoxic	Naringenin (**92**), isolated from the flowers of *M. malabathricum*, inhibited MCF-7 human breast cancer cell proliferation with an IC_{50} value of 0.28 μ*M*	[601, 604]

2.16 *Phyllanthus niruri*

Phyllanthus niruri L., commonly known as "Gale of the wind", "Chancapiedra" (stone breaker) and "Seed under a leaf" is an annual herb belonging to the Euphorbiaceae. These common names were given based on the plant's morphological and pharmacological properties [608]. Other vernacular names are "meniran" and "meniran green", which were accorded by the Javanese and Sundanese in Indonesia [609]. The plant is distributed widely in tropical and subtropical countries including Sri Lanka, the coastal regions of India, and Southeast Asia [609, 610]. This species occurs as a weed in cultivated and wasteland areas, especially during the rainy season

[611]. In Indonesia, *P. niruri* is found at an altitude of 1–1000 m above sea level in loose soils along riverbanks and beaches, and in open fields [612].

This species grows up to 25–45 cm in height each year and the leaves are sessile and alternate and 7–12 cm in size [609]. The trunk is a large, round, slippery, hairless stem and green in color. The flowers have a single form, located near the petiole, hanging and white in color. The crown is small, flat and purplish-green. The seeds are small and are hard and brownish, while the taproot is off-white [609]. The leaves and inflorescence of *P. niruri* are illustrated in Plate 16.

Plate 16 Leaves and inflorescence of *Phyllanthus niruri* (family Phyllanthaceae)

2.16.1 Ethnobotany

Since ancient times, *P. niruri* has played an important role in traditional medicine. The whole plant parts have been used traditionally to treat kidney stones in Brazil, hepatitis, dysentery, enteritis and urinary infection in China, and malaria in the Republic of the Congo [613]. This species is utilized for bronchitis, anemia, leprosy, asthma, kidney calculi, ulcers, wounds, sores, scabies, ringworm, gonorrhea, menstruation, diabetes, stimulating the liver, improving digestion, alleviating the appetite and producing laxative effects in India, and for gallstones, kidney calculi, fever, and excess uric acid in Latin America [613, 614]. The plant is used also for viral infections and hepatitis in Indonesia, and diarrhea, kidney disorders, gonorrhea, and coughs in Malaysia. Furthermore, in Thailand, the aerial parts have use as a treatment for anorexia and malaria [613]. Due to its wide-ranging ethnomedical properties, this species has been commercialized as a supplement in capsule and in powdered form.

2.16.2 Phytochemistry

Phyllanthus niruri has been reported to biosynthesize various natural products, including alkaloids, flavonoids, lignans, terpenoids and phenols. The isolation from the stems, aerial parts and roots has afforded three alkaloids, namely, 4-methoxy-*nor*-securinine (**587**), *nor*-securinine (**588**) and 4-methoxy-securinine (**590**) [614]. Flavonoids such as quercitrin (**579**), isoquercitrin (**592**), rutin (**86**), astragalin (**485**), kaempferol-4′-rhamnopyranoside (**593**) [615], eridictyol-7-rhamnopyranoside (**594**) and fisetin (**595**) [616] were isolated from the leaves, aerial parts and roots.

Several lignans were isolated from the leaves and aerial parts of *P. niruri*, including phyllanthin (**596**) [617], hypophyllanthin (**597**) [617, 618], niranthin (**598**), nirtetralin (**599**), phyltetralin (**600**) [619, 620], hinokinin (**601**), isolintetralin (**602**) [621], *seco*-isolariciresinol trimethyl ether (**603**), hydroxyniranthin (**604**), *seco*-4-hydroxylintetralin (**605**), dibenzylbutyrolactone (**606**) [622], nirphyllin (**607**), phylnirurin (**608**) [623], lintetralin (**609**) [623], 2,3-desmethoxy-*seco*-isolintetralin (**610**), 2,3-desmethoxy-*seco*-isolintetralin diacetate (**611**), and demethylenedioxy niranthin (**612**) [622].

In addition, ricinoleic acid (**613**), estradiol (**614**), β-sitosterol (**479**) and 24-isopropyl-cholesterol (**615**) are among the miscellaneous constituents reported from the seeds, leaves and aerial parts [614]. Other isolated phenols were identified as catechin (**90**), epicatechin (**616**), gallocatechin (**617**), epigallocatechin (**91**), epicatechin-3-*O*-gallate (**618**), and epigallocatechin-3-*O*-gallate (**619**). These compounds were isolated from the plant roots and root cultures [624]. Phytochemical constitiuents isolated from *Phyllanthus niruri* are shown in formula drawings **589–618**.

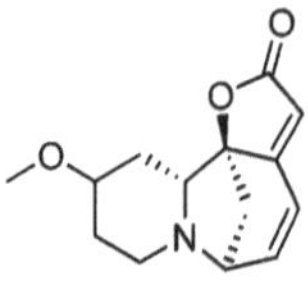

590 (4-methoxy-securinine)

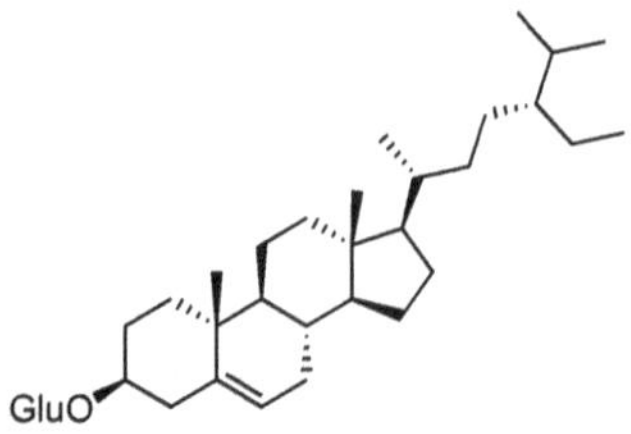

591 (β-sitosterol-3-*O*-β-D-glucopyranoside)

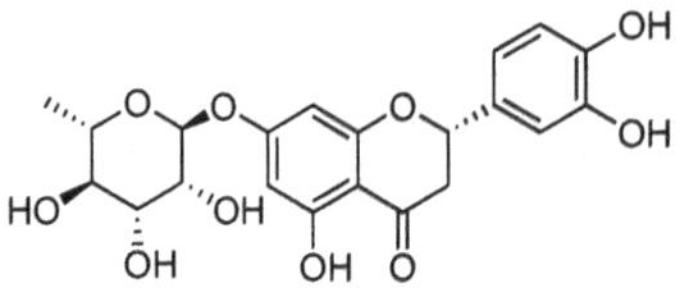

594 (eriodictyol-7-rhamnopyranoside)

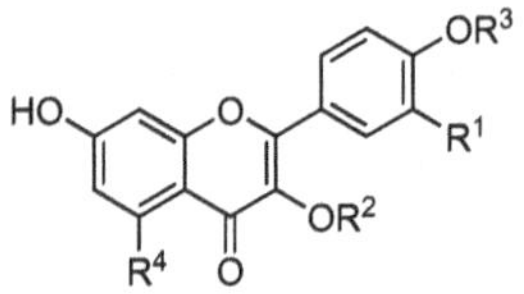

592 (isoquercitrin) R^1 = OH, R^2 = glc, R^3 = H, R^4 = OH
593 (kaempferol-4'-rhamnopyranoside) R^1 = R^2 = H, R^3 = rha, R^4 = OH
595 (fisetin) R^1 = OH, R^2 = R^3 = R^4 = H

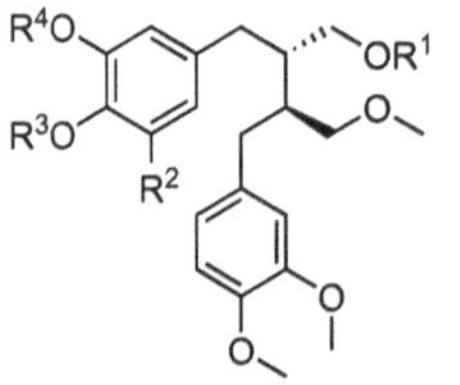

596 (phyllanthin) R^1 = R^3 = R^4 = CH_3, R^2 = H
603 (*seco*-isolariceresinol trimethyl ether) R^1 = R^2 = H, R^3 = R^4 = CH_3
612 (demethylenedioxy niranthin) R^1 = R^2 = CH_3, R^3 = R^4 = H

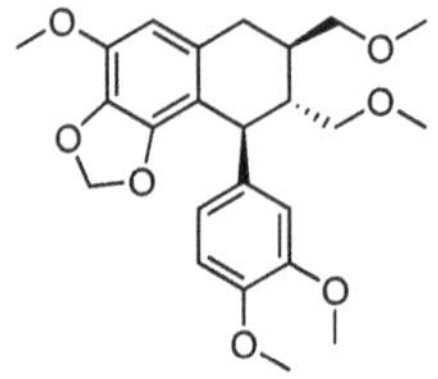

597 (hypophyllanthin)

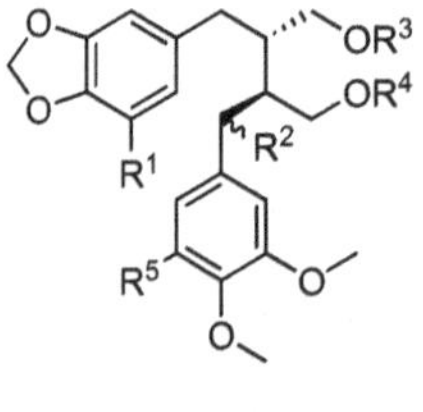

598 (niranthin) R^1 = OMe, R^2 = H, R^3 = R^4 = CH_3, R^5 = H
604 (hydroxyniranthin) R^1 = OMe, R^2 = OH, R^3 = R^4 = CH_3, R^5 = H
607 (nirphyllin) R^1 = OMe, R^2 = H, R^3 = R^4 = CH_3, R^5 = OMe
610 (2,3-desmethoxy-*seco*-isolintetralin) R^1 = R^2 = R^3 = R^4 = R^5 = H
611 (2,3-desmethoxy-*seco*-isolintetralin diacetate) R^1 = R^2 = H, R^3 = R^4 = $COCH_3$, R^5 = H

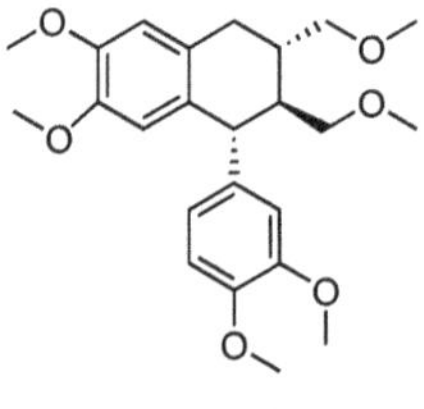

600 (phyltetralin)

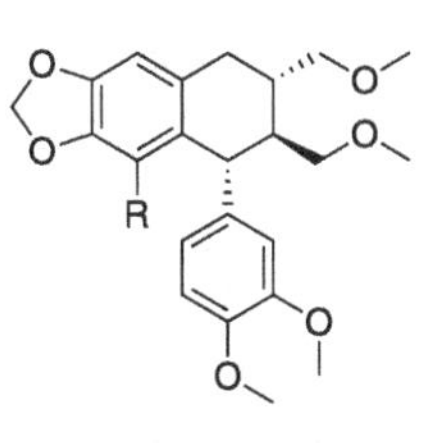

602 (isolintetralin) R = H
599 (nirtetralin) R = OMe

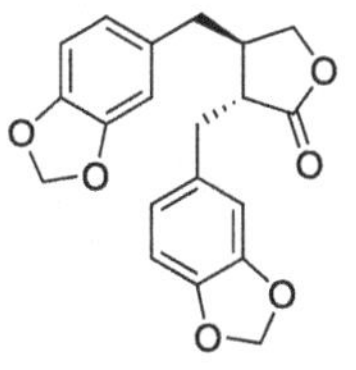

601 (hinokinin)

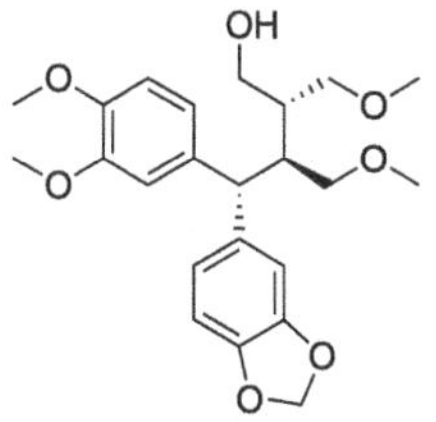

605 (*seco*-4-hydroxylintetralin)

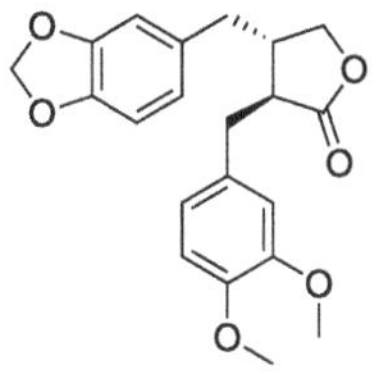

606 (dibenzylbutyrolactone)

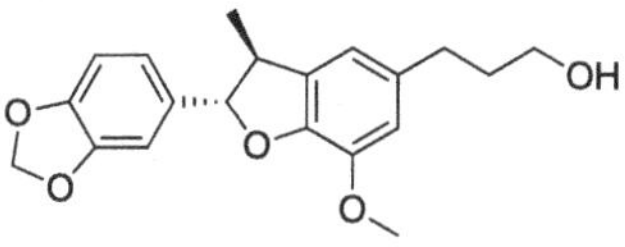

608 (phyllniruin)

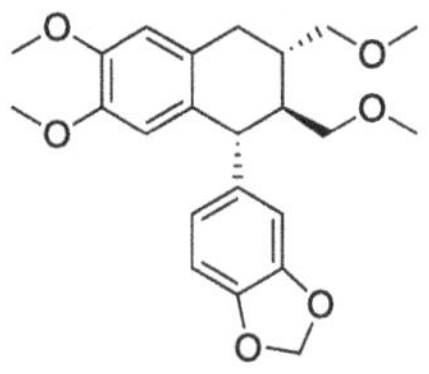

609 (lintetralin)

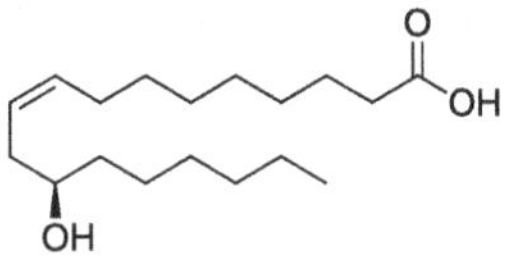

613 (ricinoleic acid)

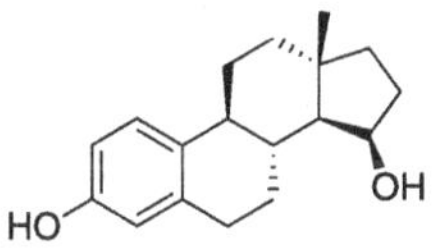

614 (estradiol)

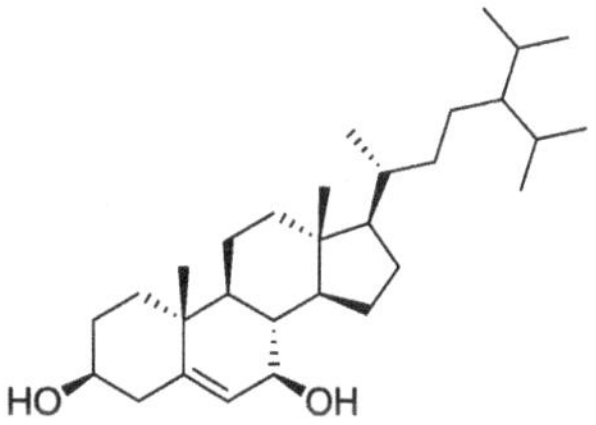

615 (24-isopropyl cholesterol)

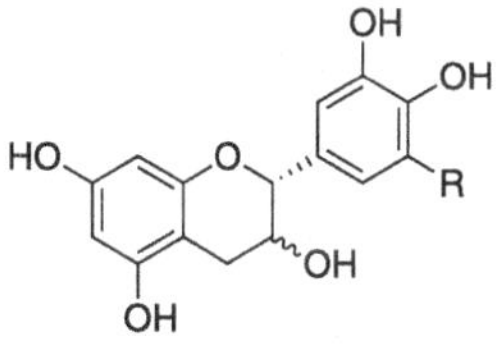

616 (epicatechin) OH *cis*, R = H
617 (gallocatechin) OH *trans*, R = OH

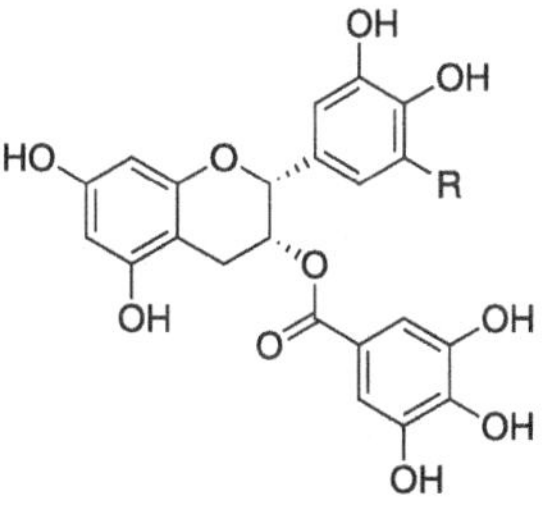

618 (epicatechin-3-*O*-gallate) R = H
619 (epigalocatechin-3-*O*-gallate) R = OH

2.16.3 Biological Activities

Phyllanthus niruri has been reported to possess various biological properties. An extract of this species had an effect against hepatitis B and other viral infections. Ellagic acid (**487**) from this plant demonstrated cytotoxicity against HepG2/C3A hepatocellular carcinoma cells [625]. Furthermore, phyllanthin (**596**) was reported to inhibit the increased levels of serum alanine transaminase and aspartate aminotransferase at doses of 120 and 180 mg/kg, respectively, in a hepatotoxicity mouse model caused by CCl_4. Their serum levels were found to decrease in the ranges of 74–83% and 62–63%, respectively, which were comparable with the standard used, silymarin, with a 66% decrease [626].

An aqueous extract of 400 and 800 mg/kg of *P. niruri* was given to rats to study the effect of chronic granuloma and anti-inflammatory activity in carrageenan and chronic granuloma models was observed [627]. Furthermore, a spray-dried powder of *P. niruri* at doses of 100, 200, 800, or 1600 mg/kg exhibited inhibitory effects on carrageenan-induced paw edema in a rodent model when compared to the positive control used. Additionally, a methanol extract of the leaves of *P. niruri* at a doses of 100, 200, and 400 mg/kg demonstrated reductions in inflammation of 46.8, 55.3, and 69.1%, respectively, in carrageenan-induced paw edema of Swiss albino rats. The standard drug ibuprofen was used at a dose of 20 mg/kg. It was also reported that the spray-dried powder of the plant inhibited leukocyte migration at 100 or 200 mg/kg, again when compared to a positive control [628] The aqueous and methanol extracts of *P. niruri* inhibited microsomal lipid peroxidation induced by Fe^{2+} and ascorbate. The leaf and fruit extracts demonstrated antioxidant activity and inhibited superoxide ions in vitro [629]. Moreover, an aqueous extract from the leaves reduced oxidative stress in the kidneys through the prevention of the decrease of endogenous antioxidant enzyme levels [630]. Jose and co-workers reported in 2014 activity in DPPH and hydroxyl radical-scavenging assays by the flavonoids of this plant [631].

2.17 *Ficus deltoidea*

Ficus deltoidea Jack, commonly known as "mistletoe fig", belongs to the Moraceae family [632]. This species is native to the Malay Archipelago and is widely distributed in Malaysia, Thailand, and Indonesia [633]. This understory plant commonly occcurs on lowlands and mountains at elevations up to 3,200 m [634]. The species is called "Mas Cotek", "Telinga Beruk", or "Serapat Angin" by the Malays, while the Sabahan, Sarawakian, and Kalimantan Island inhabitants call it "Sempit-Sempit" and "Agoluran". Other vernacular names are "Tabat Barito" (Indonesia) and "Kangkalibang" (Africa) [635]. This species is one of the most popular plants in Malaysia and is used widely for medicinal purposes [632].

Ficus deltoidea usually starts as an epiphyte, which grows on a large tree, hence the name "mistletoe fig". It is an evergreen shrub that grows up to 4.5–6.7 m high with 0.9–3 m wide zigzagged branches [632]. The trunk is slender and the bark is whitish-grey. The leaves are dark, leathery, succulent, and spoon-shaped to obovate in shape, and 4–8 cm in length [631, 634]. The male flowers have small leaves with purple spots, while the female flowers have thick, large leaves with black spots [635]. The characteristics of the leaves aid in the identification of male and female plants [632] (Plate 17).

Plate 17 Leaves and inflorescence of *Ficus deltoidea* (family Moraceae)

2.17.1 Ethnobotany

Traditionally, *Ficus deltoidea* has been used widely by the elderly to treat various illnesses. In Peninsular Malaysia, the roots and leaves when powdered have been used to treat wounds, rheumatism and sores [634, 636]. A decoction of the leaves is used as an antidiabetic, as a tonic after childbirth, and to treat irregular menstrual cycles and leukorrhea, and to boost the libido [636–639]. The fruits are chewed to relieve toothache, colds, fevers, and headaches [638, 640]. The plant is used also as a treatment for gout, high blood pressure, pneumonia, diarrhea, and skin infections [641]. Moreover, women in Indonesia consume the whole plant as a health-giving and aphrodisiac tonic [635].

In Malay traditional medical practice, *F. deltoidea* is often consumed as a tea [642]. This plant is formulated in the form of capsules and a tonic tea and is marketed throughout Malaysia [643]. It is commercialized in a sachet containing 1–2 g of dried leaves to be infused in 1 dm^3 of hot water [637]. The plant juice from this species is believed to enhance beauty and health [644].

2.17.2 Phytochemistry

Ficus deltoidea is known to contain various classes of constituents, such as members of the flavonoid and triterpene structural classes [645]. The leaves of *F. deltoidea* afforded moretenol (**620**) [646] and later, lupeol (**621**) [647], and two flavonoids, namely, vitexin (**622**) and isovitexin (**623**), were isolated from this same plant part [648].

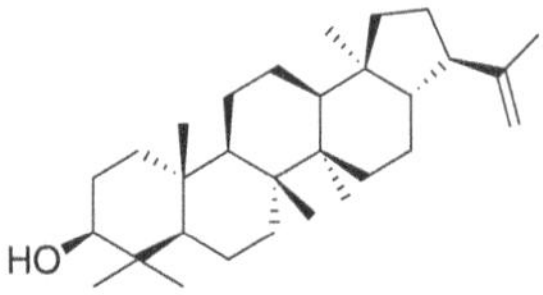

620 (moretenol)

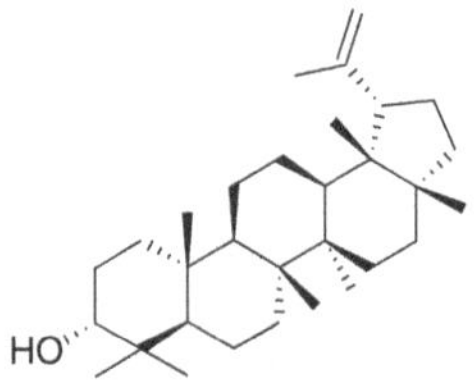

621 (lupeol)

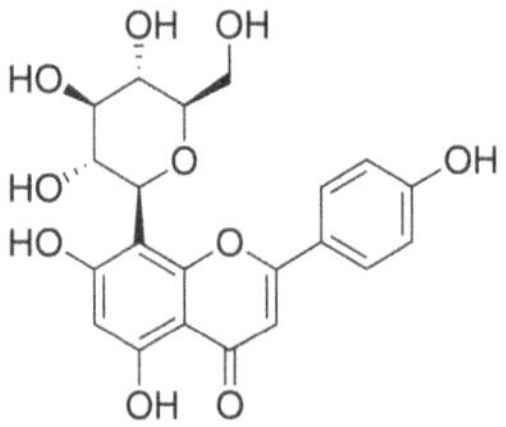

622 (vitexin)

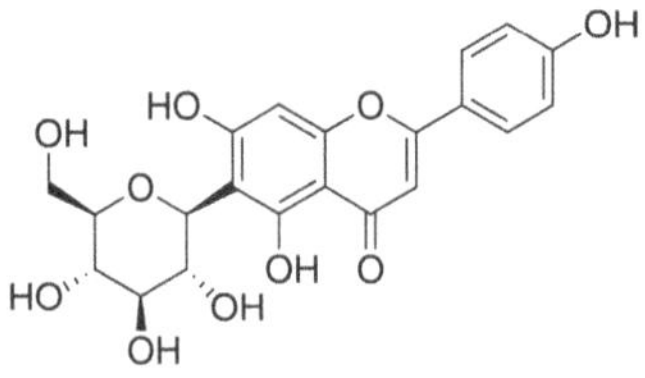

623 (isovitexin)

2.17.3 Biological Activities

Ficus deltoidea has a broad spectrum of bioactivities, such as antibacterial, antidiabetic, antioxidant, anti-inflammatory, antiulcerogenic, cytotoxic, and wound-healing effects [631, 634]. Farsi and co-workers reported in 2011 that a *n*-butanol extract from the leaves of *F. deltoidea* possesses appreciable α-glucosidase and α-amylase inhibitory effects in controlling postprandial hyperglycemia [649]. A 2012 study by Adam and co-workers revealed that an ethanol extract reduced fasting blood sugar while a hot aqueous extract was reported to stimulate the secretion of insulin and induced intracellular Ca^{2+} ions to initiate the release of insulin [650]. Furthermore, vitexin (**622**) and isovitexin (**623**) reduced postprandial blood glucose levels [648].

A methanol extract of *F. deltoidea* inhibited the growth of *S. aureus* with an *MIC* value of 3.1 mg/cm^3 [651]. A broth microdilution study for antibacterial activity against *Enterococcus faecalis*, *Streptococcus mutans*, *S. mitis*, *Aggregatibacter actinomycetemcomitans*, *Porphyromonas gingivalis*, and *Fusobacterium nucleatum* showed weak to notable inhibition, with *MIC* and *MBC* values in the range 0.6–2.5 mg/cm^3 [651, 652]. Additionally, Uyub and co-workers found in 2010 that leaf extracts of *F. deltoidea* exhibited antimicrobial activity against *Helicobacter pylori*, a Gram-negative bacterium that plays a major role in chronic gastritis and gastric ulcers [653]. The methanol extract showed the highest activity with an inhibition zone diameter of 12.0 $\pm$ 0.6 mm followed by the petroleum ether (10.0 $\pm$ 0.6 mm) and chloroform (8.0 $\pm$ 0.1 mm) extracts. Additionally, Jamal and co-researchers reported in 2011 that ethanol and methanol extracts of the leaves of *F. deltoidea* displayed inhibitory effects on the growth of *Bacillus subtilis* [654]. The research described in Ref. [652] supported the previous findings, which reported that a methanol extract of the whole plant of *F. deltoidea* demonstrated antibacterial activity towards *Staphylooccus aureus*, *B. subtilis*, *Escherichia coli*, and *Pseudomonas aeruginosa*, and antifungal activity against *Candida albicans*.

The whole plant extract of *F. deltoidea* has potential as a healing agent for gastric ulcers. An experiment was conducted using four groups of rats treated with either a negative control (distilled water), a positive control (omeprazole), and 250 or 500 mg/kg of the extract. The results showed that the higher dose was more effective in gastric protection compared to the lower dose of the extract. The study suggested also that the *F. deltoidea* extract had an antiulcerogenic effect due to the inhibition of submucosal edema as well as from reduction in ulcer areas and leukocyte infiltration [655].

In traditional medicine, *Ficus deltoides* has been used widely as a postpartum tonic, antidiabetic, wound reliever, antirheumatic, and for irregular menstrual cycles. Other uses are enhancing beauty as well as being a libido booster. Due to its therapeutic properties established to date, this species has been formulated and commercialized as capsules, a tonic tea, and an herbal juice.

2.18 Clinacanthus nutans

Clinacanthus nutans (Burm. f.) Lindau (Plate 18) is a popular herb of the family Acanthaceae. This species is distributed in China, Thailand, Vietnam, Malaysia, and Indonesia and occurs on grassland and scrubland locations [656]. It is known vernacularly as "Pokok stawa ular", "Belalai gajah" and "Sabahan snake grass" in Malaysia and Brunei [657]. In Thailand, "Phaya Plongtong", "Phaya Yo" and "Saled Pangpon Tua Mea" are among the popular names for this species [658, 659]. In China, it is referred to as "E zui hua", "Sha be she cao" and "You dun cao" [658, 660, 661], and, in Indonesia, as "Ki tajam", "Kijatan", and "Daun dandang gendis" [657, 662, 663].

Plate 18 Leaves of *Clinacanthus nutans* (family Acanthaceae)

This species is 1–3 m in height [656] with a green cylindrical, striated, and glabrescent stem [664] and the leaves are simple, opposite, and lanceolately arranged (2.5–13.0 cm long and 0.5–1.5 cm wide) [656, 665]. The leaf bases are rounded and pubescent on the nerves and the petiole is 0.3–2.0 cm in length. The flowers are simple, dull yellow or greenish-yellow in color with a green base borne in dense terminal racemes (4–6 cm wide) [656, 666]. Its calyx has a length of 1 cm and is glandular-pubescent. The fruits are produced in the form of a capsule about 2 cm long and with short hairs [667].

2.18.1 Ethnobotany

Clinacanthus nutans has multiple folkloric uses due to its known biological effects [667]. In Malaysia, the fresh leaves are used as a herbal tea by boiling in water. In Thailand, an alcohol extract of the fresh leaves is utilized as an external treatment for skin rashes, snake and insect bites, and *Herpes simplex* and *Varicella zoster* viral lesions [668]. Furthermore, the leaves have been used also to produce beverages by mixing with apple juice, sugarcane, or green tea. The Chinese use the whole plant to treat hematomas, contusions, strains, sprains, rheumatism, and other maladies [669, 670]. In Indonesia, the fresh leaves have use to treat diabetes, fever, and dysentery [671].

This species has also been commercialized in the forms of a liquid beverage, tea, soap, massage oil, ointment, balm, cream, lotion, capsule, and powder. The liquid beverage, tea, capsule, and powder formulations are used ethnomedically to treat cancer, as a detoxification method, in diabetes, for hypertension, and to regulate the menstrual cycle and protect the kidneys. Soap, cream, and lotion formulations, are utilized for the treatment of skin problems such as relieving rashes, urticaria, itching, and the symptoms of *Herpes genitalis* and *H. zoster*. The balm and ointment are beneficial in relieving inflammation and muscular and joint pain as well as for colds and influenza [666].

2.18.2 Phytochemistry

Several isolation studies have been conducted on various parts of *C. nutans* to afford flavonoids, sulfur-containing compounds, and chlorophyll, isoprenoid, and miscellaneous phenolic derivatives. A methanol extract of *C. nutans* collected in Thailand afforded seven *C*-glycosyl flavones, namely, vitexin (**622**), isovitexin (**623**), orientin (**624**), isoorientin (**625**), shaftoside (**626**), isomollupentin-7-*O*-*β*-glucopyranoside (**627**), and apigenin 6,8-*C*-*α*-L-pyranoarabinoside (**628**) [672, 673]. In 1998, two sulfur-containing compounds were characterized, namely, clinacosides A (**629**) and B (**630**) [674]. Almost two decades later, seven further sulfur-containing compounds were isolated from an ethanol extract of the aerial parts, and elucidated as (*E*)-3-methylsulfinyl-2-propenol (**631**), (*E*)-entadamide A (**632**), and (*E*)-entadamide C (**633**), as well as clinamides A–C (**634**–**636**) [675].

In addition, chlorophylls a and b and six derivatives were reported from a chloroform extract of the leaves, which were identified as purpurin-18-phytyl ester (**637**), 13^2-hydroxy-(13^2R)-phaeophytin b (**638**), 13^2-hydroxy-(13^2S)-phaeophytin a (**639**), 13^2-hydroxy-(13^2R)-phaeophytin a (**640**), 13^2-hydroxy-(13^2S)-phaeophytin b (**641**), and phaeophorbide-a (**642**) [676, 678]. Stigmasterol (**109**) and β-sitosterol (**479**) are among the phytosterols isolated from this species [678]. Other reported compounds are lupeol (**620**) [679, 680], 3*β*-lup-20(29)-en-3-ol (**643**), and sinapic acid (**489**) [680].

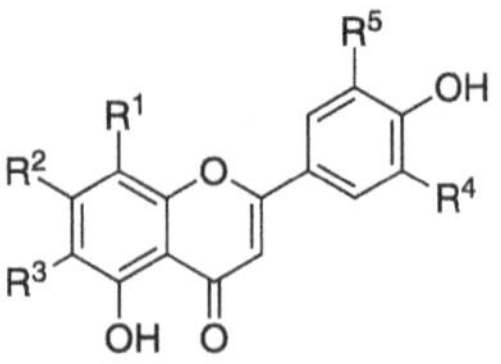

624 (orientin) R^1 = glc, R^2 = OH, R^3 = R^4 = H, R^5 = OH
625 (isoorientin) R^1 = H, R^2 = OH, R^2 = glc, R^4 = OH, R^5 = H
626 (schaftoside) R^1 = glc,R^2 = OH, R^3 = ara, R^4 = R^5 = H
627 (isomollupentin-7-*O*-*β*-glucopyranoside) R^1 = H, R^2 = glu, R^3 = ara, R^4 = R^5 = H
628 (apigenin 6,8-*C*-*α*-L-pyranarabinoside) R^1 = ara, R^2 = OH, R^3 = ara, R^4 = R^5 = H

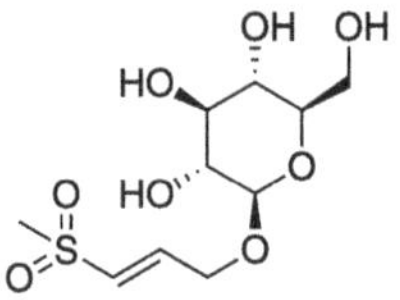

629 (clinacoside A)

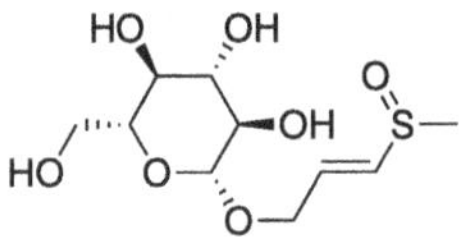

630 (clinacoside B)

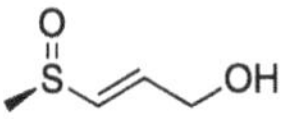

631 ((*E*)-3-methylsulfinyl-2-propenol)

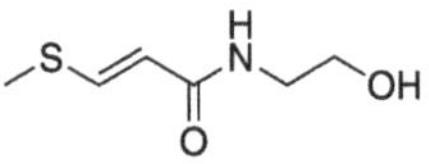

632 ((*E*)-entadamide A)

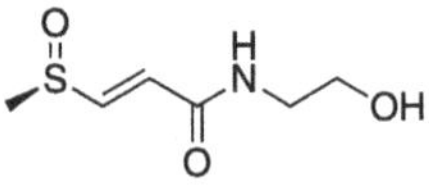

633 ((*E*)-entadamide C)

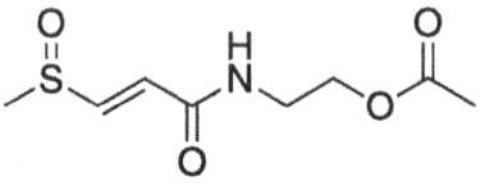

634 (clinamide A)

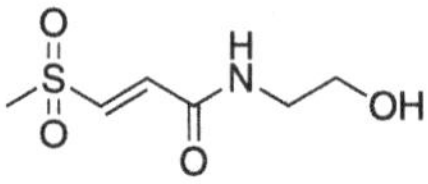

635 (clinamide B)

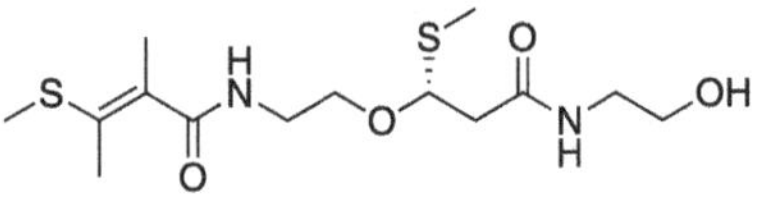

636 (clinamide C)

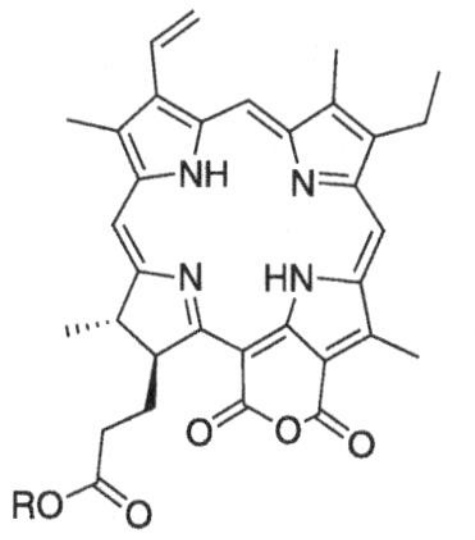

637 (purpurin-18 phytyl ester)

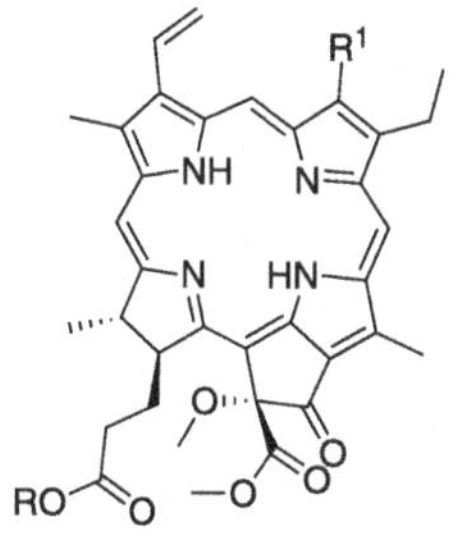

638 (13^2-hydroxy-(13^2R)-phaeophytin b) R^1 = CHO
639 (13^2-hydroxy-(13^2R)-phaeophytin a) R^1 = CH_3

R = 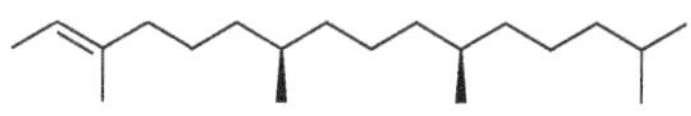

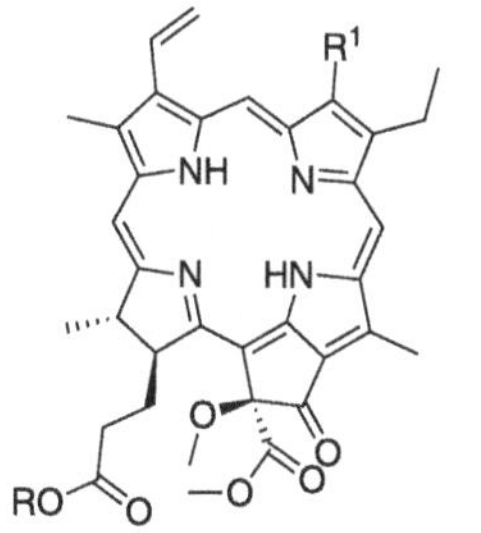

640 (13^2-hydroxy-(13^2S)-phaeophytin a) R^1 = CH_3
641(13^2-hydroxy-(13^2S)-phaeophytin b) R^1 = CHO

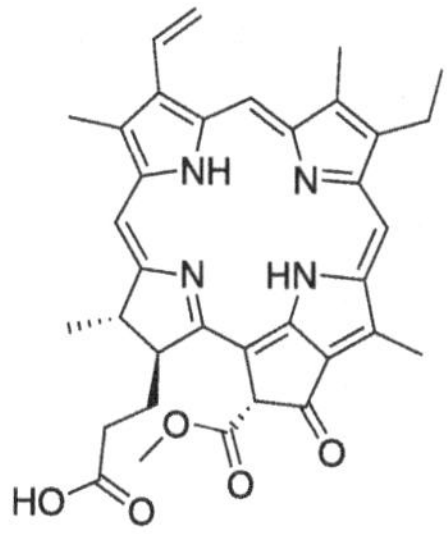

642 (phaeophorbide-a)

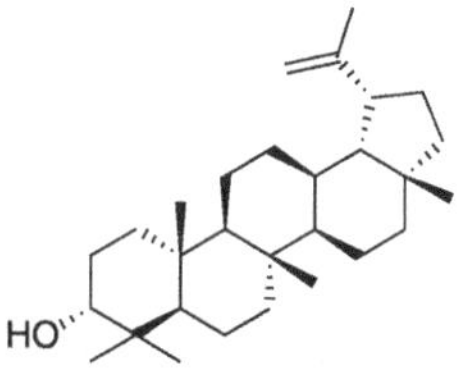

643 (3β-lup-20(29)-en-3-ol)

2.18.3 Biological Activities

Antioxidant activities of *C. nutans* have been evaluated using several types of in vitro assays. Using a 2,2-diphenyl-1-picrylhydrazyl (DPPH) assay, the highest percentage inhibition (82%) was demonstrated by a 4 mg/cm^3 concentration level of a petroleum ether leaf extract. Another parameter was employed by comparing the ages of the leaves and buds of *C. nutans*. Of the 1, 6, and 12 months-old leaves and buds tested, a methanol extract of the 12-month-old buds showed the highest inhibition with an IC_{50} of 64.6 mg/cm^3 followed by the 12-month-old leaves, having an IC_{50} value of 112.1 mg/cm^3. Using different treatment conditions of hot water leaf extracts, the unfermented leaf extract exhibited a higher inhibition value than the fermented extract. Additionally, Yong and co-workers investigated the effect of extraction solvents on DPPH activity by testing water, methanol, and chloroform leaf extracts. It was found that the highest inhibition was from the chloroform extract followed by those extracted with methanol and water. Furthermore, DPPH inhibitory activity was reported to be reduced by 70% from day 1 to day 4 of the storage duration of an aqueous ethanolic extract [657].

Crude extracts of *C. nutans* of different polarities were tested against the Gram-positive bacteria *Bacillus subtilis* and *B. cereus*, *Staphylococcus aureus*, methicillin-resistant *S. aureus* (MRSA), methicillin-sensitive *S. aureus* (MSSA), *Micrococcus luteus*, *Propionibacterium acnes*, *Staphylococcus epidermidis*, *Streptococcus* sp., *Aeromonas hydrophila*, *Escherichia coli*, *Neisseria gonorrhoeae*, *Pseudomonas aeruginosa*, *Salmonella enterica*, and *Vibrio harveyi* and *V. parahaemolyticus*. Among the solvents used, the water, methanol, ethanol, chloroform, and ethyl acetate extracts all showed low antibacterial inhibition levels [657].

2.19 *Cananga odorata*

Cananga odorata Hook. f. & Thomson (Plate 19) is a medium-sized evergreen tree belonging to the family Annonaceae and has two varieties, namely, *C. odorata* var. *odorata* and *C. odorata* var. *fruticosa* [681, 682]. Commonly known as "Ylang-ylang" in English, this plant is native to the tropical Indo-Pacific region, including Malaysia, Cambodia, Laos, Myanmar, Indonesia, Papua New Guinea, the Solomon Islands, and the Philippines [683]. Typically, it grows wild along forest edges in the northern region of Peninsular Malaysia, extending to Negeri Sembilan, and is a popular ornamental plant in streets, parks, and large gardens due to its pleasant fragrance, which enhances landscaping [683, 684]. Also, this species is frequently found in Sabah but is much rarer in Sarawak. It inhabits lowland forests, both primary and secondary, up to an altitude of 200 m [685]. The Malay name "Kenanga" (or "Cananga" in pre-standardized spelling) is the basis for the genus name, while the species name "*odorata*" is derived from the Latin word "odor", meaning fragrant or sweet-scented [686].

Plate 19 Flower of *Cananga odorata* (family Annonaceae). Photograph courtesy Prenn (https://commons.wikimedia.org/wiki/File:Cananga_odorata_02.JPG)

Cananga odorata reaches heights of up to 30 m and exhibits a rapid growth rate of over two meters annually in its early years. It features long, drooping twigs that extend three to six meters, with a smooth, greyish-white to silvery bark. The leaves are dark, glossy green, simple, alternate, and ovate-oblong in shape [686]. Its characteristic aggregate fruits are olive-like, greenish-black, and consist of 8 to 15 individually stalked fruits that ripen from dark green to black. The seeds are pale brown, flattened, ellipsoid, smooth, and irregularly pitted. The flowers have three sepals and six petals, measuring up to 8 cm in length; the petals are twisted when young and become limp and drooping as they mature. The dwarf variety, *C. odorata* var. *fruticosa*, grows to only 1.5 m and features curled petals. This variant flowers year-round but does not produce fruit [685].

2.19.1 Ethnobotany

In Malay traditional medicine, the flowers of *C. odorata* are boiled to produce an infusion that is consumed for the treatment of asthma and malaria. Furthermore, these flowers are utilized in postpartum baths for new mothers. A handful of flower buds is crushed along with other ingredients, such as *Parameria polyneura* (serapat root) and turmeric, and the resulting juice is ingested. An infusion derived from the bark of the ylang-ylang tree is used to reduce body odors. Additionally, the leaves are pounded and applied topically to affected areas to alleviate itching [687]. The dried flowers of *C. odorata* are worn in women's hair and also placed between clothes due to their fragrant scent.

The plant has long been cultivated for its highly valued flowers in perfumery and aromatherapy, and is known for its exotic floral fragrance reminiscent of jasmine. It is also grown in plantations for the production of its essential oil, which is extracted from the flowers and used in the manufacturing of numerous beauty products, including hair oils, soaps, perfumes, and shampoos. Additionally, the oil is utilized as a flavoring agent in foods and beverages [688, 689].

2.19.2 Phytochemistry

The phytochemistry of *C. odorata* has been studied extensively and is thus well documented, especially regarding the essential oil of its flowers [689–691]. In the earliest reports by Gaydou et al. in 1986, the essential oil was found to contain 52 compounds, constituted mainly by monoterpenes, sesquiterpenes, and phenols [692]. With advancements in analytical methods applied, additional compounds have since been identified in the flower's essential oil [692–695]. Bioinformatics advancements have further improved the production and quality of the essential oil, revealing several genes responsible for the biosynthesis pathway of volatile compounds through high-throughput RNA sequencing technology [696]. The main components of this oil include *p*-methylanisole (**644**), methyl benzoate (**645**), benzyl benzoate (**646**), benzyl acetate (**647**), geranyl acetate (**648**), cinnamyl acetate (**649**), (*E*,*E*)-farnesyl acetate (**650**), linalool (**401**), geraniol (**651**), benzyl salicylate (**652**), germacrene D (**653**), and caryophyllene (**654**). The floral scent of *C. odorata* is primarily attributed to terpenes **401**, **644**, and **654**, which were shown to constitute 8.7, 0.6, and 26.8% of the oil, respectively [695].

In addition to studies on the essential oil's chemical components, research on the non-volatile constituents of *C. odorata* has revealed a diverse array of bioactive compounds from various parts of the plant over the years. This work began in 1986, when the alkaloid sampangine (**655**) was first extracted from the chloroform extract of the stem bark [697]. Following this, liriodenine (**656**), an oxoaporphine alkaloid, was identified as a potent inhibitor of topoisomerase II in both in vitro and in vivo studies [698]. In 1999, a comprehensive study of the seeds of *C. odorata* resulted in the isolation of sixteen compounds. This included a new alkaloid, ushinsunine-β-*N*-oxide

(**657**), along with the known compounds cleistopholine (**658**), liriodenine (**656**), anonaine (**659**), nornuciferine (**660**), *N*-acetylnornuciferine (**661**), ushinsunine (**662**), norushinsunine (**663**), asimilobine (**664**), reticuline (**665**), *N*-(*E*)-feruloyltyramine (**666**), β-sitosterol (**667**), stigmasterol (**109**), lyscamine (**668**), anaxagoreine (**669**), and (*E*)-cinnamic acid (**670**) [699].

In a 2001 study, four compounds were purified from the fruits of *C. odorata*, inclusive of cananodine (**671**) and γ-eudesmol (**672**), along with two new eudesmane sesquiterpenes, cryptomeridiol-11-*α*-L-rhamnoside (**673**) and γ-eudesmol-11-*α*-L-rhamnoside (**674**). All four compounds exhibited cytotoxic effects against a hepatocarcinoma cell line, with sesquiterpenes **672** and **673** demonstrating the most potent cytotoxic effects [700]. Compound **675** contained in the dichloromethane extract of the flowers showed antibacterial and antifungal properties [701]. Also, Caloprisco and co-workers reported two lactones, isosiphonodin (**676**) and a new spirolactone, canangone (**677**), from an acetone extract of the dried leaves and stems [702].

Additionally, a new megastigmane glucoside, canangaionoside (**678**), was characterized from a methanol extract of dried leaves [703]. Further work by Matsumoto and colleagues in 2014 revealed two new lignan dicarboxylates and seven terpenoid derivatives from a methanol extract of the flower buds. The lignans were designated as canangalignans I (**679**) and II (**680**), while the terpenoids were named canangaterpenes I through VI (**681–686**), and these were obtained along with a known terpenoid, (*E*)-[(1*R*,3*R*,5*S*,6S,8*S*)-6-hydroxy-1,3-dimethoxy-2-oxaspiro[4.5]decan-8-yl]methyl caffeate (**687**) [704, 705]. Nagashima and colleagues isolated five new monoterpene glucosides, canangafruticosides A–E (**688–692**) [706]. Moreover, a bioassay-guided isolation on a bark ethanol extract of the plant revealed that three constituents, *O*-methylmoschatoline (**693**), liriodenine (**656**), and 3,4-dihydroxybenzoic acid (**694**), exhibited potent antibacterial and antifungal activities, with the oxoaporphine alkaloid **656** being the most active [707]. The structural drawings **644–694** depict the molecular structures of the constituents isolated from various parts of the *C. odorata* to date.

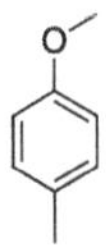

644 (*p*-methylanisole)

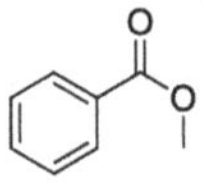

645 (methyl benzoate)

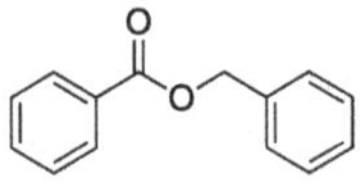

646 (benzyl benzoate)

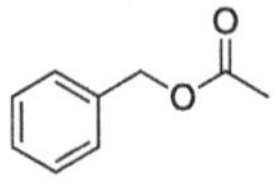

647 (benzyl acetate)

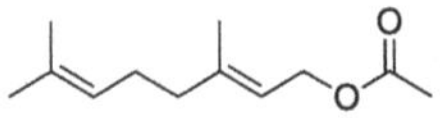

648 (geranyl acetate)

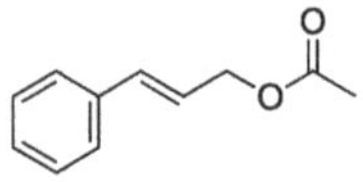

649 (cinnamyl acetate)

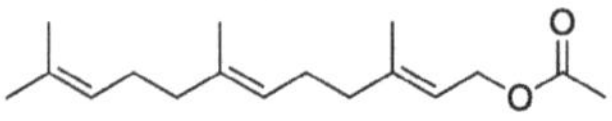

650 ((*E*,*E*)-farnesyl acetate)

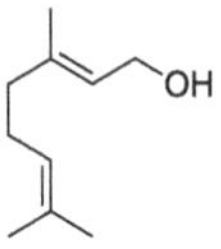

651 (geraniol)

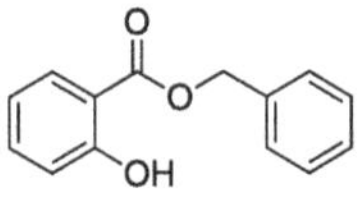

652 (benzyl salicylate)

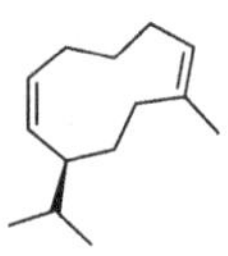

653 (germacrene D)

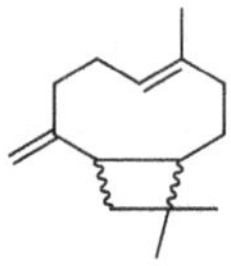

654 (caryophyllene)

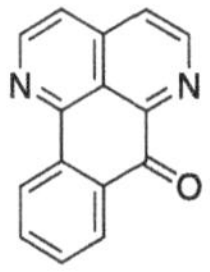

655 (sampangine)

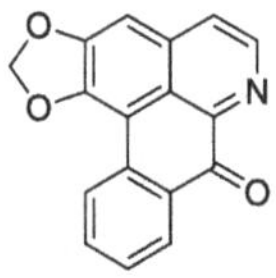

656 (liriodenine)

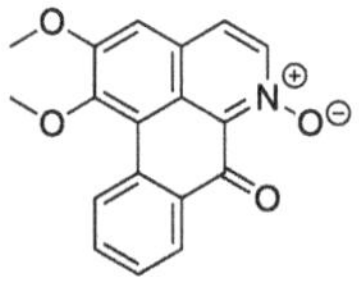

657 (ushinsunine-β-*N*-oxide)

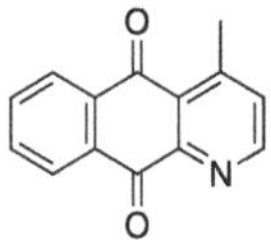

658 (cleistopholine)

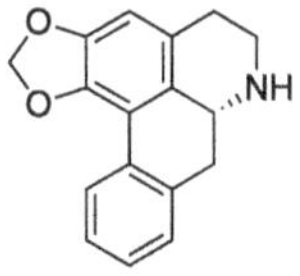

659 (anonaine)

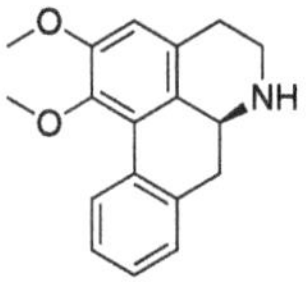

660 (nornuciferine)

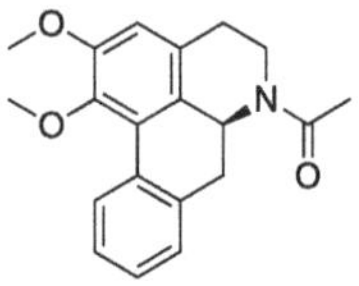

661 (*N*-acetylnornuciferine)

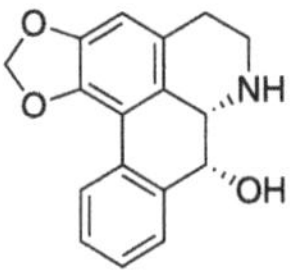

662 (ushinsunine)

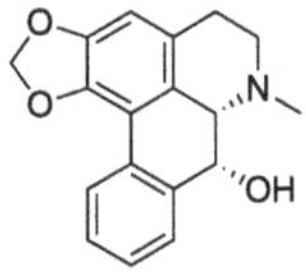

663 (norushinsunine)

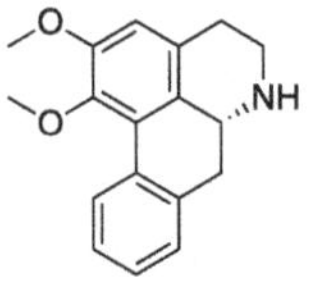

664 (asimilobine)

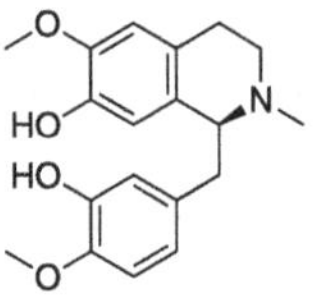

665 (reticuline)

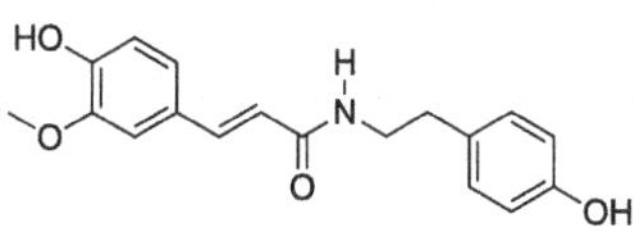

666 (*N*-(*E*)-feruloyltyramine)

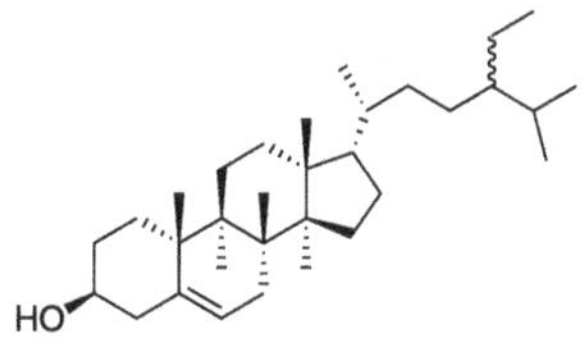

667 (β-sitosterol)

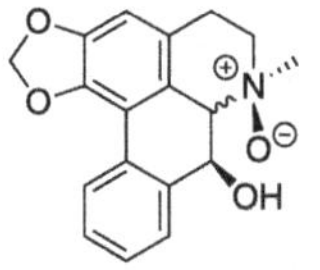

668 (lyscamine)

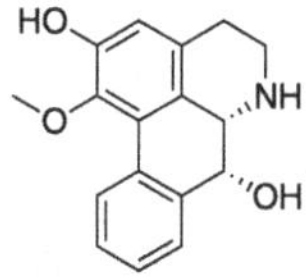

669 (anaxagoreine)

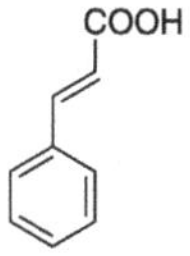

670 ((*E*)-cinnamic acid)

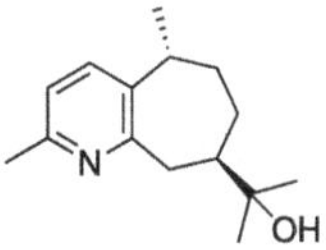

671 (cananodine)

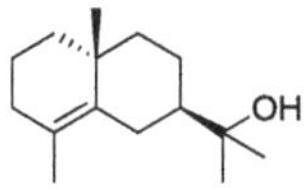

672 (γ-eudesmol)

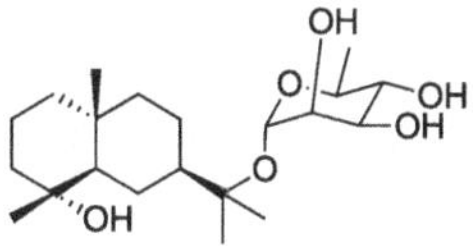

673 (cryptomeridiol 11-α-L-rhamnoside)

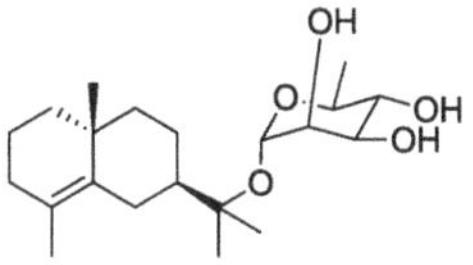

674 (γ-eudesmol 11-α-L-rhamnoside)

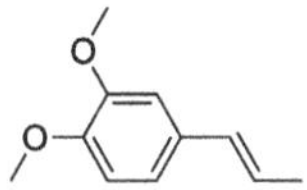

675 (methyl isoeugenol)

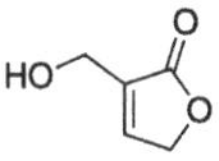

676 (isosiphonidin)

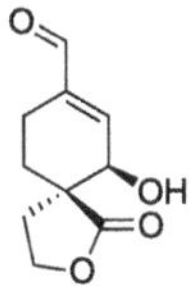

677 (canangone)

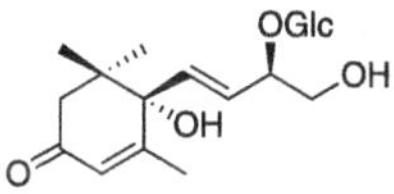

678 (canangaionoside)

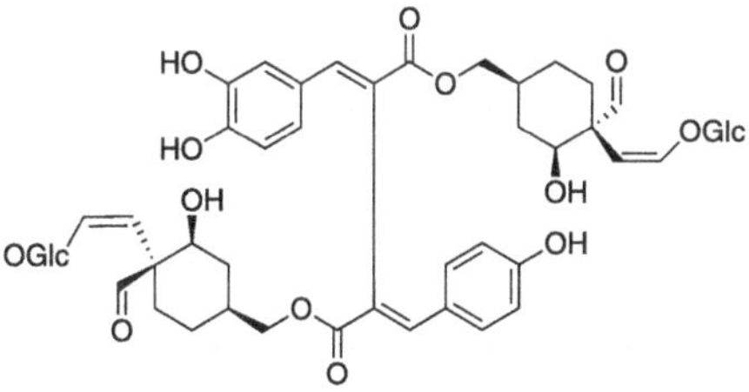

679 (canangalignan I)

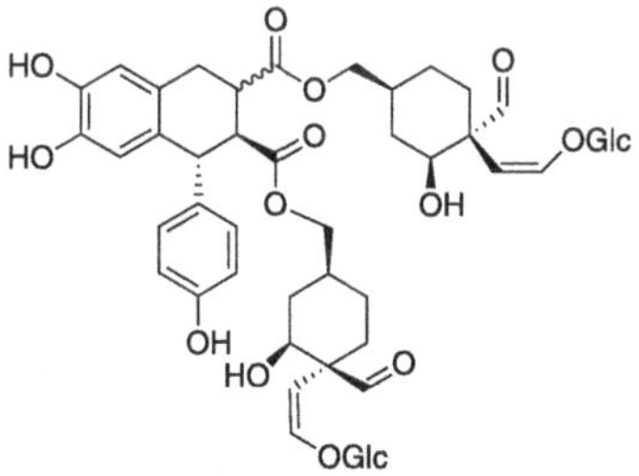

680 (canangalignan II)

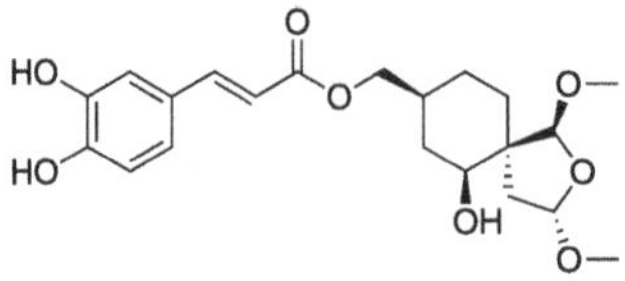

681 (canangaterpene I)

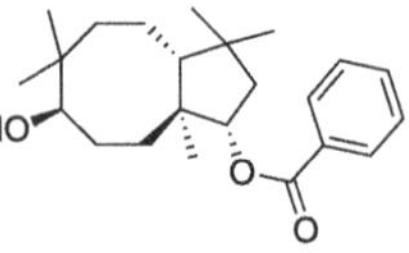

682 (canangaterpene II)

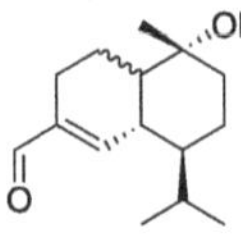

683 (canangaterpene III)

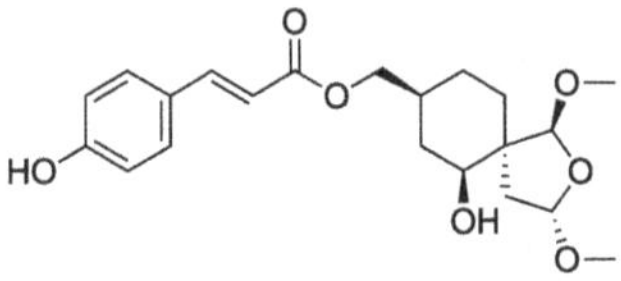

684 (canangaterpene IV)

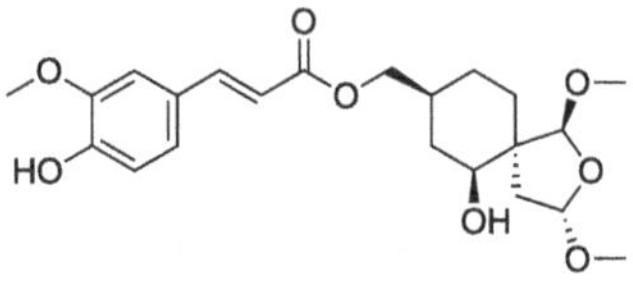

685 (canangaterpene V)

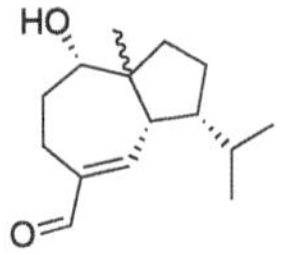

686 (canangaterpene VI)

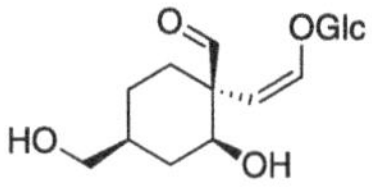

688 (canangafruticoside A)

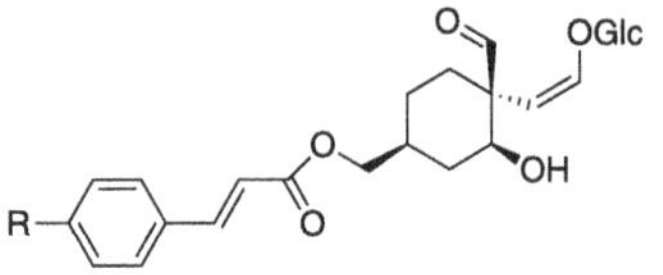

689 (canangafruticoside B) R = H
690 (canangafruticoside C) R = OH

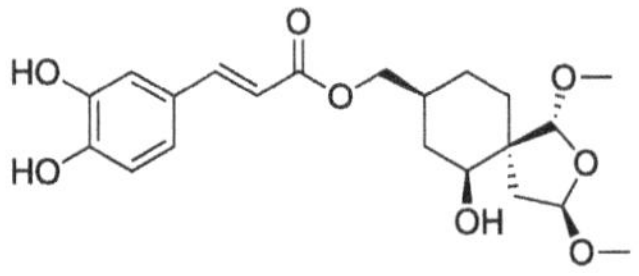

687 ((*E*)-[(1*R*,3*R*,5*S*,6*S*,8*S*)-6-hydroxy-1,3-dimethoxy-2-oxaspiro[4.5]decan-8-yl]methyl caffeate)

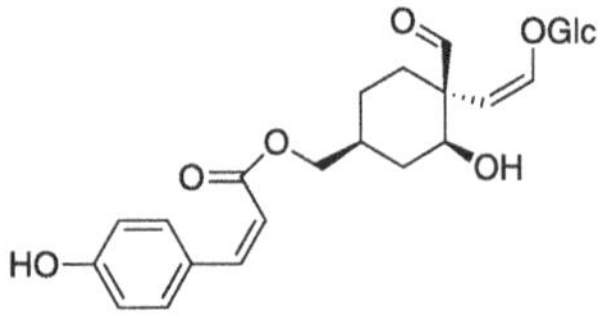

691 (canangafruticoside D)

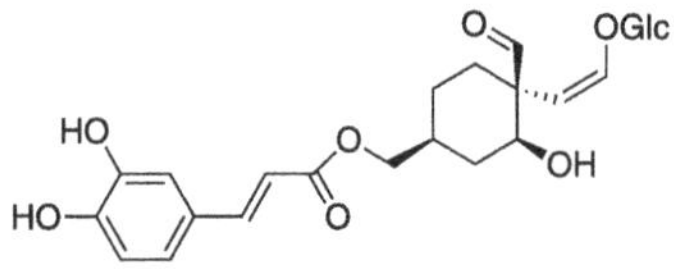

692 (canangafruticoside E)

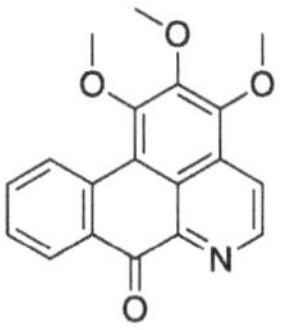

693 (*O*-methylmoschatoline)

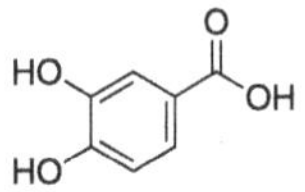

694 (3,4-dihydroxybenzoic acid)

2.19.3 Biological Activities

Cananga odorata has a wide range of biological properties, including antimicrobial, antioxidant, anti-inflammatory, antidiabetic, cytotoxic, hepatoprotective, anxiolytic, and sedative effects. Studies have highlighted the plant's potential in skincare, relaxation therapy, and insect control, particularly through its essential oil. Extracts from the leaves, stems, and flower buds have activity against various pathogens, with the essential oil showing insecticidal and repellent effects. Over the past few decades, numerous studies have explored the antimicrobial potential of *C. odorata*, particularly extracts produced from the stem bark and leaves. The use of various assays, including agar well and disk diffusion methods, has demonstrated potent antimicrobial effects. Indonesian *C. odorata* stem bark and flower extracts showed growth inhibition activities against *Propionibacterium acnes* and *Candida albicans* [708]. The stem bark ethanol extract (400 μg/well) produced a 19 ± 1.6 mm inhibition zone against *P. acnes* (activity index 0.63), while the *n*-hexane extract (100 μg/well) inhibited *C. albicans* by 17 ± 1.6 mm (activity index 0.56), when compared to standard antibiotics [708]. Evaluation of the isolated compounds revealed that *O*-methylmoschatoline (**693**), liriodenine (**656**), and 3,4-dihydroxybenzoic acid (**694**) were the key contributors to the antibacterial and antifungal activities observed, with liriodenine being the most potent [708, 709].

In a separate study, a methanol extract of *C. odorata* leaves showed discernible antimicrobial activities against Gram-positive and Gram-negative bacteria, with the Gram-negative strains tested exhibiting less susceptibility [710]. The essential oil of *C. odorata* demonstrated antimicrobial effects against *Staphylococcus aureus* ATCC 25,923, with a MIC_{90} value of 0.2 mg/cm^3, but showed no observed inhibitory effects on *Escherichia coli* or *Pseudomonas aeruginosa* even at high concentrations [711]. Another study reported weak antibacterial activity of the essential oil against *P. acnes*, with inhibition zones of 8.8–9.5 mm, and no effect against *Malassezia furfur* [712].

Interestingly, combining *C. odorata* essential oil with other essential oils enhanced its antimicrobial properties. A blend with thyme oil showed a 38.4% increase in the inhibition zone against *S. aureus* compared to thyme alone, while a slight antagonistic effect was observed against *E. coli* [713]. Similarly, a combination of lavender, clary sage, and *C. odorata* oils exhibited potent antibacterial and antifungal effects, highlighting the potential for synergistic interactions [713].

Further research on the antibiofilm activity of the essential oil of *C. odorata* revealed a significant dose-dependent inhibition of *S. aureus* biofilm formation (more than 80% at 0.01% *v*/*v*), with no effect on bacterial growth. The study identified (6*E*)- and (6*Z*)-nerolidol as key compounds responsible for this activity. Another study incorporated the essential oil with magnetic iron oxide nanoparticles as a coating on implant surfaces, significantly reducing biofilm formation by *S. aureus* and *Klebsiella pneumoniae*, with over a 2-log reduction in biofilm on coated catheter specimens. These findings suggest that *C. odorata* oil, especially in combination with nanomaterials, holds a potential for antibiofilm applications in biomedical devices [714].

The antioxidant potential of *C. odorata* has been investigated. An ethyl acetate extract of the stem bark showed DPPH inhibition (79%) among the various plant extracts evaluated [709]. In another study, the *C. odorata* essential oil exhibited free radical-scavenging effects, with 63.8 ± 0.5% DPPH inhibition, which was greater than that of trolox, a well-known antioxidant [715]. This sample also demonstrated a potent lipid peroxidation inhibitory effect (75.5 ± 0.5%) in the β-carotene bleaching test and an effective superoxide radical-scavenging effect in the luminol-photochemiluminescence assay [715]. Additionally, the essential oil from flowers of *C. odorata* collected in Madagascar, exhibited notable DPPH-scavenging activity (80.0 ± 0.02%) [715]. A methanol extract of the leaves also showed significant ferric ion reducing power, with a 290.0 ± 13.1% reduction at 0.5 μg/cm^3 [717].

Studies have shown that the *C. odorata* flower essential oil possesses repellent, oviposition-deterrent, and insecticidal properties against various mosquito species. In 2011, the weak insecticidal activity of *C. odorata* essential oil in soybean oil was demonstrated against adult mosquitoes (*Aedes aegypti*, *Culex quinquefasciatus*, and *Anopheles dirus*), with LC_{50} values of 9.8%, 8.8, and 5.0%, respectively [718]. The essential oil also showed oviposition-deterrent and ovicidal effects against these species, with a 10% concentration level in soybean oil being able to reduce egg laying and hatching. However, further studies were recommended, as some mosquito eggs showed a lack of susceptibility to this insecticidal action [719].

This essential oil has also been evaluated for its mosquito-repellent activity. One study found that *C. odorata* essential oil prepared in soybean oil exhibited repellent activities against *A. aegypti*, *A. dirus*, and *C. quinquefasciatus*, with ED_{50} values of 0.045, 2.15, and <0.003 mg/cm^2, respectively. This essential oil also provided moderate protection against these mosquitoes, with protection times of 8.4 min for *A. aegypti*, 24 min for *A. dirus*, and 60 min for *C. quinquefasciatus*, respectively [720]. In comparison, DEET, a positive control and gold standard of mosquito repellents, showed complete protection for five hours against *A. aegypti* at a 23.8% concentration [721]. In a further study, *C. odorata* essential oil prepared in ethyl alcohol at 0.33 mm^3/cm^2 provided 86.7 $\pm$ 10.4 min of protection against *A. aegypti* and 126.0 $\pm$ 15.8 min against *C. quinquefasciatus* [722]. Another research group reported that *C. odorata* essential oil prepared in coconut oil at 0.33 mm^3/cm^2 demonstrated even better protection, with 98.9% effectiveness against *A. aegypti* and an improved protection time of 88.7 $\pm$ 10.4 min [723].

Beyond mosquito control, *C. odorata* essential oil has shown repellent activity against other pests. The oil demonstrated strong repellent effects against *Tribolium castaneum*, a pest of stored products. In one study, the oil at a concentration of 5 mm^3/g of oats exhibited the most potent repellent effect compared to plant other essential oils [724]. Another study by the same group showed that *C. odorata* essential oil at 0.2 μg/cm^2 resulted in 98% repellency after two and four hours of exposure, making it an excellent candidate for use as a natural repellent against *T. castaneum* [725]. The larvicidal and pupicidal effects of *C. odorata* essential oil were also evaluated, where it exhibited higher activities against the larvae and pupae of *C. quinquefasciatus* and *Anopheles dirus* than against *A. aegypti* [726]. Moderate insecticidal activity, with a LD_{50} of 53 ppm, was also observed against *A. aegypti* larvae. Benzyl acetate (**647**), linalool (**401**), and benzyl benzoate (**646**) were identified as the major active compounds [727]. In addition to its mosquito control potential, *C. odorata* essential oil also has larvicidal and oviposition-deterrent effects against the house fly, *Musca domestica*, a major pest and disease vector. The oil exhibited a median lethal time (LT_{50}) of 52.08 h and an LC_{50} of 29.4%, compared to the chemical insecticide cypermethrin ($LT_{50} = 31.63$ h, $LC_{50} = 11.5\%$). It also demonstrated 100% oviposition-deterrent activity at both 1.65 and 3.30 mm^3/cm^2 concentrations [723].

Further research has shown that *C. odorata* leaf essential oil can also control termites and other pests. This essential oil showed toxicity against *Sitophilus zeamais*, a pest of corn storage, with an LD_{50} of 33.1 μg/adult in topical application [727]. The oil exhibited fumigant activity against *S. zeamais*, with an LD_{50} of 14.8 mg/

dm^3, and again linalool (**401**) was suggested as the key component responsible for the insecticidal effects [727].

A methanol extract of *C. odorata* flower buds has been shown to have significant inhibitory effects on melanogenesis. In a study using B16 melanoma 4A5 cells, compounds from the extract were found to reduce melanin production without cytotoxicity. Two terpenoid derivatives, canangaterpene I (**681**) ($IC_{50} = 3.6\ \mu M$) and canangafruticoside E (**692**) ($IC_{50} = 2.5\ \mu M$), exhibited more potent inhibitory activities than the positive control used, arbutin ($IC_{50} = 174\ \mu M$) [704]. These findings highlight the potential of *C. odorata* flower buds in treating skin disorders or for cosmetic applications targeting hyperpigmentation. Additionally, the phenylpropanoid *N*-(*E*)-feruloyltyramine (**666**) from *C. odorata* seeds produced a more potent inhibition of tyrosinase expression in B16 melanoma cells than kojic acid, a standard tyrosinase inhibitor [728]. However, an aqueous extract of *C. odorata* did not exhibit any observable antityrosinase activity, as shown by the lack of inhibition of dopachrome formation ($-19.8 \pm 0.7\%$) [729].

Studies supporting the traditional use of this plant for treating fever, asthma, and pain have confirmed its anti-inflammatory properties. Specifically, the essential oil of *C. odorata* inhibited lipoxygenase activity by 80% at a concentration of 0.5 $\mu g/cm^3$, a result comparable to that of the known inhibitor, nordihydroguaiaretic acid [730]. This effect was attributed to compounds inclusive of linalool (**715**), which is known to be anti-inflammatory [730, 731]. Additionally, a leaf methanol leaf extract inhibited nitric oxide release by $97.9 \pm 14.6\%$ in RAW264.7 macrophage cells at 50 $\mu g/cm^3$, indicating its potential as an anti-inflammatory agent, since nitric oxide is involved in inflammation [717]. An ethanol extract of the fruits was found to reduce paw edema in Wistar albino rats in a carrageenan-induced model, but only with a very high LD_{50} value of > 2 g/kg. This extract inhibited paw volume by 62.9% at 100 mg/kg, compared to 60.14% inhibition by aspirin at 300 mg/kg. This effect may be due to flavonoid and other phenolic constituents that inhibit both cyclooxygenase and lipoxygenase pathways [732].

The essential oil of *C. odorata* also has shown various physiological effects on humans. In one study, inhalation of this oil very slightly reduced both systolic (from 106.4 ± 11.5 mmHg to 105.2 ± 10.7 mm Hg) and diastolic blood pressure (from 70.6 ± 10.5 mm Hg to 69.2 ± 11.7 mm Hg), and also lowered the pulse rate (73.4 ± 7.4 bpm vs. 75.3 ± 7.5 bpm in the control group). It also decreased stress levels, as reflected in a reduction of the stress index (from 73.3 to 49.5 KU/dm^3) and an increase in alpha brain waves [733]. Similar findings were reported from inhaling the essential oil, reducing both blood pressure and pulse rate, suggesting a decrease in autonomic nervous system arousal. Interestingly, this study also found that inhalation improved alertness and attentiveness, suggesting a harmonization effect where physiological relaxation was paired with increased behavioral activation. In contrast, a transdermal application of *C. odorata* essential oil had a calming effect, reducing physiological arousal without enhancing alertness, emphasizing that the route of administration can influence the oil's effects [734]. Jung and colleagues further confirmed the sedative effect using a sphygmomanometer and from electrocardiogram measurements, showing a significant decrease in heart rate following inhalation [735].

In addition to its sedative properties, the essential oil of *C. odorata* has been shown to affect cognitive performance and brain activity. One investigation found that its inhalation reduced alertness and increased calmness in healthy volunteers, although it did not enhance cognitive performance or processing speed, in contrast to a peppermint aroma [736]. Ishiguchi and colleagues observed that inhaling the essential oil increased alpha 1 brain waves (8–9.9 Hz), associated with deep relaxation, while also reducing alertness, supporting its relaxant effects [737]. Further analysis revealed a reduction in the amplitude of auditory P300 waves, linked to cognitive processing, suggesting that the essential oil has a relaxing influence on cognitive function. Watanabe and colleagues extended these findings by showing that exposure to *C. odorata* essential oil prolonged P300 latency in both healthy individuals and patients with temporal lobe epilepsy (TLE). While P300 amplitude was reduced in the healthy subjects, the TLE patients showed no such reduction, suggesting that their information-processing ability remained unchanged, possibly due to already lower baseline P300 levels [738].

Canaga odorata also has spermicidal activity in both in vitro and in vivo studies. In an in vitro investigation, a 50% ethanol extract of the root bark immobilized sperm from healthy male rats within seconds, and in vivo administration of the extract (50 mg/100 g body weight/day) reduced sperm motility ($5 \pm 0.4\%$ vs. $30 \pm 2\%$ in the control). Higher doses (100 mg/100 g body weight/day) also led to a reduced sperm count and 94% abnormal sperm morphology. Biochemical analysis revealed that the extract decreased testosterone production, disrupted sperm metabolism in the testes, and impaired sperm maturation [739]. Comparison with the antifertility agent gossypol showed that the *C. odorata* extract might be more effective, as sperm motility partially recovered after discontinuing the extract. The active component responsible for this effect was identified as a 52 kD protein, which immobilized sperm in vitro within seconds [740].

Extracts from the leaves and stems of *C. odorata* exhibited α-amylase inhibitory effects, with inhibition rates of $22.6 \pm 1.3\%$ and $25.3 \pm 3.3\%$, respectively, at a concentration of 7.8 $\mu g/cm^3$ against porcine pancreatic α-amylase. However, no inhibitory effect was observed on α-glucosidase [741]. These findings suggested that *C. odorata* extracts may have potential for managing postprandial hyperglycemia. Additionally, the terpenoid derivatives canangaterpene I (**681**), canangafruticoside D (**691**) and (*E*)-[(1*R*,3*R*,5*S*,6*S*,8*S*)-6-hydroxy-1,3-dimethoxy-2-oxaspiro[4.5]decan-8-yl]methyl caffeate (**687**), from *C. odorata* flower buds, were found to inhibit aldose reductase with IC_{50} values of 1.2, 1.5, and 0.8 μM, respectively, and were compared to the reference compound chlorogenic acid ($IC_{50} = 0.7\ \mu M$) [705].

A methanol leaf extract of *C. odorata* showed anti-hepatitis B virus (HBV) activity, with an IC_{50} of 56.5 $\mu g/cm^3$ and a CC_{50} of 540.2 $\mu g/cm^3$ for Hep38.7-Tet cells, affording a Selectivity Index of 9.6. In HepG2-NTCP cells, the extract reduced heptatis B surface antigen production by 58% at 100 $\mu g/cm^3$ [742]. The essential oil of *C. odorata* has also been investigated for its anxiolytic effects.

A randomized aromatherapy controlled trial evaluating *C. odorata* essential oil was conducted in patients undergoing their first interventional neuroradiology procedure, and it was found that the group exposed to the essential oil experienced a significant reduction in salivary α-amylase activity and a greater decrease in State and Trait Anxiety, when compared to the placebo group. Both groups showed a reduction in State and Trait Anxiety, but the essential oil-treated group demonstrated a more significant improvement. No significant changes were found in salivary cortisol, blood pressure, or heart rate, suggesting that aromatherapy with *C. odorata* essential oil may reduce anxiety effectively during stressful medical procedures [743].

A recent study explored the potential adjunct use of *C. odorata* essential oil in cancer chemotherapy, by developing pH-responsive niosomes co-loaded with oxaliplatin and *C. odorata* essential oil (Ox-Y@NSs) for breast cancer treatment [744]. In MDA-MB-231 breast cancer cells, Ox-Y@NSs exhibited a low IC_{50} (0.0002 μg/cm^3) and significantly increased the percentages of late apoptotic (8.4%) and necrotic (5.1%) cells. Gene and protein assays confirmed that Ox-Y@NSs induced apoptosis by upregulating three pro-apoptotic markers (Tp53, Bax, and caspase-7) and downregulating the anti-apoptotic protein Bcl2, thereby activating the intrinsic apoptosis pathway [744].

2.20 *Citrus hystrix*

Citrus hystrix DC. is a species of citrus belonging the family Rutaceae. It is native to southern China and tropical Southeast Asia, including Malaysia. The plant grows up to 30 m in height, although it can be pruned regularly to maintain a more manageable size for easier harvesting of both the leaves and fruits. The leaves are distinctive, with winged petioles that create the appearance of two leaves fused together. They are dark green, glossy, and typically range from 5 to 8 cm in length. The fruits, about the size of Western limes, are dark green with a bumpy texture (Plate 20). However, it yields little to no juice [745].

Plate 20 Fruits and leaves of *Citrus hystrix* (family Rutaceae). Photograph courtesy Jpbrigand2 (https://commons.wikimedia.org/wiki/File:C_hystrix_kaffir_lime_leaves.jpg)

2.20.1 Ethnobotany

Each part of *C. hystrix* has distinct uses. In Malaysia, this plant is known commonly as "Limau Purut" or "Kafir lime". The fruits are often used in shampoos to reduce dandruff and promote hair growth [746]. Both the leaves and fruits are utilized for their ethnomedical properties, including enhancing sexual performance and treating conditions such as hypertension, heart disease, and diarrhea [747]. The fruit juice is also believed to promote gum health and is used as a natural teeth cleaner. The fresh leaves are a key ingredient in a variety of dishes, such as green and red curries, soups, Tom Yam, masak lemak (rich coconut milk-based dishes), asam pedas (sour and spicy dishes), singgang (sour soup), stir-fries, kerabu (Malaysian salads), and other culinary preparations. Additionally, the essential oil extracted from the leaves is used as a flavoring agent, in aromatherapy, and is incorporated into cosmeceutical products, including soaps and lotions [746].

2.20.2 Phytochemistry

Studies on the chemical constituents of *C. hystrix* have revealed several key classes of compounds, including coumarin, flavonoid, phenolic, and terpenoid derivatives. The volatile terpenoids in *C. hystrix* contribute to its aromatic properties, particularly in the essential oil constituents. Research has shown that the major compounds in the fresh leaf essential oil are citronellal (**695**), β-citronellol (**696**), limonene (**402**), and citronellyl acetate (**697**). In turn, the major components in the fruit peel essential oil include β-pinene (**698**), sabinene (**412**), citronellol (**695**), limonene (**402**), and terpinene-4-ol (**699**) [748]. Thus far, the non-volatile isoprenoids isolated from the plant include the tetracyclic triterpenoids lupeol (**620**) and agrostophillinol (**700**), as well as the sterols β-sitosterol (**667**) and sitosteryl-*β*-D-glucopyranoside (**701**) [749, 750].

Among the primary non-volatile constituents, coumarins are the predominant phytochemical class in *C. hystrix*, with 28 different compounds isolated to date [749, 751–753]. These can be categorized into three major types: (a) the furocoumarins, bergaptol (**702**), bergamottin (**703**), oxypeucedanin (**704**), 6′,7′-dihydroxybergamottin (**705**), citrusosides B–D (**706–708**), isoimperatorin (**538**), oxypeucedanin hydrate (**709**), oxypeucedanin (**704**), (+)-(*R*)-6′-hydroxy-7′-methoxybergamottin (**710**), (+)-(*R*)-oxypeucedaninmethanolate (**711**), pangelin (**712**), 2′-methoxyoxypeucedanin hydrate (**713**), and xanthotoxol (**714**); (b) the simple coumarins, umbelliferone (**715**), osthol (**716**), seselin (**717**), suberosin (**718**), suberenol (**719**), 7-demethylsuberosin (**720**), tamarin (**721**), murraol (**722**), and (*E*)-dehydroosthol (**723**); and (c) the pyranocoumarins, xanthoxyletin (**724**), xanthyletin (**725**), dentatin (**726**), and 5-methoxyseselin (**727**). Substituents such as methyl, ethyl, methoxy, isopentenyl, and glucosyl groups occur in the structures of these coumarins, contributing to the chemical diversity of *C. hystrix*, and leading to differences in the biological activities reported for these compounds.

Citrus hystrix also biosynthesizes 25 different reported flavonoids featuring two main types: (a) the flavonols, quercetin (**85**), isorhamnetin (**482**), kaempferol (**88**), myricetin (**89**), tamarixetin (**728**), and natsudaidain (**729**), and (b) the flavone type, including luteolin (**389**), apigenin (**87**), nobiletin (**367**), tangeretin (**730**), 5,7,8,4′-tetramethoxyflavone (**731**), diosmin (**732**), cosmosiin (**174**), saponarin (**733**), and apiin (**734**) [753–755]. Additionally, the flavanones hesperetin (**735**), naringin (**736**), and hesperidin (**737**) were also identified in the plant. Other flavonoid constituents of *C. hystrix* are the anthocyanin derivative peonidin (**738**), and the flavans epicatechin (**615**), epigallocatechin (**91**), and epigallocatechin 3-gallate (**618**). Anthocyanins such as peonidin (**738**) are significant due to their role as natural pigments in plants [753, 754].

Several miscellaneous types of constituents have been isolated from *C. hystrix*, including the phenolic acids, vanillic acid (**534**), 4,6,7-trimethoxy-5-chlorogenic acid

(**739**), caffeic acid (**537**), and ferulic acid (**489**) [754]. In addition, two glyceroglycolipids have been isolated from the leaves of *C. hystrix*, namely, 1,2-di-*O*-linolenoyl-3-*O*-galactopyranosyl-*sn*-glycerol (**585**) and 1-*O*-α-linolenoyl-2-*O*-palmitoyl-3-*O*-β-galactopyranosyl-*sn*-glycerol (**740**) [756]. Other isolated compounds include 11-hydroxynoracronycine (**741**) and (+)-syringaresinol (**742**) [757]. The structures of these chemical constituents are shown in the structural formulas.

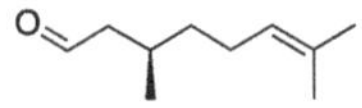
695 (citronellal)

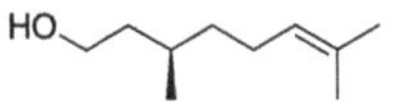
696 (β-citronellol)

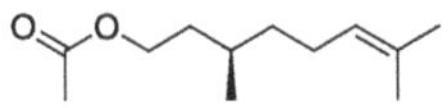
697 (citronellyl acetate)

698 (β-pinene)

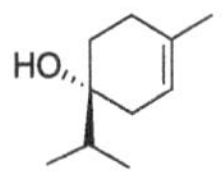
769 (terpinen-4-ol)

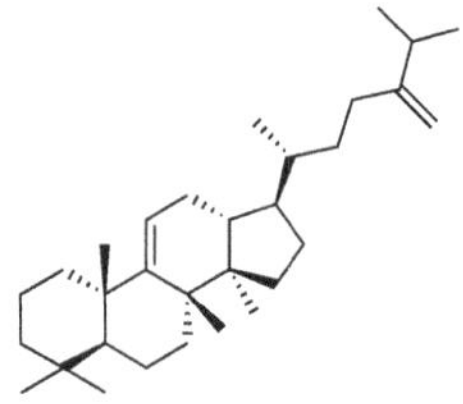
700 (agrostophillinol)

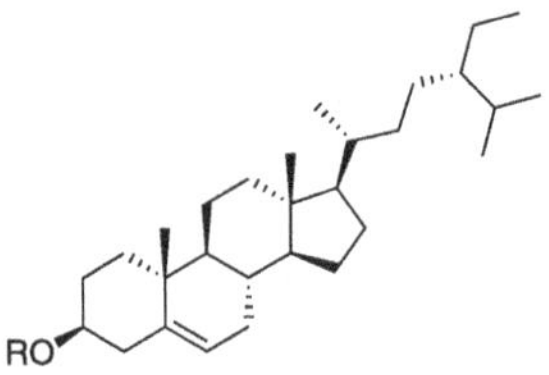
701 (sitosteryl-β-D-glucopyranoside) R = Glc

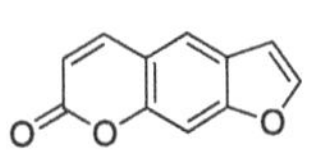
702 (bergaptol)

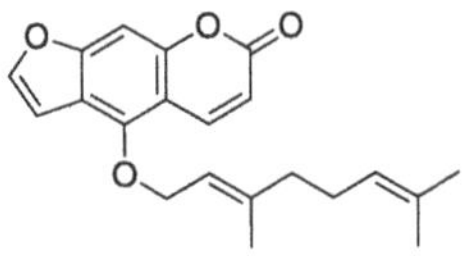
703 (bergamottin)

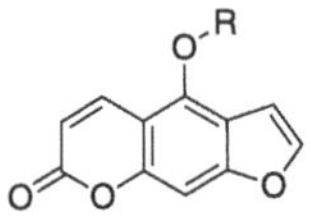
704 (oxypeucedanin) R = S_1
705 (6′,7′-dihydroxybergamottin) R = S_2
706 (citrusoside B) R = S_3
707 (citrusoside C) R = S_4
708 (citrusoside D) R = S_5
709 (oxypeucedanin hydrate) R = S_6
710 ((+)-(*R*)-6′-hydroxy-7′-methoxybergamottin) R = S_7
711 ((+)-(*R*)-6′-oxypeucedaninmethanolate) R = S_8
712 (pangelin) R = S_9
713 (2′-methoxyoxypeucedanin hydrate) R = S_{10}
714 (xanthotoxol) R = H

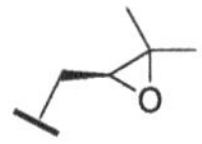
S_1

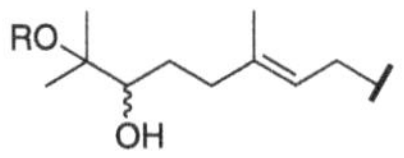
S_2 R = H
S_7 R = Me

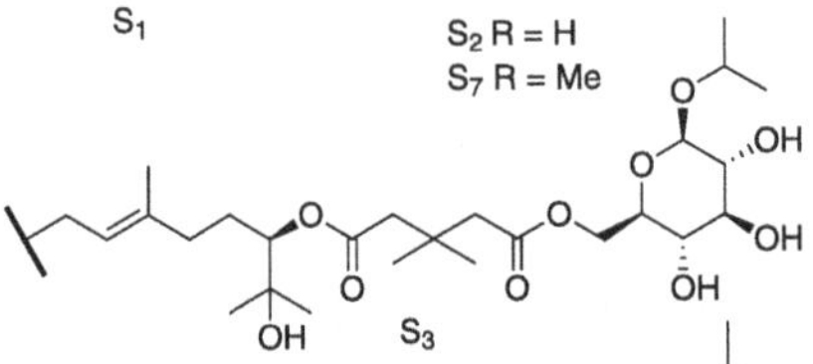
S_3

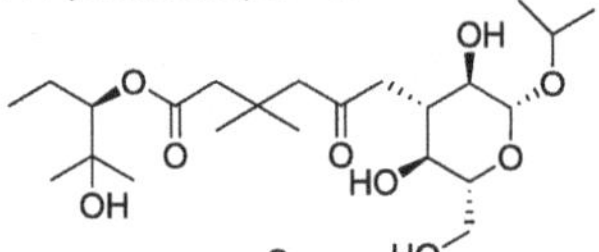
S_5

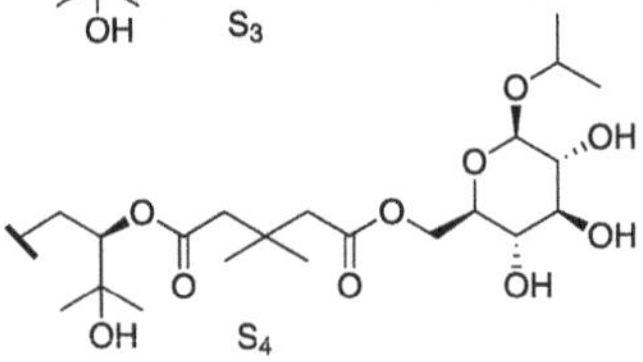
S_4

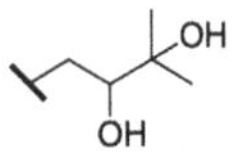
S_6

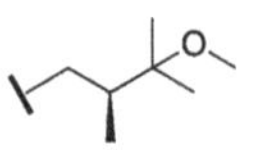
S_8

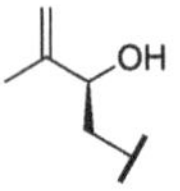
S_9

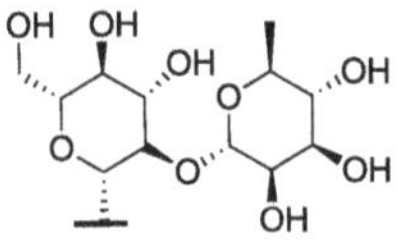
S_{10}

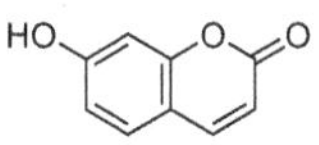

715 (umbelliferone)

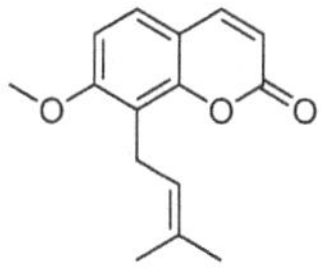

716 (osthol)

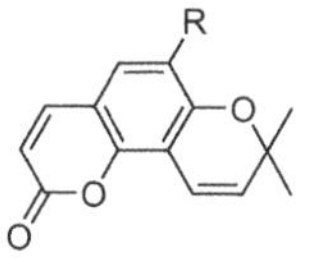

717 (seselin) R = H
727 (5-methoxyseselin) R = OMe

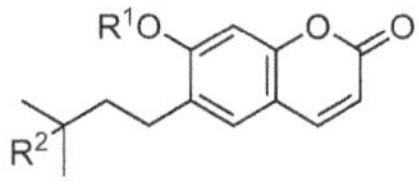

718 (suberosin) R^1 = Me, R^2 = H
719 (suberenol) R^1 = Me, R^2 = OH
720 (7-demethylsuberosin) R^1 = R^2 = H

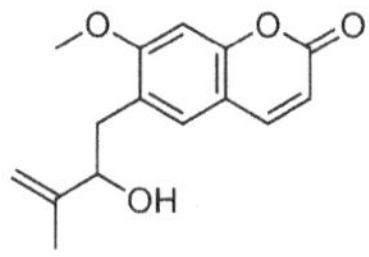

721 (tamarin)

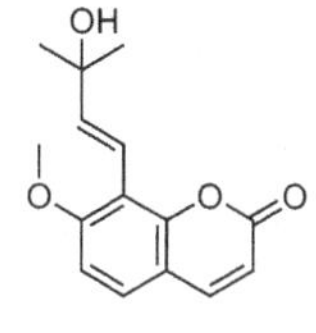

722 (murraol)

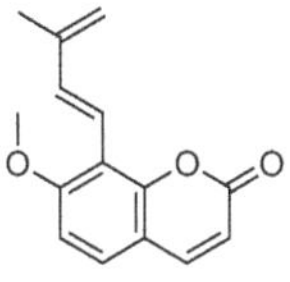

723 ((*E*)-dehydroosthol)

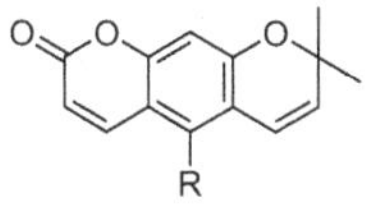

724 (xanthoxyletin) R = OMe
725 (xanthyletin) R = H

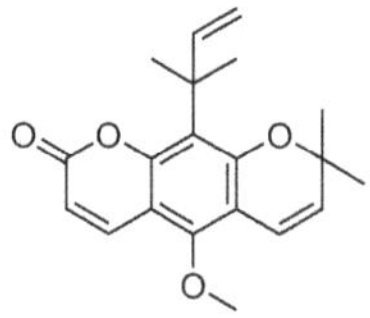

726 (dentatin)

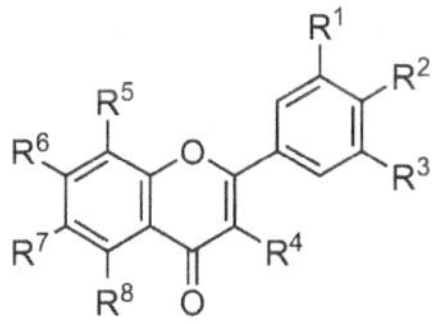

728 (tamarixetin) $R^3 = R^5 = R^7$ = H, $R^1 = R^4 = R^6 = R^8$ = OH, = R^2 = OMe
729 (natsudaidain) R^3 = H, R^4 = OH, = $R^2 = R^3 = R^5 = R^6 = R^7 = R^8$ = OMe
730 (tangeretin) $R^1 = R^2 = R^4$ = H, $R^3 = R^5 = R^6 = R^7 = R^8$ = OMe
731 (5,7,8,4'-tetramethoxyflavone) $R^1 = R^2 = R^4 = R^7$ = H, $R^3 = R^5 = R^6 = R^8$ = OMe
732 (diosmin) $R^3 = R^4 = R^5 = R^7$ = H, $R^1 = R^8$ = OH, R^6 = O-S_{12}, R^2 = OMe
733 (saponarin) $R^1 = R^4 = R^5$ = H, R^7 = Glc, $R^2 = R^3 = R^8$ = OH, R^6 = O-Glc
734 (apiin) $R^1 = R^3 = R^4 = R^5 = R^7$ = H, $R^2 = R^8$ = OH, R^6 = O-S_{13}

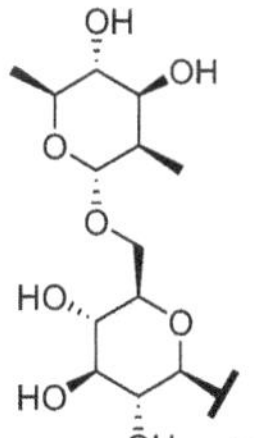

S_{12}

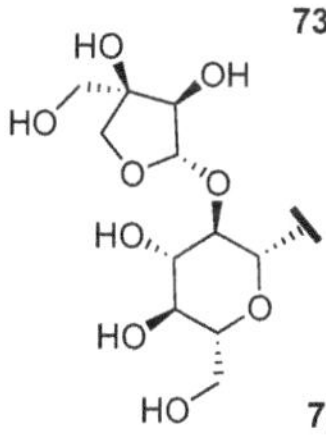

S_{13}

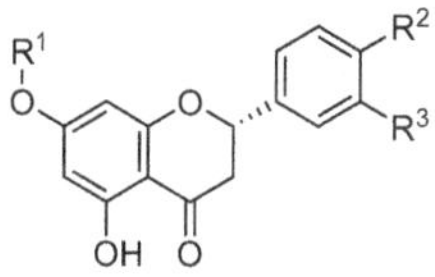

735 (hesperetin) R^1 = H, $R^2 = R^3$ = OH
736 (naringin) R^1 = S_{12}, R^2 = OH, R^3 = H
737 (hesperidin) R^1 = S_{12}, R^2 = OMe, R^3 = OH

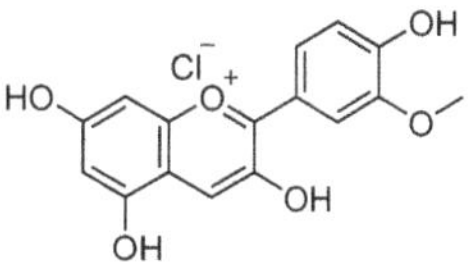

738 (peonidin)

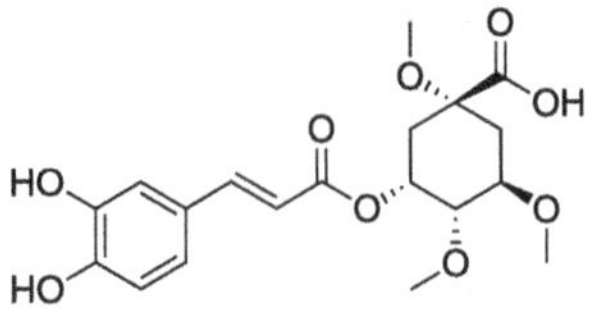

739 (4,6,7-trimethoxy-5-chlorogenic acid)

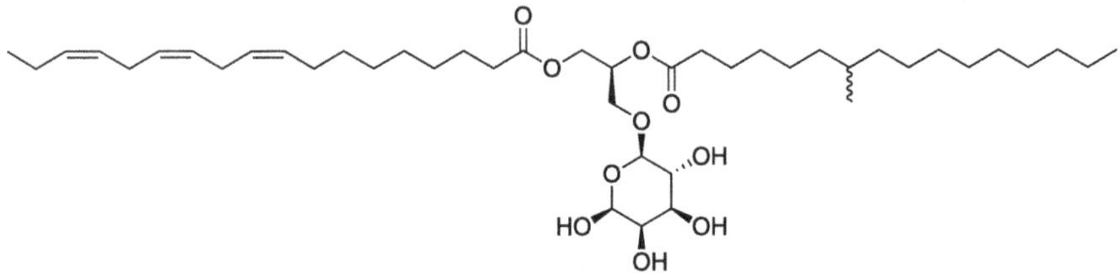

740 (1-*O*-α-linolenoyl-2-*O*-palmitoyl-3-*O*-β-galactopyranosyl-*sn*-glycerol)

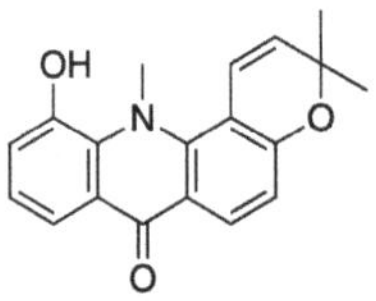

741 (11-hydroxynoracronycine)

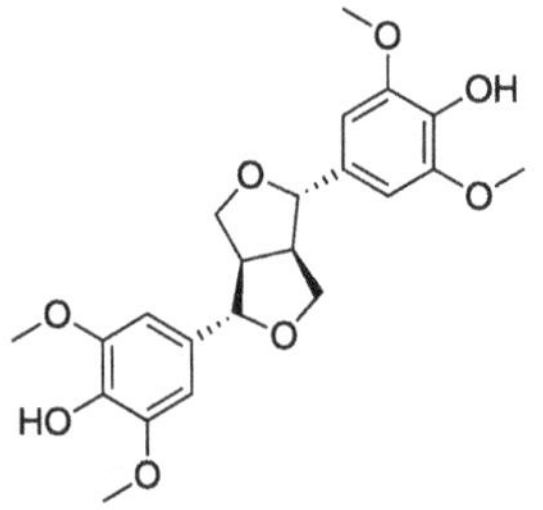

742 ((+)-syringaresinol)

2.20.3 Biological Activities

Citrus hystrix has garnered significant attention for its diverse biological properties that have been reported to date. Research has demonstrated its broad-spectrum antimicrobial activity, including antibacterial, antifungal, and antiviral effects, as well as anti-inflammatory and antioxidant properties that may help combat oxidative stress and chronic inflammation. Additionally, several bioactive constituents, such as coumarins, flavonoids, and terpenoids, have shown cytotoxic and neuroprotective potential, with growth inhibitory effects against various cancer cell lines and potentially beneficial effects on neurodegenerative diseases like Alzheimer's. The plant's ability to modulate key enzymes, promote wound healing, and support cardiovascular health underscore its ethnomedical value as a functional food and natural therapeutic agent.

Citrus hystrix is known also for its antimicrobial properties, and exerts antibacterial, antifungal, and anti-*Vibrio* activities. A number of studies have demonstrated the broad-spectrum efficacy of its essential oil and various types of extracts against a range of pathogens. In a study by Waikedre et al., the essential oil from *C. hystrix* showed antimicrobial effects against several microbial strains [758]. Hence, this oil inhibited *Staphylococcus epidermidis* (zone of inhibition: 7 mm), *Candida albicans* (*MIC*: 75 $\mu g/cm^3$), *Cryptococcus neoformans* (*MIC*: 50 $\mu g/cm^3$), and *Saccharomyces cerevisiae* (*MIC*: 50 $\mu g/cm^3$). Additionally, the antibacterial properties of compounds isolated from the plant's roots were assessed by Panthong and co-workers. The coumarin, tamarin (**721**) and the lignan, syringaresinol (**742**) were found to exhibit activity against multidrug-resistant *Acinetobacter baumannii* with a *MIC* of 100 $\mu g/cm^3$, while limonene (**402**) demonstrated a more potent effect with a *MIC* of 50 $\mu g/cm^3$ [757]. A further investigation on *C. hystrix* leaf essential oil by Wongsariya et al. revealed its antibacterial potency with *MIC*s of 1.1 mg/cm^3 against both *Porphyromonas gingivalis* and *Streptococcus mutans*, and 2.1 mg/cm^3 for *Streptococcus sanguinis* [759]. At a concentration of 4.3 mg/cm^3, the oil also exhibited a 99% inhibition of biofilm formation. Moreover, when applied at 4 × *MIC* and 2 × *MIC*, *C. hystrix* leaf oil showed lethal effects on *P. gingivalis* within two to four hours. Interestingly, a synergistic effect was observed when combined with chlorhexidine, as the *MIC* values of the strains decreased four-fold. The active compound in the leaf essential oil was identified as citronellal (**695**). Borusiewicz et al. reported similar antibacterial activities from this essential oil, which was tested against 50 strains of multidrug-resistant *A. baumannii*. The *MIC* values ranged from 0.125 to 1 mm^3/cm^3, confirming the oil's broad-spectrum antibacterial potential [760].

In 2020, an antifungal activity evaluation of *C. hystrix* leaf essential oil and its microemulsion against *Trichophyton mentagrophytes* was reported, in which both samples exhibited antifungal activity with *MIC*s of 1.1 mg/cm^3, and the microemulsion showed good physical and chemical stability [761]. In addition, the ethanol extract of *C. hystrix* peel was shown to inhibit *Salmonella* sp. with an *MIC* of 0.6%, demonstrating its potential as a natural antimicrobial agent. In an in vivo murine model, daily administration of 16 mg of the extract for three consecutive days, reduced bacterial loads in the spleen, liver, and ileum [762]. The anti-*Vibrio* activity of *C. hystrix* was also investigated by Singhapol and Tinrat, who demonstrated that the ethanol fraction of its peel exhibited antimicrobial effects against *Vibrio parahaemolyticus*, a marine bacterium causing foodborne gastroenteritis [763]. At a concentration of 50 mg/cm^3, this extract showed more potent antimicrobial activity when compared to other citrus extracts, suggesting its potential for use in controlling a disease that spreads in shrimp. In 2021, Chit-aree and co-workers explored the efficacy of *C. hystrix* essential oil against *Colletotrichum gloeosporioides*, the causative agent of anthracnose disease in mango fruits. The oil exhibited inhibitory effects on mycelial growth with very high *MIC* values ranging from 1,500 to 50,000 ppm and effectively reduced disease progression in vivo, indicating some potential for use in agricultural applications [764].

Recent research has focused on the antibacterial efficacy of *C. hystrix* essential oil against multidrug-resistant (MDR) *S. aureus*, including both methicillin-susceptible

(MSSA) and methicillin-resistant (MRSA) strains [765]. The essential oil demonstrated weak *MIC* values of 18.3 mg/cm^3 for MSSA and 17.9 mg/cm^3 for MRSA. Time-kill studies revealed that the essential oil was capable of completely killing both MSSA and MRSA within 12 h at 1 × *MIC*.

Citrus hystrix also has repellent and larvicidal activities against various mosquito species, including *A. aegypti* and *A. minimus*. Nararak et al. demonstrated the spatial repellent efficacy of *C. hystrix* leaf essential oil, with the highest activity observed at concentrations of 1 and 2%. The oil also exhibited significant combined irritant and repellent effects at concentrations ranging from 1– 5%, with greater inhibition observed against *A. minimus* mosquitoes compared to *A. aegypti* [766]. In addition to repellent activity, *C. hystrix* possesses larvicidal properties. Nishan et al. investigated the ethanol extracts of the pulps of *C. hystrix* and *Citrus aurantifolia* against 3rd and 4th instar larvae of *A. aegypti*. This *C. hystrix* extract at a concentration of 3.85 mg/cm^3 resulted in 66.7% larval mortality over 24 h. This larvicidal effect was comparable to that of synthetic controls at the same concentration over a 72-h period, further supporting the potential of *C. hystrix* as a natural larvicidal agent [767].

Similar to other functional food plant sources, *C. hystrix* has been investigated for its potent antioxidant properties. In 2008, Butryee and Kupradinum reported that a plant leaf extract demonstrated antioxidant effects both in vitro and in vivo. Using the erythrocyte micronucleus test in mice, the leaf extract tested exhibited clastogenic and anticlastogenic effects [768]. The total antioxidant capacity of aqueous and lipid extracts of *C. hystrix* was evaluated using the in vitro oxygen radical absorbance capacity (ORAC) and ferric reducing antioxidant power (FRAP) assays, affording values of 433 μM Trolox equivalent/g and 95 μM Fe^{2+} equivalent/g, respectively. In 2009, Butryee and colleagues further explored the impact of boiling on *C. hystrix* antioxidant capacity, showing that boiling reduced the total antioxidant capacity, with the order of antioxidant potency being: fresh > deep-fat frying > boiling [754]. The antioxidant potential of 33 known compounds isolated from the roots of *C. hystrix*, including coumarin, lignan and other derivatives, has also been assessed [757]. Among these, the quinoline, 11-hydroxy-noracronycine (**741**) and the lignan, (+)-syringaresinol (**742**) exhibited notable antioxidant activity, with IC_{50} values of 0.2 mg/cm^3 and 0.032 mg/cm^3, respectively, in the DPPH assay. (+)-Syringaresinol (**742**) also displayed potent superoxide scavenging ability (IC_{50} 1.5 mg/cm^3), although its crude extract of origin showed a superior antioxidant activity, suggesting a synergistic effect between the compounds present. Another research group also studied the in vitro antioxidant effects of *C. hystrix* leaf extracts in terms of their DPPH, cupric reducing antioxidant capacity, and nitric oxide scavenging activities. The ethanol extract used demonstrated the most potent free radical-scavenging activity as evident through its highest total antioxidant capacity, followed by the chloroform and methanol extracts [769]. Ratanachamnong et al. have investigated the wound-healing potential of *C. hystrix* water extract, noting antioxidant effects on human keratinocytes and fibroblasts. This extract demonstrated free-radical scavenging in the DPPH (IC_{50} 14.9 mg/cm^3) and NO assays (IC_{50} 4.5 mg/cm^3).

Additionally, at a concentration of 50 μg/cm^3, this *C. hystrix* extract reduced reactive oxygen species (ROS) generated by H_2O_2, and dose-dependently promoted cell migration, supporting its potential as a wound-healing accelerator [770].

Citrus hystrix and its constituents have also been investigated for their cancer-related biological effects [771]. In 1995, Murakami et al. isolated 1,2-di-*O*-linolenoyl-3-*O*-galactopyranosyl-*sn*-glycerol (**585**) and 1-*O*-*α*-linolenoyl-2-*O*-palmitoyl-3-*O*-*β*-galactopyranosyl-*sn*-glycerol (**740**) from *C. hystrix*, both of which demonstrated tumor-promoting inhibition and potent activity against Epstein-Barr virus activation, with lower IC_{50} values compared to several control agents [756]. In a two-stage carcinogenesis experiment involving ICR mice, compound **585** exhibited antitumor-promoting activity at a dose ten times lower than that of α-linolenic acid. These compounds showed anti-inflammatory properties, specifically in inhibiting edema formation in a 12-*O*-tetradecanoylphorbol-13-acetate (TPA) promoted mouse model, suggesting that the anti-inflammatory actions of *C. hystrix* may play a role in its cancer-related attributes by modulating the phospholipase A2 pathway involved in inflammation [756]. In 2014, Tunjung et al. evaluated the cytotoxic effects of a *C. hystrix* leaf ethyl acetate extract using a variety of cancer cell lines, including cervical cancer (HeLa) and neuroblastoma (UKF-NB3). The study revealed that this ethyl acetate had weak cytotoxic effects, with IC_{50} values ranging from 14.1 to 40.7 μg/cm^3 [771]. In 2017, Borusiewicz et al. examined *C. hystrix* peel essential oil and its ability to inhibit cancer cell proliferation. The essential oil displayed anti-proliferative and cytotoxic effects against melanoma cells (WM793 and A375) while showing lower cytotoxicity to normal skin fibroblasts [760].

In 2018, Sun et al. identified bergamottin (**703**) as a cytotoxic coumarin from *C. hystrix*, showing inhibitory effects against pancreatic cancer cells (PANC-1). Bergamottin (**703**) induced cell shrinkage, membrane blebbing, and organelle disintegration while suppressing cell migration and colony formation. Additionally, a molecular investigation suggested that bergamottin downregulated proteins in the Akt/mTOR signaling pathway, making it a candidate for additional laboratory investigation. A further study explored the activity of two components, β-citronellol (**696**) and citronellal (**695**), of a *C. hystrix* hexane extract against a breast cancer cell line [772]. Ho and co-workers showed that these two compounds significantly inhibited MDA-MB-231 breast cancer cell proliferation, colony formation, and migration. The mechanism of action involved induction of cell cycle arrest and apoptosis through the caspase-3 dependent pathway [772]. In addition, a hexane fraction of *C. hystrix* leaves was reported to inhibit the proliferation of a leukemia cell line, speficially by reducing the expression of the WT1 protein in K562 cells. The study identified lupeol (**621**) as an active compound responsible for this anti-proliferative effect. Furthermore, agrostophillinol (**700**), isolated from *C. hystrix* leaves, exhibited selective cytotoxicity against leukemia cells (EoL-1 and HL-60) without affecting normal peripheral blood mononuclear cells (PBMCs) [750, 773].

Several constituents have demonstrated anti-inflammatory and enzyme-inhibitory activities. Among these, a group of coumarins, including bergamottin (**703**), was shown to have anti-inflammatory effects. Bergamottin (**703**) inhibited inflammation with an IC_{50} value of 14 μM, and was compared to the reference compound *N*-(iminoethyl)-ornithine (IC_{50}: 7.9 μM). Other derivatives with structural differences in their side chains exhibited less potent activities in this investigation [756]. In addition, agrostophillinol (**700**) was also reported to have promising anti-inflammatory effects. At 2.7 $\mu g/cm^3$, **700** significantly inhibited IL-6 production, in a more potent manner than the reference anti-inflammatory drug used, dexamethasone. This suggests that *C. hystrix* has some potential for modulating inflammation [750, 773]. In 2010, furanocoumarins isolated from *C. hystrix* were evaluated for their cholinesterase inhibitory activity. Among these, (+)-*R*-6′-hydroxy-7′-methoxybergamottin (**710**) demonstrated the most potent acetylcholinesterase (AChE) inhibition with an IC_{50} value of 11.2 μM [749]. This finding was supported by further research in 2016, where Sammi et al. showed that the monoterpene, citronellal (**695**) inhibited AChE activity at both the biochemical and genomic levels, while also modulating the expression of the choline transporter and acetyltransferase genes [774].

Moreover, *C. hystrix* extracts has been evaluated for their enzyme-inhibitory potential. Abirami et al. demonstrated in 2014 that fresh juice of *C. hystrix* fruits inhibited a range of enzymes, including α-amylase, α-glucosidase, tyrosinase, AChE, and β-glucuronidase, with inhibition rates ranging from 60 to 80% at the concentration levels used [747].

In addition, *C. hystrix* has neuroprotective effects. In 2020, Pattarachotanant and Tencomnao investigated the anti-senescent mechanisms of *C. hystrix* extracts in SH-SY5Y human neuroblastoma cells. The extracts mitigated high glucose-induced cytotoxicity, reduced reactive oxygen species (ROS) generation, and promoted cell cycle progression through the activation of the SIRT1/GAPDH pathway [775].

Further expanding on the therapeutic potential of *C. hystrix*, Ali et al. evaluated its thrombolytic and membrane-stabilizing activities. An ethanol leaf extract exhibited thrombolytic activity with a clot-lysing value of 13.7%, compared to 37.3% for the standard streptokinase. In terms of membrane-stabilizing effects, the extract had a 74.4% hemolytic activity against human erythrocytes, significantly less than the 93.2% level observed with acetylsalicylic acid, but still indicating that *C. hystrix* may afford therapeutic benefits for cardiovascular conditions [769]. However, despite its potential benefits, *C. hystrix* has also been associated with skin-inflammatory effects. In 1999, Koh and Ong reported a case of phytophotodermatitis in a human subject exposed to the juice of *C. hystrix*, which contained high levels of furanocoumarins and hence may cause photosensitivity [776]. While *C. hystrix* oil was shown to promote blood circulation and reduce skin temperature, as demonstrated by Hongratanaworakit and Buchbauer in 2007, its potential for topical applications should be considered with caution due to its photosensitizing properties [777].

2.21 *Polygonum minus*

Polygonum minus Huds., also known by its synonym *Persicaria minor* (Huds.) Opiz., is a herbaceous plant in the family Polygonaceae. Native to Southeast Asia, it is found primarily in Malaysia, Indonesia, Thailand, Laos, and Vietnam, but has also become widely distributed in parts of Europe and Australia. It grows naturally in damp areas like ditches, river banks, and lakes, thriving in cool, hilly regions [778]. The plant can be cultivated in a range of soils, from mineral to peat, provided there is adequate water. It requires an annual rainfall of 2,500–3,000 mm and is unsuitable for areas having dry spells exceeding one month unless it is irrigated. In Malaysia, *P. minus* (commonly known as "Kesum") reaches 1.0 m in lowland areas and up to 1.5 m in hilly regions. Its dark green, aromatic leaves are lanceolate, measuring 5–7 cm in length and 0.5–2.0 cm in width, arranged alternately along a cylindrical, slightly reddish stem with short internodes and easily rooting nodes (Plate 21). The small white purple-colored flowers are found at the end of the shoot. There are two common growth forms of this species, namely, creeping and vertical. The vertical-growing type is preferred for commercial production due to its benefits in field maintenance and harvesting efficiency [778, 779].

Plate 21 *Polygonum minus* leaves (family Polygonaceae). Photograph courtesy Penarc (https://commons.wikimedia.org/wiki/File:Persicaria_8_11_2010_014.jpg)

2.21.1 Ethnobotany

Polygonum minus is a popular ingredient in many Malaysian dishes, including "asam laksa" (spicy and sour fish noodle soup), "nasi kerabu" (herb rice salad with vegetables and grilled meat), and "asam pedas" (spicy and sour fish stew), and the plant is valued for its distinctive aroma that adds a unique flavor to these dishes. Traditionally, a decoction of *P. minus* leaves is used to cure digestive disorders and as a tonic taken after birth [265]. A paste made from the crushed leaves mixed with a small amount of kerosene is applied to the skin to treat fungal infections [780]. Its essential oil is applied to the scalp to treat dandruff and is also used for aromatherapy in folk medicine. The essential oil of *P. minus* is believed to have high potential for use in the flavor and fragrance industries [778, 781].

2.21.2 Phytochemistry

Polygonum minus has been listed in the National Agro-Food Policy by the Malaysian government to promote its production as a means of boosting the agricultural economy. It is also recognized in the Herbal Product Blueprint as an important crop due to its high essential oil content (72.5%) [782]. The major constituents of its essential oil include the aldehydes decanal (**743**) and dodecanal (**744**), as well as the sesquiterpenes cyclolongifolene oxide (**745**), β-caryophyllene (**404**), seychellene (**746**), α-bergamotene (**747**), valencene (**748**), and isolongifolol (**749**) [783]. In addition to the volatile terpenoids, *P. minus* also produces other non-volatile phytochemicals. Previous studies have led to the isolation of flavonoids, terpenoids, and phenolic compounds. Among the flavonoids, rutin (**86**), catechin (**90**), quercetin (**85**), kaempferol (**88**), isorhamnetin (**484**), myricetin (**89**), apigetrin (**750**), isoquercetin (**751**), hyperoside (**752**), and astragalin (**485**) have been reported [784–786]. Furthermore, two new flavonoids were successfully isolated from *P. minus* namely, 6,7,4′,5′-dimethylenedioxy-3,5,3′-trimethoxyflavone (**753**) and 6,7-methylenedioxy-5,3′,4′,5′-tetramethoxyflavone (**754**) [787]. In 2018, a phytochemical study of the stem methanol extract of the plant led to the purification and characterization of a new compound, polygonumin A (**755**), along with its known C-3″ diastereomer, vanicoside A (**759**) [788]. Subsquently, further polygonumin derivatives, specifically polygonumins B (**756**), C (**757**), and D (**758**), were obtained [789]. Among other chemical constituents of *P. minus*, the sesquiterpenoid xanthorrhizol (**760**) and simple acids such as oxalic (**761**), gallic (**95**), and (*E*)-coumaric acid (**538**) were reported [790].

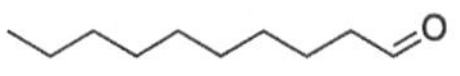

743 (decanal)

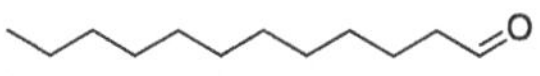

744 (dodecanal)

745 (cyclolongifolene oxide)

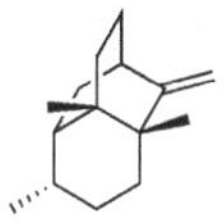

746 (seychellene)

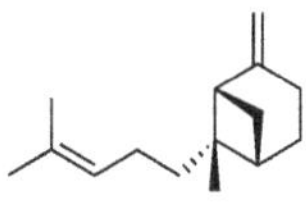

747 (α-bergamotene)

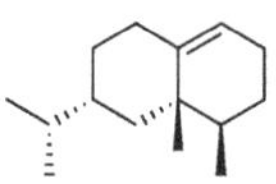

748 (valencene)

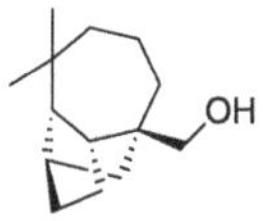

749 (isolongifolol)

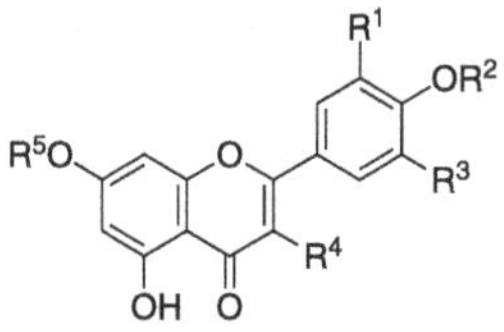

750 (apigetrin) $R^1 = R^3 = R^2 = R^4 = H$, $R^5 =$ β-D-Glc
751 (isoquercetin) $R^1 = OH$, $R^3 = R^2 = R^5 = H$, $R^4 =$ *O*-β-D-Glc
752 (hyperoside) $R^1 = OH$ $R^2 = R^3 = R^5 = H$,
$R^4 =$ *O*-β-D-Gal

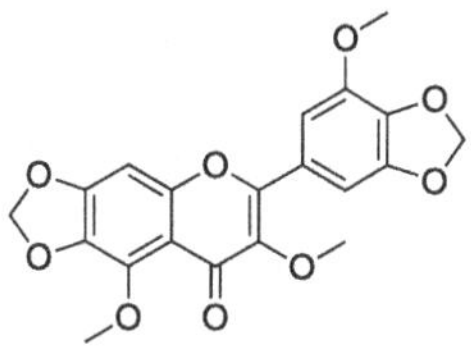

753 (6,7,4′,5′-dimethylenedioxy-3,5,3′-trimethoxyflavone)

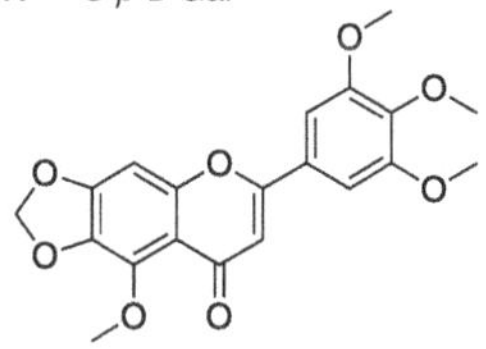

754 (6,7-methylenedioxy-5,3′,4′,5′-tetramethoxyflavone)

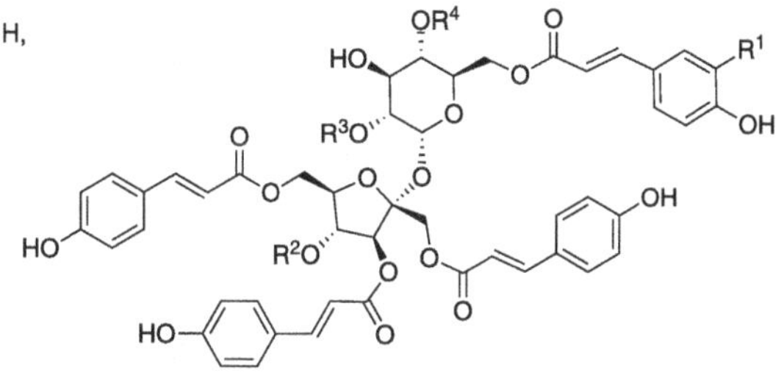

755 (polygonumin A) $R^1 = R^3 = H$, $R^4 = R^2 = COCH_3$
756 (polygonumin B) $R^1 = OMe$, $R^3 = H$, $R^2 = R^4 = COCH_3$
757 (polygonumin C) $R^1 = R^2 = R^3 = H$, $R^4 = COCH_3$
758 (polygonumin D) $R^1 = R^2 = R^3 = R^4 = H$
759 (vanicoside A) $R^1 = CH_3$, $R^2 = R^4 = H$, $R^3 = COCH_3$

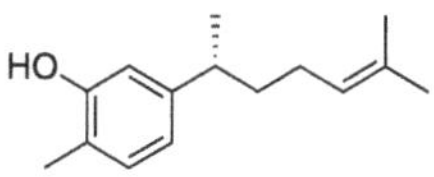

760 (xanthorrizhol)

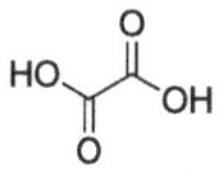

761 (oxalic acid)

2.21.3 Biological Activities

Polygonum minus has a variety of pharmacological and biological activities in both in vivo and in vitro experimental models, and exhibits antiviral, antioxidant, antigenotoxicity, antiulcer, antifungal, and cytotoxic effects. Studies on its toxicity suggest that it is safe for further pharmacological investigation. An acute toxicity study of an aqueous extract showed no signs of hepatotoxicity, nephrotoxicity, or adverse effects on food intake, growth, organ weights, hematological parameters, or mortality in experimental animals [791]. In a study on genotoxicity, *P. minus* was evaluated for its potential to cause genetic damage in human lymphocytes. The aqueous extract was found to be non-genotoxic, supporting its safety for daily consumption [792]. Extracts of the plant, including the aqueous extract, were tested against HeLa cervical cancer cells, showing no significant inhibition of cell viability, indicating accordingly a low cytotoxicity to normal cells [793].

Several studies have demonstrated the potential of *Polygonum minus* leaves as a cytotoxic agent against cancer cells. Korulkin and Muzychkina showed that flavonoids from *P. minus* and other *Polygonum* species induced apoptosis in human oral cavity carcinoma cells (HSC-2), salivary gland tumor cells (HSG), and normal human fibroblasts from the gums (HGF). The flavonoid constituents of the plant, including myricetin (**89**), and quercetin (**85**), are believed to play a key role in mediating its cancer cell line cytotoxicity [794]. Other studies have shown that methanol, hexane, and ethyl acetate extracts of *P. minus* exhibit antiproliferative activity against colon cancer cells (HCT116), while methanol and aqueous extracts encouraged the growth of normal colon cells (CCD841) in a dose-dependent manner, supporting a healing effect [786]. However, the authors also reported that the same aqueous extract had a lack of cancer cell line cytotoxic selectivity, by stimulating the growth of both colon cancer cells (HCT116) and normal colon cells (CCD841), with cell viability increasing to 134 and 157% at the high concentrations of 167 and 333 μg/cm^3, respectively. These results remain unclear and require further investigation.

The efficacy of *P. minus* constituents in exhibiting antioxidant, anti-cholinesterase (AChE), anti-butyrylcholinesterase (BChE), anti-HIV-1 protease and cytotoxic activities was further supported by a study conducted by Ahmad and colleagues. In their investigation, they identified four new compounds named polygonumins A (**755**), B (**756**), C (**757**), and D (**758**) produced by *P. minus* [788, 789]. Among these, **758** demonstrated antioxidant activity, with an IC_{50} value of 27 μg/cm^3 in a DPPH assay. Additionally, polygonumin C (**757**) exhibited AChE activity, showing 71.8 ± 0.6% inhibition. Polygonumins A (**755**) and D (**758**) showed inhibition of HIV-1 protease activity, surpassing in each case the activity of the positive control, pepstatin A, with inhibition rates of 91.6 ± 0.2% and 91.4 ± 0.1%, respectively. Earlier work from the same group showed polygonumin A (**755**) to have low micromolar antiproliferative effects against human leukemia (K562), breast adenocarcinoma (MCF-7), and colorectal cancer (HCT116) cell lines, with statistical significance ($p < 0.05$). Notably, **755** also exhibited selective cytotoxicity toward the cancer cells used, demonstrating cytotoxicity to V79–4 cells at a high concentration of 50 μg/cm^3 [788].

A number of studies have explored the potential of *P. minus* to protect normal cells from drug-induced toxicity. For example, Rashid et al. demonstrated that the essential oil of *P. minus* can protect hepatocytes from cisplatin overdose [783]. Thus, a high dose of *P. minus* ethanol extract (400 mg/kg) reduced the serum AST, ALT, and bilirubin levels in Sprague–Dawley rats compared to a cisplatin-induced group [806].

Further studies have shown that the aqueous extract of the plant exhibited hepatoprotective activity against paracetamol-induced liver toxicity, while the methanol extract demonstrated similar protection against carbon tetrachloride-induced liver toxicity [795]. These findings were supported by another study showing that a methanol extract helped to prevent liver damage caused by these same inducers [796]. Previously, the same group showed that a methanol extract protected stomach cells from ulcer formation [785]. The extract was able to reduce stomach pH in Wistar albino mice to 1.8 ± 0.1. Additionally, the total acidity of the gastric juice in the treated rats was significantly lower (29.8 ± 3.0 mEq/dm^3) compared to the control group (73.8 ± 6.4 mEq/dm^3).

Numerous studies have confirmed that *P. minus* exhibits strong antibacterial activity against both Gram-positive and Gram-negative bacteria. Methanol, *n*-hexane, and aqueous extracts inhibited the growth of key pathogens, including *Staphyloccus aureus*, *Escherichia coli*, *Bacillus subtilis*, and *Salmonella typhimurium* [794, 797]. Hence, the methanol extract produced large inhibition zones against *S. aureus* and *B. subtilis* [798], while the aqueous extract was effective against *S. typhimurium* and *Serratia marcescens*, two common causes of gastroenteritis and urinary tract infections [797]. These findings suggest that *P. minus* could serve as a natural remedy to treat bacterial infections, especially those caused by multidrug-resistant strains like methicillin-resistant *S. aureus* (MRSA). In addition to its antibacterial effects, *P. minus* has reported antifungal properties. Methanol, chloroform, and acetone extracts have some activity against plant pathogens such as *Colletotrichum capsici*, with *MIC* values greater than 20 µg/cm^3 [799]. The essential oil of the plant was found to inhibit the growth of *Candida albicans*, a common cause of candidiasis and dermatitis, with a clear inhibition zone of 12 ± 0.2 mm [794]. Moreover, *P. minus* has been shown to inhibit the growth of *Trichophyton rubrum*, the causative agent of dermatophytosis and ringworm. An ethanol extract of *P. minus* exhibited a clear inhibition zone of 14 ± 1.6 cm at a 20% concentration, indicating that the fungus is highly sensitive to the plant's antifungal compounds [799].

Polygonum minus also has demonstrated antiviral activity, including against *Herpes simplex* virus type 1 (HSV-1) and vesicular stomatitis virus (VSV) [786, 800]. These findings broaden the therapeutic potential of *P. minus* to include the treatment of viral infections. Beyond its antimicrobial properties, the *P. minus* ethanol extract reduced ulcer area, when compared with a carboxymethyl cellulose negative control group, in an ethanol-induced ulcer model [791]. Another study on gastroprotective activity against oxidative stress using an ethanol-induced model suggested that *P. minus* could be beneficial in managing gastric ulcers and oxidative damage in the stomach. Gallic acid (**95**) has been identified as a main active compound, followed by (*E*)-coumaric acid (**538**) in manifesting gastroprotective activity. This activity was evaluated through antioxidant assays, mucus barrier formation, and the maintenance of prostaglandin E2 (PGE2) and superoxide dismutase levels [801]. In addition to its antimicrobial and gastroprotective properties, an aqueous extract of *P. minus* was more effective than butylated hydroxytoluene in preventing microbial growth in refrigerated duck meatballs stored at 4°C for 21 days [802]. This highlights the potential of *P. minus* as a natural preservative, possibly offering an alternative to synthetic antimicrobial agents in food products.

Polygonum minus extracts have shown antioxidant properties in both in vitro and in vivo studies. One such investigation demonstrated that an extract of the plant exhibited DPPH radical-scavenging activity, with an IC_{50} value of 84.5 ± 7.3% [803]. This finding was further supported by other studies, which demonstrated that the methanol extract of *P. minus* produced significant DPPH radical-scavenging effects [786, 798]. In addition to DPPH scavenging, extracts have ferric reducing antioxidant power (FRAP) activities, which were measured to assess the reduction of iron ions. Abdullah et al. (2017) found that a methanol extract of *P. minus* had a high FRAP activity of 1728.3 ± 1 μmol Fe^{2+}/g extract [786], while Khalid and Babji (2018) also reported significant ferric reducing activity (63.6 ± 0.7 μmol Fe^{2+}/g) [804]. In vivo studies on hepatotoxicity-induced Sprague Dawley rats further supported the antioxidant potential of *P. minor* in reducing the levels of oxidative stress markers, such as superoxide dismutase (SOD), glutathione peroxidase (GPx), and catalase, all of which are critical enzymes involved in the body's defense against oxidative damage. Additionally, a plant extract protected red blood cells from destruction induced by hydrogen peroxide, a reactive oxygen species that contributes to oxidative stress in RBCs [783, 804].

Research by Hussin et al. and co-workers explored the anti-aging potential of *P. minor* by evaluating its ability to inhibit elastase, an enzyme that breaks down the protein elastin, which is responsible for the elasticity of connective tissues [805]. A plant extract inhibited elastase activity by 57.6 ± 1.0%, which was comparable with the control compound quercetin (52.9 ± 1.2%), a well-known antioxidant and anti-aging agent. This anti-aging effect was attributed to the presence of several bioactive compounds in *P. minus*, including, myricetin (**89**), catechin (**90**), and isorhamnetin (**484**), as well as quercetin (**85**) itself [806].

3 Key Biologically Active Malaysian Medicinal and Aromatic Plants—Outlook

Malaysian medicinal plants are a rich and diverse source of bioactive compounds, each offering a wide array of potential therapeutic benefits. These plants, deeply rooted in traditional medicine, have attracted increasing attention from modern scientific researchers due to their notable biological properties. From antimicrobial and anti-inflammatory effects to neuroprotective and antitumor properties, the therapeutic potential of these plants is vast and holds promise for the development of novel treatments in modern medicine. Below are some of the key bioactive compounds found in notable Malaysian medicinal and aromatic plants, highlighting their wide-ranging applications in health and wellness.

One of the most well-known and extensively studied plants in Malaysia is *Eurycoma longifolia* (see Sect. 2.1), commonly known as "Tongkat Ali". This plant is renowned for its aphrodisiac properties and testosterone-enhancing effects, and biosynthesizes as key bioactive compounds eurycomanone (**1**) and other quassinoids. These compounds exhibit a broad spectrum of activities, including anti-inflammatory, antimicrobial, antioxidant, and cytotoxic effects. Eurycomanone (**1**), in particular, has been shown to stimulate testosterone production, supporting *E. longifolia* as an important natural remedy for sexual dysfunction. Its ability to combat chronic diseases further underscores its ethnomedical value. However, gaps in understanding the synergistic effects of its compounds and limited mechanistic insights into its broader pharmacological activities remain. Exploring these areas through advanced omics approaches could unlock new therapeutic applications for *E. longifolia*.

Another well-researched plant is *Andrographis paniculata* (see Sect. 2.3), known for its potential medicinal properties and bitter taste. The bioactive compounds in this plant, including andrographolide (**117**), and other labdane diterpenoid lactones, contribute to its demonstrated anti-inflammatory, immunomodulatory, and cytotoxic activities. Andrographolide (**117**) has been shown to inhibit the NF-κB signaling pathway and induce apoptosis in cancer cells, making this plant a potential treatment for conditions like rheumatoid arthritis and being worthy of further laboratory investigation for its antitumor effects in animal models. Its broad pharmacological effects add to the growing list of Malaysian plants with clinical potential. However, the lack of clinical trials validating these effects in human populations limits its application in modern medicine. Future research could prioritize clinical studies and explore its potential for targeted drug development.

In terms of women's health, *Marantodes pumilum* (see Sect. 2.2) (formally known as *Labdisia pumila*) is a plant used traditionally for addressing hormonal imbalances. In containing flavonoids such as quercetin (**85**) and kaempferol (**88**), *M. pumilum* plays a useful role in managing menopausal symptoms, preventing osteoporosis, and helping to control obesity. These compounds work in part by modulating estrogenic pathways, offering a natural approach to hormonal balance. Additionally, the plant's antibacterial, hepatoprotective, and other properties contribute to its overall therapeutic scope, making it a valuable resource in holistic healthcare. However, further research is needed to standardize its bioactive compounds for therapeutic use and to address knowledge gaps regarding its long-term safety and optimal dosages.

Similarly, *Morinda citrifolia* (see Sect. 2.4), commonly known as "noni", is renowned for its diverse medicinal uses. The plant contains bioactive compounds like anthraquinones, iridoids, and phenolic acids, which exhibit significant antidiabetic, antihypertensive, and hepatoprotective effects. However, caution is advised when using noni, as prolonged use may lead to nephrotoxicity, highlighting the need for careful, regulated consumption. In addition, the mechanisms driving its nephrotoxicity are poorly understood, necessitating further toxicological studies to ensure its safe therapeutic use.

In the realm of cognitive health, *Centella asiatica* (see Sect. 2.5), or "Gotu Kola", has gained prominence for its neuroprotective properties. Triterpene constituents such as asiatic acid (**248**) and madecassic acid (**249**) have shown antihyperglycemic and neuroprotective effects. The potential ability of Gotu Kola to support cognitive function while regulating blood glucose is of significance. However, current research efforts have often neglected the synergistic potential of its bioactive compounds. Molecular networking approaches could provide valuable insights into these interactions and their clinical implications.

Another plant with potential applications in metabolic and cardiovascular health is *Orthosiphon stamineus* (see Sect. 2.6), or "Java tea". This plant contains diterpenes and flavonoids, including sinensetin (**348**), which are responsible for its antioxidant, antidiabetic, and antihypertensive effects. This plant has been shown to improve lipid profiles, reduce blood glucose levels, and help manage hypertension in laboratory studies, making it a potential valuable resource in the management of certain cardiovascular diseases and metabolic disorders like type 2 diabetes. To advance its therapeutic applications, further studies are required to standardize its extracts and validate these effects through human trials.

In addition to these, *Cananga odorata* (see Sect. 2.19) ("Ylang-ylang") also contributes significantly to the pharmacological knowledge of Malaysian plants. The bioactive compounds in *C. odorata*, including *O*-methylmoschatoline (**693**), liriodenine (**656**), and 3,4-dihydroxybenzoic acid (**694**), demonstrate potent antibacterial and antifungal properties. The essential oil derived from the plant has insecticidal effects, and its extracts have shown promise in skin whitening and anti-aging applications. Constituents like canangaterpene I (**681**) inhibit melanogenesis, offering potential for managing hyperpigmentation and promoting skin health. However, the precise molecular mechanisms underpinning its dermatological benefits remain poorly understood. Future studies should explore its bioactivity using cutting-edge metabolomics and bioinformatics approaches to unlock its full potential.

Citrus hystrix (see Sect. 2.20), or "Kaffir lime", is another important plant in traditional medicine, containing a range of bioactive compounds, such as tamarin (**721**) and (+)-syringaresinol (**742**), and limonin (**762**), which exhibit antibacterial, antioxidant, and cytotoxic properties. The essential oil from its leaves, rich in citronellal (**695**), has demonstrated antibacterial activity against oral pathogens like *Porphyromonas gingivalis* and *Streptococcus mutans*, while also inhibiting biofilm formation. Furthermore, these compounds have shown potential neuroprotective effects, suggesting their use in treating neurodegenerative disorders like Alzheimer's disease. Nonetheless, the lack of large-scale preclinical and clinical studies hinders its advancement into pharmaceutical applications. Strategic partnerships with biotechnology firms could facilitate the development of standardized extracts and clinical-grade formulations.

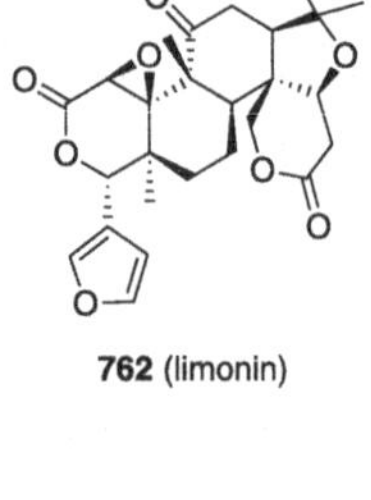

762 (limonin)

Finally, *Piper sarmentosum* (see Sect. 2.11), is valued for its content of flavonoids such as myricetin (**89**) and quercetin (**85**), which exhibit potent antioxidant effects. These compounds induce apoptosis in cancer cells, inclusive of those from human oral carcinomas and salivary gland tumors. Moreover, *P. sarmentosum* has been associated with some therapeutic benefits, including as a muscle-relaxant.

In conclusion, the bioactive compounds found in Malaysian medicinal and aromatic plants collectively represent a valuable and underexplored resource for further modern pharmacological investigations. Constituents such as andrographolide (**117**), eurycomanone (**1**), asiatic acid (**248**), and myricetin (**89**) exemplify the therapeutic potential of these plants in possibly aiding in the future therapy of chronic diseases, metabolic disorders, cancer, and neurological conditions. However, significant gaps remain, particularly in understanding the precise mechanisms of action, the standardization of bioactive extracts, and in conducting large-scale preclinical and clinical studies to validate their efficacy and safety. Addressing these gaps requires multidisciplinary collaboration, integrating advanced omics technologies, bioinformatics, and innovative cultivation and processing methods to ensure consistency and reproducibility. As research continues to uncover the full range of benefits these plants may offer, it is becoming increasingly clear that Malaysian flora could play a central role in the development of natural, sustainable therapies for a wide variety of health challenges. The potential of these plants extends beyond pharmacological applications, offering opportunities for partnerships between researchers, policymakers, and industry stakeholders to create standardized formulations and clinical-grade products. By leveraging these synergies, Malaysian medicinal plants could not only preserve traditional knowledge but also drive breakthroughs in drug discovery, therapeutic innovation, and sustainable economic development.

Acknowledgments The authors would like to express their sincere gratitude to Professor Dr. Khozirah Shaari of Universiti Putra Malaysia for her trust, encouragement, and invaluable support in the development of this chapter. We also extend our heartfelt thanks to Professor Heinz Falk for his assistance with the preparation of the manuscript. Additional appreciation is due to Dr. Nik Fatini Azmin and Dr. Isna Athirah Othman for their help in drawing some of the molecular structures using ChemDraw.

References

1. Abuga I, Sulaiman SF, Abdul Wahab R, Ooi KL, Rasad MSBA (2022) Phytochemical constituents and antibacterial activities of 45 Malay traditional medicinal plants. J Herb Med 32:100496
2. Ministry of Finance Malaysia (2015) Eleventh Malaysia plan 2016–2020. Speech by the Prime Minister in the Dewan Rakyat. Kuala Lumpur, Malaysia 32
3. Li X, Yuan W, Wu J, Zhen J, Sun Q, Yu M (2022) Andrographolide, a natural anti-inflammatory agent: an update. Front Pharmacol 13:920435
4. Ismail SB, Wan Mohammad WMZ, George A, Hussain NHN, Kamal ZMM, Liske E (2012) Randomized clinical trial on the use of PHYSTA freeze-dried water extract of *Eurycoma*

longifolia for the improvement of quality of life and sexual well-being in men. Evid-Based Complt Altern Med 2012:429268
5. Gohil KJ, Patel JA, Gajjar AK (2010) Pharmacological review on *Centella asiatica*: a potential herbal cure-all. Indian J Pharm Sci 72:546
6. Patwardhan B, Vaidya ADB, Chorghade M (2014) Ayurveda and natural products discovery. Curr Sci 86:789
7. Bhat R, Karim AA (2010) Tongkat Ali (*Eurycoma longifolia* Jack): a review on its ethnobotany and pharmacological importance. Fitoterapia 81:669
8. Kassim AR (2002) Species assembly and site preference of tree species in a primary Seraya-ridge forest of peninsular Malaysia. J Trop Forest Sci 14:287
9. Chua GLS, Koh BL, Lau S, Lee SC, Mathias S, Turner IM, Yong JWH, Heah HH (1995) The nutrient status of the plateau heath forest on Gunung Keriong, Pahang, Peninsular Malaysia. J Trop Forest Sci 8:240
10. Chua LSL, Kamarudin S, Markandian M, Hamidah M (2005) A preliminary checklist of vascular plants from the Machinchang Range, Pulau Langkawi, Peninsular Malaysia. Malay Nat J 57:155
11. Ang HH, Ikeda S, Gan EK (2001) Evaluation of the potency activity of aphrodisiac in *Eurycoma longifolia* Jack. Phytother Res 15:435
12. Kulip J (2009) Medicinal plants of Sabah, Malaysia: potential for agroforestry. JIRKAS Working Report 60:47
13. Adenan M (1998) Opportunities on the planting of medicinal and herbal plants in Malaysia. Published privately
14. Hsuan K (1990) The concise flora of Singapore, gymnosperms and dicotyledons. Singapore University Press, Singapore
15. Ulbricht C, Conquer J, Flanagan K, Isaac R, Rusie E, Windsor RC (2013) An evidence-based systematic review of Tongkat Ali (*Eurycoma longifolia*) by the natural standard research collaboration. J Diet Suppl 10:54
16. Jiwajinda S, Santisopasri V, Murakami A, Hirai N, Ohigashi H (2001) Quassinoids from *Eurycoma longifolia* as plant growth inhibitors. Phytochemistry 58:959
17. Perry LM (1980) Medicinal plants of East and Southeast Asia: attributed properties. Econ Bot 34:361
18. Darise M, Kohda H, Mizutani K, Tanaka O (1982) Eurycomanone and eurycomanol, quassinoids from the roots of *Eurycoma longifolia*. Phytochemistry 21:2091
19. Darise M, Kohda H, Mizijtani K, Tanaka O (1983) Revision of configuration of the 12-hydroxy group of eurycomanone and eurycomanol, quassinoids from *Eurycoma longifolia*. Phytochemistry 22:1514
20. Ismail Z, Ismail N, Lassa J (2009) Malaysian herbal monograph. Forest Research Institute Malaysia, Kuala Lumpur
21. Kuo PC, Shi LS, Damu AG, Su C-R, Huang C-H, Ke C-H, Wu J-B, Lin A-J, Bastow KF, Lee K-H (2003) Cytotoxic and antimalarial β-carboline alkaloids from the roots of *Eurycoma longifolia*. J Nat Prod 66:1324
22. Kuo PC, Damu AG, Lee KH, Wu TS (2004) Cytotoxic and antimalarial constituents from the roots of *Eurycoma longifolia*. Bioorg Med Chem 12:537
23. Samy J, Sugumaran Manickam, Lee KLW, Wong KM (2005) Herbs of Malaysia: an introduction to the medicinal, culinary, aromatic and cosmetic use of herbs. Times Editions, Shah Alam, Selingor, Malaysia
24. Maneenoon K, Khuniad C, Teanuan Y, Saedan N, Prom-in S, Rukleng N, Kongpool W, Pinsook P, Wongwiwat W (2015) Ethnomedicinal plants used by traditional healers in Phatthalung Province, Peninsular Malaysia. J Ethnobiol Ethnomed 11:1
25. Chai PPK (2006) Medicinal plants of Sarawak. Lee Ming Press, Kuching, Sarawak, Malaysia
26. George A, Chinnapan S, Biggins J (2018) Antiviral activity of a standardized root water extract of *Eurycoma longifoliaa* (Physta) against dengue virus. In: Proceedings of the 34th international conference on natural products, Batu Ferringhi, Penang, Malaysia

27. George A, Suzuki N, Abas AB, Mohri K, Utsuyama M, Hirokawa K, Takara T (2016) Immunomodulation in middle-aged humans via the ingestion of Physta® standardized root water extract of *Eurycoma longifolia* Jack—a randomized, double-blind, placebo-controlled, parallel study. Phytother Res 30:627
28. Byrne M (2023) The ultimate Tongkat Ali supplement guide 2023: benefits, side effects & how to use. Published privately
29. Jagananth IB, Ng LT (2000) Herbs: the green pharmacy of Malaysia, 1st edn. MARDI, Kuala Lumpur
30. Chan KL, Lee SP, Sam TW, Han BH (1989) A quassinoid glycoside from the roots of *Eurycoma longifolia*. Phytochemistry 28:2857
31. Chan KL, Lee SP, Sam TW, Tan SC, Noguchi H, Sankawa U (1991) 13*β*,18-Dihydroeurycomanol, a quassinoid from *Eurycoma longifolia*. Phytochemistry 30:3138
32. Tada H, Yasuda F, Otani K, Doteuchi M, Ishihara Y, Shiro M (1991) New antiulcer quassinoids from *Eurycoma longifolia*. Eur J Med Chem 26:345
33. Park S, Nhiem NX, Kiem PV, Minh CV, Tai BH, Kim N, Yoo J-H, Ko H-J, Kim SH (2014) Five new quassinoids and cytotoxic constituents from the roots of *Eurycoma longifolia*. Bioorg Med Chem Lett 24:3835
34. Meng D, Li X, Han L, Zhang L, An W, Li X (2014) Four new quassinoids from the roots of *Eurycoma longifolia* Jack. Fitoterapia 92:105
35. Chan KL, Iitaka Y, Noguchi H, Sugiyama H, Saito I, Sankawa U (1992) 6*α*-Hydroxyeurycomalactone, a quassinoid from *Eurycoma longifolia*. Phytochemistry 31:4295
36. Fiaschetti G, Grotzer MA, Shalaby T, Castelletti D, Arcaro A (2011) Quassinoids: from traditional drugs to new cancer therapeutics. Curr Med Chem 18:316
37. Morre JD, Grieco PA, Morre DM (1998) Mode of action of the anticancer quassinoids — inhibition of the plasma membrane NADH oxidase. Life Sci 63:595
38. Tran TVA, Malainer C, Schwaiger S, Atanasov AG, Heiss EH, Dirsch VM, Stuppner H (2014) NF-κB inhibitors from *Eurycoma longifolia*. J Nat Prod 77:483
39. Mitsunaga K, Zuo K, Tanaka T, Ohkawa Y, Kobayashi Y, Sawaguchi T, Ohmoto T (1994) Canthin-6-one alkaloids from *Eurycoma longifolia*. Phytochemistry 35:795
40. Choo CY, Chan KL (2002) High performance liquid chromatography analysis of canthinone alkaloids from *Eurycoma longifolia*. Planta Med 64:382
41. Talbott SM, Talbott JA, George A, Pugh M (2013) Effect of Tongkat Ali on stress hormones and psychological mood state in moderately stressed subjects. J Int Soc Sports Nutr 10:28
42. Danial M, Chan LK, Syed Alwee SSR, Subramaniam S (2011) Seed histology of recalcitrant *Eurycoma longifolia* plants during germination and its beneficial attribute for hairy roots production. J Med Plants Res 5:93
43. Varghese CP, Ambrose C, Jin SC, Lim YJ, Keisaban T (2013) Antioxidant and anti-inflammatory activity of *Eurycoma longifolia* Jack, a traditional medicinal plant in Malaysia. Int J Pharm Sci Nanotech 5:1875
44. Wahab NA, Mokhtar NM, Halim WNHA, Das S (2010) The effect of *Eurycoma longifolia* Jack on spermatogenesis in estrogen-treated rats. Clinics 65:93
45. Panjaitan RGP, Handharyani E, Chairul MW (2013) Hepatoprotective activity of *Eurycoma longifolia* Jack roots. Indian J Trad Knowl 12:225
46. Al-Salahi OSA, Kit-Lam C, Majid AMSA, Al-Sueade FSR, Saghir SAM, Abdullah WZ, Ahamed MBK, Yusoff NM (2013) Anti-angiogenic quassinoid-rich fraction from *Eurycoma longifolia* modulates endothelial cell function. Microvasc Res 90:30
47. Husen R, Pihie AHL, Nallappan M (2004) Screening for antihyperglycaemic activity in several local herbs of Malaysia. J Ethnopharmacol 95:205
48. Ngoc PB, Pham TB, Nguyen HD, Tran TT, Chu HH, Chau VM, Lee J-H, Nguyen TD (2016) A new anti-inflammatory β-carboline alkaloid from the hairy-root cultures of *Eurycoma longifolia*. Nat Prod Res 30:1360
49. Han YM, Woo SU, Choi MS, Park YN, Kim SH, Yim H, Yoo HH (2016) Antiinflammatory and analgesic effects of *Eurycoma longifolia* extracts. Arch Pharm Res 39:421

50. Ang HH, Chan KL, Mak JW (1995) Effect of 7-day daily replacement of culture medium containing *Eurycoma longifolia* Jack constituents on the Malaysian *Plasmodium falciparum* isolates. J Ethnopharmacol 49:171
51. Mohd Abd Razak MR, Noor Rain A, Ismail Z, Zakiah I (2005) Effect of *Eurycoma longifolia* extract on the glutathione level in *Plasmodium falciparum* infected erythrocytes in vitro. Trop Biomed 22:155
52. Mohd Abd Ridzuan MR, Sow A, Noor Rain A, Mohd Ilham A, Zakiah I (2007) *Eurycoma longifolia* extract-artemisinin combination: parasitemia suppression of *Plasmodium yoelii*-infected mice. Trop Biomed 24:111
53. Farouk A-E, Benafri A (2007) Antibacterial activity of *Eurycoma longifolia* Jack. A Malaysian medicinal plant. Saudi Med J 28:1422
54. Farouk A-E, Mohd Nawi MN, Hassan S (2008) Antibacterial peptides from *Eurycoma longifolia* (Tongkat Ali) and *Labisia pumila* (Kacip Fatimah) leaves in Malaysia. Sci Broneiana 9:55
55. Kong C, Yehye WA, Rahman NA, Tan M-W, Nathan S (2014) Discovery of potential anti-infectives against *Staphylococcus aureus* using a *Caenorhabditis elegans* infection model. BMC Compl Altern Med 14:1
56. Kavitha N, Noordin R, Chan KL, Sasidharan S (2012) In vitro anti-*Toxoplasma gondii* activity of root extract/fractions of *Eurycoma longifolia* Jack. BMC Compl Altern Med 12:91
57. Girish S, Kumar S, Aminudin N (2015) Tongkat Ali (*Eurycoma longifolia*): a possible therapeutic candidate against *Blastocystis* sp. Parasit Vectors 8:332
58. Jantan I, Zaki ZM, Ahmad AR, Ahmad R (1999) Evaluation of smoke from mosquito coils containing Malaysian plants against *Aedes aegypti*. Fitoterapia 70:237
59. Ang HH, Cheang HS (1999) Studies on the anxiolytic activity of *Eurycoma longifolia* Jack roots in mice. Jpn J Pharmacol 79:497
60. Ang HH, Lee KL (2002) Effect of *Eurycoma longifolia* Jack on orientation activities in middle-aged male rats. Fund Clin Pharmacol 16:479
61. Ang HH, Ngai TH (2001) Aphrodisiac evaluation in non-copulator male rats after chronic administration of *Eurycoma longifolia* Jack. Fund Clin Pharmacol 15:265
62. Ang HH, Sim MK (1997) *Eurycoma longifolia* Jack enhances libido in sexually experienced male rats. Exp Anim 46:287
63. Ang HH, Lee KL, Kiyoshi M (2003) *Eurycoma longifolia* Jack enhances sexual motivation in middle-aged male mice. J Basic Clin Physiol Pharmacol 14:301
64. Ang HH, Lee KL, Kiyoshi M (2004) Sexual arousal in sexually sluggish old male rats after oral administration of *Eurycoma longifolia* Jack. J Bas Clin Physiol Pharmacol 15:3
65. Ang HH, Ngai TH, Tan TH (2003) Effects of *Eurycoma longifolia* Jack on sexual qualities in middle aged male rats. Phytomedicine 10:590
66. Abd Razak M-F, Aidoo KE, Candlish AGG (2007) Mutagenic and cytotoxic properties of three herbal plants from Southeast Asia. Trop Biomed 24:49
67. Al-Salahi OSA, Ji D, Abdul Majid AMS, Kit-Lam C, Abdullah WZ, Zaki A, Din SKKJ, Mohd Yusoff N, Majid AAS (2014) Anti-tumor activity of *Eurycoma longifolia* root extracts against K-562 cell line: in vitro and in vivo study. PLoS One 9:e83818
68. Hajjouli S, Chateauvieux S, Teiten MH, Orlikova B, Schumacher M, Dicato M, Choo CY, Diederich M (2014) Eurycomanone and eurycomanol from *Eurycoma longifolia* Jack as regulators of signaling pathways involved in proliferation, cell death and inflammation. Molecules 19:14649
69. Nurhanan MY, Hawariah LPA, Ilham AM, Shukri MAM (2005) Cytotoxic effects of the root extracts of *Eurycoma longifolia* Jack. Phytother Res 19:994
70. Ohishi K, Toume K, Arai MA, Koyano T, Kowithayakorn T, Mizoguchi T, Itoh M, Ishibashi M (2015) 9-Hydroxycanthin-6-one, a β-carboline alkaloid from *Eurycoma longifolia*, is the first Wnt signal inhibitor through activation of glycogen synthase kinase 3β without depending on casein kinase 1α. J Nat Prod 78:1139
71. Mohamad M (2002) Compendium of medicinal plants used in Malaysia. Herbal Medicine Research Centre, Institute for Medical Research, Kuala Lumpur

72. Tee TT, Lope Pihie AH (2005) Induction of apoptosis by *Eurycoma longifolia* Jack extracts. Anticancer Res 25:2205
73. Tee TT, Cheah YH, Lope Pihie AH (2007) F16, a fraction from *Eurycoma longifolia* Jack extract, induces apoptosis via a caspase-9-independent manner in MCF-7 cells. Anticancer Res 27:3425
74. Tong KL, Chan KL, AbuBakar S, Low BS, Ma HQ, Wong PF (2015) The in vitro and in vivo anti-cancer activities of a standardized quassinoids composition from *Eurycoma longifolia* on LNCaP human prostate cancer cells. PLoS One 10:1
75. Zakaria Y, Rahmat A, Pihie AHL, Abdullah NR, Houghton PJ (2009) Eurycomanone induce apoptosis in HepG2 cells via up-regulation of p53. Cancer Cell Int 9:16
76. Ang HH, Sim MK (1998) *Eurycoma longifolia* increases sexual motivation in sexually naive male rats. Arch Pharm Res 21:779
77. Chan K-L, Low B-S, Teh C-H, Das PK (2009) The effect of *Eurycoma longifolia* on sperm quality of male rats. Nat Prod Commun 4:1331
78. Chen CK, Wan Mohamad WMZ, Ooi FK, Ismail SB, Abdullah MR, George A (2014) Supplementation of *Eurycoma longifolia* Jack extract for 6 weeks does not affect urinary testosterone: epitestosterone ratio, liver and renal functions in male recreational athletes. Int J Prev Med 5:728
79. Frydrychová S, Opletal L, Macáková K, Lustyková A, Rozkot M, Lipenský J (2011) Effects of herbal preparation on libido and semen quality in boars. Reprod Domest Anim 46:573
80. George A, Henkel R (2014) Phytoandrogenic properties of *Eurycoma longifolia* as natural alternative to testosterone replacement therapy. Andrologia 46:708
81. Henkel RR, Wang R, Bassett SH, Chen T, Liu N, Zhu Y, Tambi MI (2014) Tongkat Ali as a potential herbal supplement for physically active male and female seniors—a pilot study. Phytother Res 28:544
82. Kotirum S, Ismail SB, Chaiyakunapruk N (2015) Efficacy of Tongkat Ali (*Eurycoma longifolia*) on erectile function improvement: systematic review and meta-analysis of randomized controlled trials. Compl Ther Med 23:693
83. Low BS, Choi SB, Abdul Wahab H, Das KP, Chan KL (2013) Eurycomanone, the major quassinoid in *Eurycoma longifolia* root extract increases spermatogenesis by inhibiting the activity of phosphodiesterase and aromatase in steroidogenesis. J Ethnopharmacol 149:201
84. Mohd Effendy N, Mohamed N, Muhammad N, Mohamad NI, Shuid AN (2012) *Eurycoma longifolia*: medicinal plant in the prevention and treatment of male osteoporosis due to androgen deficiency. Evid-Based Compl Alt Med 2012:125761
85. Mat Noor M, Mohd Nor AHS, Che Hassan L (2004) The effect of *Eurycoma longifolia* Jack Tongkat Ali on sexual behaviour and sperm quality in rats. Malays J Pharm Sci 2:53
86. Qinna N, Taha H, Matalka KZ, Badwan AA (2009) A new herbal combination, Etana, for enhancing erectile function: an efficacy and safety study in animals. Int J Impot Res 21:315
87. Solomon MC, Erasmus N, Henkel RR (2014) In vivo effects of *Eurycoma longifolia* Jack (Tongkat Ali) extract on reproductive functions in the rat. Andrologia 46:339
88. Bin Mohd Tambi MI, Imran MK (2010) *Eurycoma longifolia* Jack in managing idiopathic male infertility. Asian J Androl 12:376
89. Bin Mohd Tambi MI, Imran MK, Henkel RR (2012) Standardised water-soluble extract of *Eurycoma longifolia*, Tongkat Ali, as testosterone booster for managing men with late-onset hypogonadism? Andrologia 44:226
90. Udani JK, George AA, Musthapa M, Pakdaman MN, Abas A (2014) Effects of a proprietary freeze-dried water extract of *Eurycoma longifolia* (Physta) and *Polygonum minus* on sexual performance and well-being in men: a randomized, double-blind, placebo-controlled study. Evid-Based Compl Alt Med 2014:179529
91. Zanoli P, Zavatti M, Montanari C, Baraldi M (2009) Influence of *Eurycoma longifolia* on the copulatory activity of sexually sluggish and impotent male rats. J Ethnopharmacol 126:308
92. Abdulghani M, Hussin AH, Sulaiman SA, Chan KL (2012) The ameliorative effects of *Eurycoma longifolia* Jack on testosterone-induced reproductive disorders in female rats. Reprod Biol 12:247

93. Teh CH, Abdulghani M, Morita H, Shiro M, Hussin AH, Chan KL (2011) Comparative X-ray and conformational analysis of a new crystal of 13α,21-dihydroeurycomanone with eurycomanone from *Eurycoma longifolia* and their anti-estrogenic activity using the uterotrophic assay. Planta Med 77:128
94. Qodriyah HMS, Asmadi AY (2013) *Eurycoma longifolia* in Radix^{TM} for the treatment of ethanol-induced gastric lesion in rats. Pak J Biol Sci 16:1815
95. Moreira SG, Brannigan RE, Spitz A, Orejuela FJ, Lipshultz LI, Kim ED (2000) Side-effect profile of sildenafil citrate (Viagra) in clinical practice. Adult Urol 56:474
96. Saadiah Abdul Razak H, Shuid AN, Naina Mohamed I (2012) Combined effects of *Eurycoma longifolia* and testosterone on androgen-deficient osteoporosis in a male rat model. Evid-Based Compl Altern Med 2012:872406
97. Shuid AN, El-Arabi E, Effendy NM, Saadiah H, Razak A, Muhammad N, Mohamed N, Soelaiman IN (2012) *Eurycoma longifolia* upregulates osteoprotegerin gene expression in androgen-deficient osteoporosis rat model. BMC Compl Alt Med 12:152
98. Wauquier F, Leotoing L, Coxam V, Guicheux J, Wittrant Y (2009) Oxidative stress in bone remodelling and disease. Trends Mol Med 15:468
99. Shuid AN, Ping LL, Muhammad N, Mohamed N, Soelaiman IN (2011) The effects of *Labisia pumila* var. *alata* on bone markers and bone calcium in a rat model of post-menopausal osteoporosis. J Ethnopharmacol 133:538
100. Ibrahim IS, Mohd Said M, Mohammad Zainoor N, Jamal JA (2022) Authentication of *Marantodes pumilum* (Blume) Kuntze: a systematic review. Front Pharmacol 13:1
101. Chua LS, Latiff NA, Lee SY, Lee CT, Sarmidi MR, Aziz RA (2011) Flavonoids and phenolic acids from *Labisia pumila* (Kacip Fatimah). Food Chem 127:1186
102. Abdullah N, Chermahini SH, Suan CL, Sarmidi MR (2013) *Labisia pumila*: a review on its traditional, phytochemical and biological uses. World Appl Sci J 27:1297
103. Sunarno B (2005) Revision of the genus *Labisia* (Myrsinaceae). Blumea: J Plant Taxon Plant Geogr 50:579
104. Jamal JA, Houghton PJ, Milligan SR, Jantan I (2003) The oestrogenic and cytotoxic effect of the extracts of *Labisia pumila* var. *alata* and *Labisia pumila* var. *pumila* in vitro. Sains Kesihatan 1:53
105. Ezumi MFW, Amrah SS, Suhaimi AWM, Mohsin SSJ (2007) Evaluation of the female reproductive toxicity of aqueous extract of *Labisia pumila* var. *alata* in rats. Indian J Pharmacol 39:30
106. Burkill IH (1935) A dictionary of the economic products of the Malay Peninsula. Crown Agents, London
107. Wan Hassan WE (2007) Healing herbs of Malaysia. Federal Land Development Authority (FELDA), Kuala Lumpur
108. Fathilah SN, Abdullah S, Mohamed N, Shuid AN (2012) *Labisia pumila* prevents complications of osteoporosis by increasing bone strength in a rat model of postmenopausal osteoporosis. Evid-Based Compl Alt Med 2012:948080
109. Fathilah SN, Nazrun Shuid A, Mohamed N, Muhammad N, Soelaiman N (2012) *Labisia pumila* protects the bone of estrogen-deficient rat model: a histomorphometric study. J Ethnopharmacol 142:294
110. Fathilah SN, Mohamed N, Muhammad N, Mohamed IN, Nirwana I, Shuid S, Shuid AN (2013) *Labisia pumila* regulates bone-related gene expression in postmenopausal osteoporosis model. BMC Compl Alt Med 13:217
111. Karimi E, Jaafar HZE (2011) HPLC and GC-MS determination of bioactive compounds in microwave obtained extracts of three varieties of *Labisia pumila* Benth. Molecules 16:6791
112. Avula B, Wang YH, Smillie TJ, Fu X, Li XC, Mabusela W, Syce J, Johnson Q, Folk W, Khan IA (2010) Quantitative determination of flavonoids and cycloartanol glycosides from aerial parts of *Sutherlandia frutescens* (L.) R. Br. by using LC-UV/ELSD methods and confirmation by using LC-MS method. J Pharm Biomed Anal 52:173
113. Al-Mekhlafi NA, Shaari K, Abas F, Kneer R, Jeyaraj EJ, Stanslas J, Yamamoto N, Honda T, Lajis NH (2012) Alkenylresorcinols and cytotoxic activity of the constituents isolated from *Labisia pumila*. Phytochemistry 80:42

114. Ahmad SU, Azam A, Shuid AN, Mohamed IN (2018) Antioxidant and anti-inflammatory activities of *Marantodes pumilum* (Blume) Kuntze and their relationship with the phytochemical content. Rec Nat Prod 12:518
115. Karimi E, Jaafar HZE, Ahmad S (2011) Phytochemical analysis and antimicrobial activities of methanolic extracts of leaf, stem and root from different varieties of *Labisia pumila* Benth. Molecules 16:4438
116. Liang H, Huang Q, Zou L, Wei P, Lu J, Zhang Y (2023) Methyl gallate: review of pharmacological activity. Pharmacol Res 194:106849
117. Pandey G (2009) Some medicinal plants as natural anticancer agents. Pharmacogn Rev 3:259
118. Köpcke W, Krutmann J (2008) Protection from sunburn with β-carotene—a meta-analysis. Photochem Photobiol 84:284
119. Ali Z, Khan IA (2011) Alkyl phenols and saponins from the roots of *Labisia pumila* (Kacip Fatimah). Phytochemistry 72:2075
120. Aladdin NA, Husain K, Jalil J, Sabandar CW, Jamal JA (2020) Xanthine oxidase inhibitory activity of a new isocoumarin obtained from *Marantodes pumilum* var. *pumila* leaves. BMC Compl Med Ther 20:324
121. Karimi E, Ze Jaafar H, Ahmad S (2013) Antifungal, anti-inflammatory and cytotoxicity activities of three varieties of *Labisia pumila* Benth. from microwave obtained extracts. BMC Compl Altern Med 13:2
122. Norhaiza M, Maziah M, Hakiman M (2009) Antioxidative properties of leaf extracts of a popular Malaysian herb, *Labisia pumila*. J Med Plants Res 3:217
123. Jamal JA, Ramli N, Stanslas J, Husain K (2012) Estrogenic activity of selected Myrsinaceae species in MCF-7 human breast cancer cells. Int J Pharm Pharm Sci 4:547
124. Pihie LAH, Othman F, Zakaria ZA (2011) Anticarcinogenic activity of *Labisia pumila* against 7,12-dimethylbenz[*a*]anthracene (DMBA)/croton oil-induced mouse skin carcinogenesis. Afr J Pharm Pharmacol 5:823
125. Choi H-K, Kim D-H, Kim JW, Ngadiran S, Sarmidi MR, Park CS (2010) *Labisia pumila* extract protects skin cells from photoaging caused by UVB irradiation. J Biosci Bioeng 109:291
126. Rahmi EP, Kumolosasi E, Jalil J, Husain K, Buang F, Razak AFA, Jamal JA (2020) Anti-hyperuricemic and anti-inflammatory effects of *Marantodes pumilum* as potential treatment for gout. Front Pharmacol 11:00289
127. Hairi HA, Jamal JA, Aladdin NA, Husain K, Sofi NSM, Mohamed B, Mohamed IN, Shuid AN (2018) Demethylbelamcandaquinone B (Dmcq B) is the active compound of *Marantodes pumilum* var. *alata* (Blume) Kuntze with osteoanabolic activities. Molecules 23:1686
128. Fazliana M, Wan Nazaimoon WM, Gu HF, Östenson CG (2009) *Labisia pumila* extract regulates body weight and adipokines in ovariectomized rats. Maturitas 62:91
129. Dianita R, Jantan I, Jalil J, Amran AZ (2016) Effects of *Labisia pumila* var. *alata* extracts on the lipid profile, serum antioxidant status and abdominal aorta of high-cholesterol diet rats. Phytomedicine 23:810
130. Wan Omar WFN, Giribabu N, Karim K, Salleh N (2019) *Marantodes pumilum* (Blume) Kuntze (Kacip Fatimah) stimulates uterine contraction in rats in post-partum period. J Ethnopharmacol 245:112175
131. Karimi E, Ze Jaafar H, Ghasemzadeh A, Ebrahimi M (2015) Fatty acid composition, antioxidant and antibacterial properties of the microwave aqueous extract of three varieties of *Labisia pumila* Benth. Biol Res 48:1
132. Mamat N, Azdina Jamal J, Jantan I, Husain K (2014) Xanthine oxidase inhibitory and DPPH radical scavenging activities of some Primulaceae species. Sains Malays 43:1827
133. Sanusi RAM, Shukor NAA, Sulaiman MR (2013) Anti-inflammatory effects of *Labisia pumila* (Blume) F. Vill-Naves aqueous extract. Sains Malay 42:1511
134. Al-Wahaibi A, Wan Nazaimoon WN, Norsyam WNHS, Azlan AL (2008) Effect of water extract of *Labisia pumila* var. *alata* on aorta of ovariectomized Sprague Dawley rats. Pak J Nutr 7:208

135. Boutouyrie P, Tropeano AI, Asmar R, Gautier I, Benetos A, Lacolley P, Laurent S (2002) Aortic stiffness is an independent predictor of primary coronary events in hypertensive patients: a longitudinal study. Hypertension 39:10
136. Abdul Kadir A, Nik Hussain NH, Wan Bebakar WM, Mohd DM, Mohammad WWMZ, Hassan II, Shukor N, Kamaruddin NA, Mohamud WWN (2012) The effect of *Labisia pumila* var. *alata* on postmenopausal women: a pilot study. Evid-Based Compl Altern Med 2012:216525
137. Singh GD, Ganjoo M, Youssouf MS, Koul A, Sharma R, Singh S, Sangwan PL, Koul S, Ahamad DB, Johri RK (2009) Sub-acute toxicity evaluation of an aqueous extract of *Labisia pumila*, a Malaysian herb. Food Chem Toxicol 47:2661
138. Shahrim Z, Noor M, Baharuddin PJ, Yahya A, Muhammad H, Bakar RA, Ismail Z (2006) The in vivo rodent micronucleus assay of Kacip Fatimah (*Labisia pumila*) extract. Trop Biomed 23:214
139. Mohd Fuad WE, Sulaiman SA, Islam MN, Wahab AMS, Jamalullail SMS (2005) Evaluation of the teratogenicity of aqueous extract of *Labisia pumila* var. *alata* in rats. Malays J Med Sci 12:13
140. Mohd Effendy N, Abdullah S, Yunoh MFM, Shuid AN (2015) Time and dose-dependent effects of *Labisia pumila* on the bone strength of postmenopausal osteoporosis rat model. BMC Complement Altern Med 15:0567
141. Mannerås L, Fazliana M, Wan Nazaimoon WM, Lönn M, Gu HF, Östenson CG, Stener-Victorin E (2010) Beneficial metabolic effects of the Malaysian herb *Labisia pumila* var. *alata* in a rat model of polycystic ovary syndrome. J Ethnopharmacol 127:346
142. Chua LS, Lee SY, Abdullah N, Sarmidi MR (2012) Review on *Labisia pumila* (Kacip Fatimah): bioactive phytochemicals and skin collagen synthesis promoting herb. Fitoterapia 83:1322
143. Chao W-W, Lin B-F (2010) Isolation and identification of bioactive compounds in *Andrographis paniculata* (Chuanxinlian). Chin Med 5:1
144. Raina AP, Gupta V, Sivaraj N, Dutta M (2013) *Andrographis paniculata* (Burm. f.) Wall. ex Nees (Kalmegh), a traditional hepatoprotective drug from India. Genet Resour Crop Evol 60:1181
145. Okhuarobo A, Ehizogie Falodun J, Erharuyi O, Imieje V, Falodun A, Langer P (2014) Harnessing the medicinal properties of *Andrographis paniculata* for diseases and beyond: a review of its phytochemistry and pharmacology. Asian Pac J Trop Dis 4:213
146. Kumar A, Dora J, Singh A, Tripathi R (2012) A review on king of bitter (Kalmegh). Int J Res Pharm Chem 2:116
147. Subramanian R, Asmawi MZ, Sadikun A (2012) A bitter plant with a sweet future? A comprehensive review of an oriental medicinal plant: *Andrographis paniculata*. Phytochem Rev 11:39
148. Jarukamjorn K, Nemoto N (2008) Pharmacological aspects of *Andrographis paniculata* on health and its major diterpenoid constituent andrographolide. J Health Sci 54:370
149. Mishra SK, Neelam SS, Rajender SS (2007) *Andrographis paniculata* (Kalmegh): a review. Pharmacogn Rev 1:283
150. Sujarwo W, Keim AP, Savo V, Guarrera PM, Caneva G (2015) Ethnobotanical study of Loloh: traditional herbal drinks from Bali (Indonesia). J Ethnopharmacol 169:34
151. Poolsup N, Suthisisang C, Prathanturarug S, Asawamekin A, Chanchareon U (2004) *Andrographis paniculata* in the symptomatic treatment of uncomplicated upper respiratory tract infection: systematic review of randomized controlled trials. J Clin Pharm Ther 29:37
152. Hossain MS, Urbi Z, Sule A, Rahman KMH (2014) *Andrographis paniculata* (Burm. f.) Wall. ex Nees: a review of ethnobotany, phytochemistry, and pharmacology. Sci World J 2014:1
153. Patidar S, Gontia AS, Upadhyay A, Nayak PS (2011) Biochemical constituents in Kalmegh (*Andrographis paniculata* Nees.) under various row spacing's and nitrogen levels. World Appl Sci J 15:1095
154. Balu S, Alagesaboopathi C (1993) Anti-inflammatory activities of some species of *Andrographis* Wall. (Acanthaceae). Anc Sci Life 13:180

155. Jarukamjorn K, Kondo S, Chatuphonprasert W, Sakuma T, Kawasaki Y, Nemoto N (2010) Gender-associated modulation of inducible CYP1A1 expression by andrographolide in mouse liver. Eur J Pharm Sci 39:394
156. Akbar S (2011) *Andrographis paniculata*: a review of pharmacological activities and clinical effects. Alt Med Rev 16:66
157. Zhang CY, Tan BKH (1997) Mechanisms of cardiovascular activity of *Andrographis paniculata* in the anaesthetized rat. J Ethnopharmacol 56:97
158. Huang CJ, Wu MC (2002) Differential effects of foods traditionally regarded as "heating" and "cooling" on prostaglandin E(2) production by a macrophage cell line. J Biomed Sci 9:596
159. Avani G, Rao MV (2008) In vitro cytogenetic effects of *Andrographis paniculata* (kalmegh) on arsenic. Phytomedicine 15:221
160. Lim JCW, Chan TK, Ng DS, Sagineedu SR, Stanslas J, Wong WF (2012) Andrographolide and its analogues: versatile bioactive molecules for combating inflammation and cancer. Clin Exp Pharmacol Physiol 39:300
161. Osathanunkul M, Suwannapoom C, Osathanunkul K, Madesis P, De Boer H (2016) Evaluation of DNA barcoding coupled high resolution melting for discrimination of closely related species in phytopharmaceuticals. Phytomedicine 23:156
162. Jiang C-G, Li J-B, Liu F-R, Wu T, Yu M, Xu HM (2007) Andrographolide inhibits the adhesion of gastric cancer cells to endothelial cells by blocking E-selectin expression. Anticancer Res 27:2439
163. Kumar S, Singh B, Bajpai V (2021) *Andrographis paniculata* (Burm. f.) Nees: traditional uses, phytochemistry, pharmacological properties and quality control/quality assurance. J Ethnopharmacol 275:114054
164. Vedavathy S, Rao KN (1991) Antipyretic activity of six indigenous medicinal plants of Tirumala Hills, Andhra Pradesh, India. J Ethnopharmacol 33:193
165. Koteswara Rao Y, Vimalamma G, Venkata Rao C, Tzeng YM (2004) Flavonoids and andrographolides from *Andrographis paniculata*. Phytochemistry 65:2317
166. Kishore PH, Reddya MVB, Reddy MK, Reddy KM, Gunasekara D, Caux C, Bodo B (2003) Flavonoids from *Andrographis lineata*. Phytochemistry 63:457
167. Varma A, Padh H, Shrivastava N (2011) Andrographolide: a new plant-derived antineoplastic entity on horizon. Evid-Based Compl Alt Med 2011:815390
168. Dua VK, Ojha VP, Roy R, Joshi BC, Valecha N, Devi CU, Bhatnagar MC, Sharma VP, Subbarao SK (2004) Anti-malarial activity of some xanthones isolated from the roots of *Andrographis paniculata*. J Ethnopharmacol 95:247
169. Wu TS, Chern HJ, Damu AG, Kuo PC, Su CR, Lee EJ, Teng CM (2008) Flavonoids and *ent*-labdane diterpenoids from *Andrographis paniculata* and their antiplatelet aggregatory and vasorelaxing effects. J Asian Nat Prod Res 10:17
170. Xu C, Chou GX, Wang CH, Wang ZT (2012) Rare noriridoids from the roots of *Andrographis paniculata*. Phytochemistry 77:275
171. Kuroyanagi M, Sato M, Uenoa A, Nishib K, Machi MS (1987) Flavonoids from *Andrographis paniculata*. Chem Pharm Bull 35:4429
172. Sivananthan M, Elamaran M (2013) Medicinal and pharmacological properties of *Andrographis paniculata*. Int J Biomol Biomed 3:1
173. Nyeem MA, Abdul Mannan M, Nuruzzaman M, Kamrujjaman K, Kumar Das S (2017) Indigenous king of bitter (*Andrographis paniculata*): a review. J Med Plants Stud 5:318
174. Singh RP, Banerjee S, Ramesha Rao A (2001) Modulatory influence of *Andrographis paniculata* on mouse hepatic and extrahepatic carcinogen metabolizing enzymes and antioxidant status. Phytother Res 15:382
175. Kumar RA, Sridevi K, Vijaya Kumar N, Nanduri S, Rajagopal S (2004) Anticancer and immunostimulatory compounds from *Andrographis paniculata*. J Ethnopharmacol 92:291
176. Rajagopal S, Ajaya Kumar R, Deevi DS, Satyanarayana C, Rajagopalan R (2003) Andrographolide, a potential cancer therapeutic agent isolated from *Andrographis paniculata*. J Exp Ther Oncol 3:147

177. Lee YC, Lin HH, Hsu CH, Wang CJ, Chiang TA, Chen JH (2010) Inhibitory effects of andrographolide on migration and invasion in human non-small cell lung cancer A549 cells via down-regulation of PI3K/Akt signaling pathway. Eur J Pharmacol 632:23
178. Reyes BAS, Bautista ND, Tanquilut NC, Anunciado RVP, Leung AB, Sanchez GC, Magtoto RL, Castronuevo P, Tsukamura H, Maeda K (2006) Anti-diabetic potentials of *Momordica charantia* and *Andrographis paniculata* and their effects on estrous cyclicity of alloxan-induced diabetic rats. J Ethnopharmacol 105:196
179. Verma N, Vinayak M (2008) Antioxidant action of *Andrographis paniculata* on lymphoma. Mol Biol Rep 35:535
180. Yu BC, Chang CK, Su CF, Cheng JT (2008) Mediation of β-endorphin in andrographolide-induced plasma glucose-lowering action in type I diabetes-like animals. Naunyn-Schmiedebergs Arch Pharmacol 377:529
181. Chandrasekaran CV, Gupta A, Agarwal A (2010) Effect of an extract of *Andrographis paniculata* leaves on inflammatory and allergic mediators in vitro. J Ethnopharmacol 129:203
182. Chao WW, Kuo YH, Li WC, Lin BF (2009) The production of nitric oxide and prostaglandin E2 in peritoneal macrophages is inhibited by *Andrographis paniculata, Angelica sinensis* and *Morus alba* ethyl acetate fractions. J Ethnopharmacol 122:68
183. Qin L-H, Kong L, Shi G-J, Wang ZT, Ge BX (2006) Andrographolide inhibits the production of TNF-α and interleukin-12 in lipopolysaccharide-stimulated macrophages: role of mitogen-activated protein kinases. Biol Pharmacol Bull 29:220
184. Najila MS, Rain AN, Kamel AM, Zahir SS, Khozirah S, Hakim LS, Zakiah I, Azizol A (2002) The screening of extracts from *Goniothalamus scortechinii, Aralidium pinnatifidum* and *Andrographis paniculata* for anti-malarial activity using the lactate dehydrogenase assay. J Ethnopharmacol 82:239
185. Basak A, Cooper S, Roberge G, Banik UK, Chre M, Seidahr NG (1999) Inhibition of proprotein convertases-1, -7 and furin by diterpenes of *Andrographis paniculata* and their succinoyl esters. Biochem J 338:107
186. Akowuah GA, Zhari I, Mariam A, Yam MF (2009) Absorption of andrographolides from *Andrographis paniculata* and its effect on CCl_4-induced oxidative stress in rats. Food Chem Toxicol 47:2321
187. Das S, Gautam N, Dey SK, Maiti T, Roy S (2009) Oxidative stress in the brain of nicotine-induced toxicity: protective role of *Andrographis paniculata* Nees and vitamin E. Appl Physiol Nutr Metabol 34:124
188. Lin FL, Wu SJ, Lee SC, Ng LT (2009) Antioxidant, antioedema and analgesic activities of *Andrographis paniculata* extracts and their active constituent andrographolide. Phytother Res 23:958
189. Zhang CY, Tan BKH (1998) Vasorelaxation of rat thoracic aorta caused by 14-deoxyandrographolide. Clin Exp Pharmacol Physiol 25:424
190. Mandal SC, Dhara AK, Maiti BC (2001) Studies on psychopharmacological activity of *Andrographis paniculata* extract. Phytother Res 15:253
191. Thisoda P, Rangkadilok N, Pholphana N, Worasuttayangkurn L, Ruchirawat S, Satayavivad J (2006) Inhibitory effect of *Andrographis paniculata* extract and its active diterpenoids on platelet aggregation. Eur J Pharmacol 553:39
192. Chen JH, Hsiao G, Lee AR, Wu CC, Yen MH (2004) Andrographolide suppresses endothelial cell apoptosis via activation of phosphatidyl inositol-3-kinase/Akt pathway. Biochem Pharmacol 67:1337
193. Roxas M, Jurenka J (2007) Colds and influenza: a review of diagnosis and conventional, botanical, and nutritional considerations. Alt Med Rev 12:25
194. Madav S, Tripathu HC, Mishra T, Mishra SK (1995) Analgesic, antipyretic and antiulcerogenic effects of andrographolide. Indian J Pharmacol Sci 57:121
195. Sheeja K, Kuttan G (2007) Modulation of natural killer cell activity, antibody-dependent cellular cytotoxicity, and antibody-dependent complement-mediated cytotoxicity by andrographolide in normal and Ehrlich ascites carcinoma-bearing mice. Integr Cancer Ther 6:66

196. Gupta PP, Tandon JS, Patnaik GK (1998) Antiallergic activity of andrographolides isolated from *Andrographis paniculata* (Burm. f.) Wall. Pharm Biol 36:72
197. Singh S, Mehta A, Baweja S, Akhirwal L, Mehta P (2013) Anticancer activity of *Andrographis paniculata* and *Silybum marianum* on five human cancer cell lines. J Pharmacol Toxicol 8:42
198. Harjotaruno S, Widyawaruyanti A, Sismindari ZNC (2007) Apoptosis inducing effect of andrographolide on TD-47 human breast cancer cell line. Afr J Trad Compl Alt Med 4:345
199. Chang C-C, Duann Y-F, Yen T-L, Chen Y-Y, Jayakumar T, Ong E-T, Sheu J-R (2014) Andrographolide, a novel NF-κB inhibitor, inhibits vascular smooth muscle cell proliferation and cerebral endothelial cell inflammation. Acta Cardiol Sin 30:308
200. Zhang Z, Jiang J, Yu P, Zeng X, Larrick JW, Wang Y (2009) Hypoglycemic and beta cell protective effects of andrographolide analogue for diabetes treatment. J Transl Med 7:62
201. Nugroho AE, Rais IR, Setiawan I, Pratiwi PY, Hadibarata T, Tegar M, Pramono S (2014) Pancreatic effect of andrographolide isolated from *Andrographis paniculata* (Burm. f.) Nees. Pak J Biol Sci 17:22
202. Borhanuddin M, Shamsuzzoha M, Hussain AH (1994) Hypoglycaemic effects of *Andrographis paniculata* Nees on non-diabetic rabbits. Bangladesh Med Res Counc Bull 20:24
203. Zhang X-F, Tan BK-H (2000) Anti-diabetic property of ethanolic extract of *Andrographis paniculata* in streptozotocin-diabetic rats. Acta Pharmacol Sin 21:1157
204. Kamdem RE, Sang S, Ho CT (2002) Mechanism of the superoxide scavenging activity of neoandrographolide a natural product from *Andrographis paniculata* Nees. J Agric Food Chem 50:4662
205. Misra P, Pal NL, Guru PY, Katiyar JC, Srivastava V, Tandon JS (1992) Antimalarial activity of *Andrographis paniculata* (kalmegh) against *Plasmodium berghei* NK 65 in *Mastomys natalensis*. Pharm Biol 30:263
206. Zhou J, Zhang S, Choon-Nam O, Shen HM (2006) Critical role of pro-apoptotic Bcl-2 family members in andrographolide-induced apoptosis in human cancer cells. Biochem Pharmacol 72:132
207. Amroyan E, Gabrielian E, Panossian A, Wikman G, Wagner H (1999) Inhibitory effect of andrographolide from *Andrographis paniculata* on PAF-induced platelet aggregation. Phytomedicine 6:27
208. Lingampally V, Solanki VR, Kaur A, Raja SS (2013) Andrographolide—an effective insect growth regulator of plant origin against *Tribolium confusum* (Duval). Int J Curr Res 5:22
209. Shinde N, Chauhan AS, Gupta SK, Bodakhe SH, Pandey DP (2015) Antifertility studies of curcumin and andrographolide combination in female rats. Asian Pac J Reprod 4:188
210. Akbarsha MA, Murugaian P (2000) Aspects of the male reproductive toxicity/male antifertility property of andrographolide in albino rats: effect on the testis and the cauda epididymidal spermatozoa. Phytother Res 14:432
211. Zoha MS, Hussain AH, Choudhury SA (1989) Antifertility effect of *Andrographis paniculata* in mice. Bangladesh Med Res Counc Bull 15:34
212. Sinha J, Mukhopadhyay S, Das N, Kumar Basu M (2000) Targeting of liposomal andrographolide to *L. donovani*-infected macrophages in vivo. Drug Deliv 7:209
213. Tripathi GS, Tripathi YB (1991) Choleretic action of andrographolide obtained from *Andrographis paniculata* in rats. Phytother Res 5:176
214. Low M, Khoo CS, Münch G, Govindaraghavan S, Sucher NJ (2015) An in vitro study of anti-inflammatory activity of standardised *Andrographis paniculata* extracts and pure andrographolide. BMC Compl Alt Med 15:18
215. Abu-Ghefreh AA, Canatan H, Ezeamuzie CI (2009) In vitro and in vivo anti-inflammatory effects of andrographolide. Int Immunopharmacol 9:313
216. Parichatikanond W, Suthisisang C, Dhepakson P, Herunsalee A (2010) Study of anti-inflammatory activities of the pure compounds from *Andrographis paniculata* (Burm. f.) Nees and their effects on gene expression. Int Immunopharmacol 10:1361
217. Sheeja K, Shihab PK, Kuttan G (2006) Antioxidant and anti-inflammatory activities of the plant *Andrographis paniculata* Nees. Immunopharmacol Immunotoxicol 28:129

218. Najib N, Rahman NA, Furuta T, Kojima S, Takane K, Ali Mohd M (1999) Antimalarial activity of extracts of Malaysian medicinal plants. J Ethnopharmacol 64:249
219. Chakraborty S, Biswas S, Manna SJ, Das S, Dey R (2011) Sol-gel derived silica-gel as a controlled delivery system of *Andrographis paniculata* extract and its anti-microbial efficacy. Trans Indian Inst Metals 64:189
220. Molla MTH, Ahsan MS, Alam MT, Haque ME (2010) Antibacterial activity in the leaves of seven bitter medicinal plants of Bangladesh. J Biol Sci 18:128
221. Mishra US, Mishra A, Kumari R, Murthy PN, Naik BS (2009) Antibacterial activity of ethanol extract of *Andrographis paniculata*. Indian J Pharm Sci 436
222. Mishra PK, Singh RK, Gupta A, Chaturvedi A, Pandey R, Tiwari SP, Mohapatra TM (2013) Antibacterial activity of *Andrographis paniculata* (Burm. f.) Wall. ex Nees leaves against clinical pathogens. J Pharm Res 7:459
223. Sule A, Ahmed QU, Samah OA, Omar MN (2010) Screening for antibacterial activity of *Andrographis paniculata* used in Malaysian folkloric medicine: a possible alternative for the treatment of skin infections. Ethnobotan Leafl 14:445
224. Noorkorina TT, Noorfatiehah TT (2012) Antifungal activity of *Andrographis paniculata*: an in vitro study. Intern Med J 19:382
225. Ghosh M, Thangamani D, Thapliyal M, Yasodha R, Gurumurthi K (2004) Isolation of *Andrographis paniculata* leaf protein with antifungal property. Acta Phytopathol Entomol Hung 39:377
226. Tang LIC, Ling APK, Koh RY, Chye RY, Voon KGL (2012) Screening of anti-dengue activity in methanolic extracts of medicinal plants. BMC Compl Alt Med 12:3
227. Wiart C, Kumar K, Yusof MY, Hamimah H, Fauzi ZM, Sulaiman M (2005) Antiviral properties of *ent*-labdene diterpenes of *Andrographis paniculata* Nees, inhibitors of *Herpes simplex* virus type 1. Phytother Res 19:1069
228. Akowuah GA, Zhari I, Mariam A (2008) Analysis of urinary andrographolides and antioxidant status after oral administration of *Andrographis paniculata* leaf extract in rats. Food Chem Toxicol 46:3616
229. Chen W, Su H, Feng L, Zheng X (2016) Andrographolide suppresses preadipocytes proliferation through glutathione antioxidant systems abrogation. Life Sci 156:21
230. Kurzawa M, Filipiak-Szok A, Kłodzińska E, Szłyk E (2015) Determination of phytochemicals, antioxidant activity and total phenolic content in *Andrographis paniculata* using chromatographic methods. J Chromatogr B 995–996
231. Sangeetha S, Archit R, Velu AS (2014) Phytochemical testing, antioxidant activity, HPTLC and FTIR analysis of antidiabetic plants *Nigella sativa, Eugenia jambolana, Andrographis paniculata* and *Gymnema sylvestre*. Res J Biotechnol 9:65
232. Cauley JA, John Robbins D, Chen Z, Cummings SR, Jackson RD, LaCroix AZ, LeBoff MM, Lewis CE, McGowan JM, Neuner J, Pettinger MM, Stefanick MML, Wactawski-Wende J, Watts NB (2003) Effects of estrogen plus progestin on risk of fracture and bone mineral density the women's health initiative randomized trial. J Am Med Assoc 290:1729
233. Shen Y-C, Chen C-F, Chiou W-F (2000) Suppression of rat neutrophil reactive oxygen species production and adhesion by the diterpenoid lactone andrographolide. Planta Med 66:314
234. Thakur AK, Soni UK, Rai G, Chatterjee SS, Kumar V (2014) Protective effects of *Andrographis paniculata* extract and pure andrographolide against chronic stress-triggered pathologies in rats. Cell Mol Neurobiol 34:1111
235. Batkhuu J, Hattori K, Takano F, Fushiya S, Oshiman K, Fujimiya Y (2002) Suppression of NO production in activated macrophages in vitro and ex vivo by neoandrographolide isolated from *Andrographis paniculata*. Biol Pharm Bull 25:1169
236. Gupta S, Choudhry MA, Yadava JNS, Srivastava V, Tandon JS (1990) Antidiarrhoeal activity of diterpenes of *Andrographis paniculata* (kal-megh) against *Escherichia coli* enterotoxin in in vivo models. Pharm Biol 28:273
237. Singha PK, Roy S, Dey S (2003) Antimicrobial activity of *Andrographis paniculata*. Fitoterapia 74:692

238. Calabrese C, Berman SH, Babish JG, Ma X, Shinto L, Dorr M, Wells K, Wenner CA, Standish LJ (2000) A phase I trial of andrographolide in HIV positive patients and normal volunteers. Phytother Res 14:333
239. Zou Y, Xiong H, Xiong H, Lu T, Zhu F, Luo Z, Yuan X, Wang Y (2015) A polysaccharide from *Andrographis paniculata* induces mitochondrial-mediated apoptosis in human hepatoma cell line (HepG2). Tumor Biol 36:5179
240. Agrawal RC, Agrawal N, Mishra D (2017) Chemopreventive and antiproliferative effect of *Andrographis paniculata* extract. Med One 2:e170011
241. Matsuda T, Kuroyanagi M, Suguyama S, Umehara K, Ueno A, Nishi K (1994) Cell differentiation-inducing diterpenes from *Andrographis paniculata* Nees. Chem Pharm Bull 42:1216
242. Li W, Xu X, Zhang H, Ma C, Fong H, van Breemen R, Fitzloff J (2007) Secondary metabolites from *Andrographis paniculata*. Chem Pharm Bull 55:455
243. Cáceres DD, Hancke JL, Burgos RA, Wikman GK (1997) Prevention of common colds with *Andrographis paniculata* dried extract. A pilot double blind trial. Phytomedicine 4:101
244. Pekthong D, Martin H, Abadie C, Bonet A, Heyd B, Mantion G, Richert L (2007) Differential inhibition of rat and human hepatic cytochrome P450 by *Andrographis paniculata* extract and andrographolide. J Ethnopharmacol 115:432
245. Ooi JP, Kuroyanagi M, Sulaiman SF, Muhammad TST, Tan ML (2011) Andrographolide and 14-deoxy-11,12-didehydroandrographolide inhibit cytochrome P450s in HepG2 hepatoma cells. Life Sci 88:447
246. Kondo S, Chatuphonprasert W, Jaruchotikamol A, Sakuma T, Nemoto N (2011) Cellular glutathione content modulates the effect of andrographolide on β-naphthoflavone-induced CYP1A1 mRNA expression in mouse hepatocytes. Toxicology 280:18
247. Chander R, Srivastava V, Tandon JS, Kapoor' NK (1995) Antihepatotoxic activity of diterpenes of *Andrographis paniculata* (kal-megh) against *Plasmodium berghei*-induced hepatic damage in *Mastomys natalensis*. Int J Pharmacogn 33:135
248. Rana AC, Avadhoot Y (1991) Hepatoprotective effects of *Andrographis paniculata* against carbon tetrachloride-induced liver damage. Arch Pharm Res 93:5
249. Kapil A, Koul IB, Banerjeet SK, Gupta BD (1993) Antihepatotoxic effects of major diterpenoid constituents of *Andrographis paniculata*. Biochem Pharmacol 46:182
250. Rajalakshmi G, Jothi KA, Venkatesan RS, Jegatheesan K (2012) Hepatoprotective activity of *Andrographis paniculata* on paracetamol induced liver damage in rats. J Pharm Res 5:2983
251. Trivedi NP, Rawal UM, Patel BP (2007) Hepatoprotective effect of andrographolide against hexachlorocyclohexane-induced oxidative injury. Integr Cancer Ther 6:271
252. Nagalekshmi R, Menon A, Chandrasekharan DK, Nair CKK (2011) Hepatoprotective activity of *Andrographis paniculata* and *Swertia chirayita*. Food Chem Toxicol 49:3367
253. Johri R, Johri PK (2010) Hepatoprotective activity of *Andrographis paniculata* (Nees) leaves on paracetamol induced hepatic damage in rats. J Exp Zool 13:51
254. Handa SS, Sharma A (1990) Hepatoprotective activity of andrographolide against galactosamine and paracetamol intoxication in rats. Indian J Med Res B 92:284
255. Rana AC, Avadhoot Y (1991) Hepatoprotective effects of *Andrographis paniculata* against carbon tetrachloride-induced liver damage. Arch Pharm Res 14:93
256. Subramaniam S, Khan HBH, Elumalai N, Lakshmi SYS (2015) Hepatoprotective effect of ethanolic extract of whole plant of *Andrographis paniculata* against CCl_4-induced hepatotoxicity in rats. Comp Clin Pathol 24:1245
257. Xu Y, Chen A, Fry S, Barrow RA, Marshall RL, Mukkur TKS (2007) Modulation of immune response in mice immunised with an inactivated *Salmonella* vaccine and gavaged with *Andrographis paniculata* extract or andrographolide. Int Immunopharmacol 7:515
258. Hu X, Wen Y, Liu S, Tan X, Li Z, Lu X, Long X (2015) Evaluation of the anaphylactoid potential of *Andrographis paniculata* extracts using the popliteal lymph node assay and P815 cell degranulation in vitro. J Transl Med 13:121
259. Xu Y, Chen A, Fry S, Barrow RA, Marshall RL, Mukkur TKS (2010) Immunomodulatory activity of andrographolide on macrophage activation and specific antibody response. Acta Pharmacol Sin 31:191

260. Kotewong R, Duangkaew P, Srisook E et al (2014) Structure–function relationships of inhibition of mosquito cytochrome P450 enzymes by flavonoids of *Andrographis paniculata*. Parasitol Res 113:3381
261. Lachenmeier K, Mußhoff F, Madea B, Reusch H, Lachenmeier DW (2006) Authentication of noni (*Morinda citrifolia*) juice. Dtsch Lebensmitt Rundsch 102:58
262. Burkill IH (1966) A dictionary of the economic products of the Malay peninsular. Ministry of Agriculture, Kuala Lumpur
263. Mohamad TAST, Islahudin F, Jasamai M, Jamal JA (2022) Traditional practitioner's knowledge of Malay post-partum herbal remedies in Malaysia. Arch Pharm Pract 13:11
264. Pawlus AD, Kinghorn AD (2010) Review of the ethnobotany, chemistry, biological activity and safety of the botanical dietary supplement *Morinda citrifolia* (noni). J Pharm Pharmacol 59:1587
265. Burkill IH (1936) A dictionary of the economic products of the Malay peninsula. Nature 137:255
266. Ee GCL, Wen YP, Sukari MA, Go R, Lee HL (2009) A new anthraquinone from *Morinda citrifolia* roots. Nat Prod Res 23:1322
267. Kamiya K, Hamabe W, Tokuyama S, Satake T (2009) New anthraquinone glycosides from the roots of *Morinda citrifolia*. Fitoterapia 80:196
268. Siddiqui BS, Sattar FA, Begum S, Gulzar T, Ahmad F (2006) New anthraquinones from the stem of *Morinda citrifolia* Linn. Nat Prod Res 20:1136
269. Wang J, Qin X, Chen Z, Ju Z, He W, Tan Y, Zhou X, Tu Z, Lu F, Liu Y (2016) Two new anthraquinones with antiviral activities from the barks of *Morinda citrifolia* (noni). Phytochem Lett 15:13
270. Kamiya K, Tanaka Y, Endang H, Umar M, Satake T (2005) New anthraquinone and iridoid from the fruits of *Morinda citrifolia*. Chem Pharm Bull 53:1597
271. Kim HK, Kwon MK, Kim JN, Kim CK, Lee YJ, Shin HJ, Lee J, Lee HS (2010) Identification of novel fatty acid glucosides from the tropical fruit *Morinda citrifolia* L. Phytochem Lett 3:238
272. Lee D, Yu JS, Huang P, Qader M, Manavalan A, Wu X, Kim JC, Pang C, Cao S, Kang KS, Kim KH (2020) Identification of anti-inflammatory compounds from Hawaiian noni (*Morinda citrifolia* L.) fruit juice. Molecules 25:4968
273. Siddiqui B, Sattar F, Ahmad F, Begum S (2007) Isolation and structural elucidation of chemical constituents from the fruits of *Morinda citrifolia* Linn. Arch Pharm Res 30:919
274. Samoylenko V, Zhao J, Dunbar DC, Khan IA, Rushing JW, Muhammad I (2006) New constituents from noni (*Morinda citrifolia*) fruit juice. J Agric Food Chem 54:6398
275. Siddiqui BS, Sattar FA, Ahmad F, Begum S (2008) Isolation and structure determination of two new constituents from the fruits of *Morinda citrifolia* Linn. Nat Prod Res 22:1128
276. Sang S, Cheng X, Zhu N, Stark RE, Badmaev V, Ghai G, Rosen RT, Ho CT (2001) Flavonol glycosides and novel iridoid glycoside from the leaves of *Morinda citrifolia*. J Agric Food Chem 49:4478
277. Sang S, He K, Liu G, Liu G, Zhu N, Cheng X, Wang M, Zheng Q, Dong Z, Ghai G, Rosen RT, Ho CT (2001) A new unusual iridoid with inhibition of activator protein-1 (AP-1) from the leaves of *Morinda citrifolia* L. Org Lett 3:1307
278. Sang S, Liu G, He K, Zhu N, Dong Z, Zheng Q, Rosen RT, Ho CT (2003) New unusual iridoids from the leaves of noni (*Morinda citrifolia* L.) show inhibitory effect on ultraviolet B-induced transcriptional activator protein-1 (AP-1) activity. Bioorg Med Chem 11:2499
279. Takashima J, Ikeda Y, Komiyama K, Hayashi M, Kishida A, Ohsaki A (2007) New constituents from the leaves of *Morinda citrifolia*. Chem Pharm Bull (Tokyo) 55:343
280. Han Y-S, Van der Heiden R, Verpoorte R (2001) Biosynthesis of anthraquinones in cell cultures of the Rubiaceae. Plant Cell Tissue Organ Cult 67:201
281. Osman CP, Ismail NH (2018) Antiplasmodial anthraquinones from medicinal plants: the chemistry and possible mode of actions. Nat Prod Commun 13:1591
282. Do QV, Pharm GD, Mai NT, Phan TPP, Nguyen HN, Yea YY, Ahn BZ (1999) Cytotoxicity of some anthraquinones from the stem of *Morinda citrifolia* growing in Vietnam. Tap Chi Hoa Hoc 37:94

283. Lv L, Chen H, Ho C-T, Sang S (2011) Chemical components of the roots of noni (*Morinda citrifolia*) and their cytotoxic effects. Fitoterapia 82:704
284. Kamiya K, Hamabe W, Harada S, Murakami R, Tokuyama S, Satake T (2008) Chemical constituents of *Morinda citrifolia* roots exhibit hypoglycemic effects in streptozotocin-induced diabetic mice. Biol Pharm Bull 31:935
285. Lin CF, Ni CL, Huang YL, Sheu SJ, Chen CC (2007) Lignans and anthraquinones from the fruits of *Morinda citrifolia*. Nat Prod Res 21:1199
286. Deng Y, Chin YW, Chai H, Keller WJ, Kinghorn AD (2007) Anthraquinones with quinone reductase-inducing activity and benzophenones from *Morinda citrifolia* (noni) roots. J Nat Prod 70:2049
287. Kitic D, Miladinovic B, Randjelovic M, Szopa A, Seidel V, Prasher P, Sharma M, Fatima R, Arslan Ateşşahin D, Calina D, Sharifi-Rad J (2024) Anticancer and chemopreventive potential of *Morinda citrifolia* L. bioactive compounds: a comprehensive update. Phytother Res 38:1932
288. Yamaguchi S, Ohnishi J, Sogawa M, Maru I, Ohta Y, Tsukada Y (2002) Inhibition of angiotensin I converting enzyme by noni (*Morinda citrifolia*) juice. Nippon Shokuhin Kagaku Kogaku Kaishi 49:624
289. Mandukhail SR, Aziz N, Gilani A-H (2010) Studies on antidyslipidemic effects of *Morinda citrifolia* (Noni) fruit, leaves and root extracts. Lipids Health Dis 9:88
290. Gilani AH, Mandukhail SR, Iqbal J, Yasinzai M, Aziz N, Khan A, Najeeb-ur-Rehman (2010) Antispasmodic and vasodilator activities of *Morinda citrifolia* root extract are mediated through blockade of voltage dependent calcium channels. BMC Compl Altern Med 10:2
291. Lin YL, Chou CH, Yang DJ, Chen JW, Tzang BS, Chen YC (2012) Hypolipidemic and antioxidative effects of noni (*Morinda citrifolia* L.) juice on high-fat/cholesterol-dietary hamsters. Plant Foods Hum Nutr 67:294
292. Palu AK, Brown A, Deng S, Kaluhiokalani BJNW (2021) The effects of noni (*Morinda citrifolia* L.) fruit juice on cholesterol levels: a mechanistic investigation and an open label pilot study. J Appl Pharm Sci 2:25
293. Inada AC, Silva GT, da-Silva LPR, Alves FM, de O Filiú WF, Asato MA, Junior WHK, Corsino J, de O Figueiredo P, Garcez FR, Garcez WS, de NO da Silva R, dos Santos-Eichler RA, de Guimarães R CA, de C Freitas K, Hiane PA (2020) Therapeutic effects of *Morinda citrifolia* Linn. (noni) aqueous fruit extract on the glucose and lipid metabolism in high-fat/high-fructose-fed Swiss mice. Nutrients 12:3439
294. Wigati D, Anwar K, Sudarsono, Nugroho AE (2016) Hypotensive activity of ethanolic extracts of *Morinda citrifolia* L. leaves and fruit in dexamethasone-induced hypertensive rat. J Evid-Based Compl Alt Med 22:107
295. de Oliveira Fernandes D, César FG, Melo BP, Brandão J da SF, dos Santos KJ, de Andrade MT, da Fonseca Casteluber MC, de Carvalho MV, de Barcellos LAM, Soares DD, Guimarães BJ (2023) Chronic supplementation of noni in diabetic type 1-STZ rats: effects on glycemic levels, kidney toxicity and exercise performance. Diabetol Metab Syndr 15:191
296. Sasnan GS, Hanani E, Kristianto J (2014) Effect of *Morinda citrifolia* fruit extract capsule on total cholesterol levels in patients with hypercholesterolemia. Trop J Pharm Res 13:1319
297. Shoeb A, Alwar MC, Shenoy PJ, Gokul P (2016) Effect of *Morinda citrifolia* (noni) fruit juice on high fat diet induced dyslipidemia in rats. J Clin Diagn Res 10:Ff06–10
298. Kah Hui C, Majid NI, Mohd Yusof H, Zainol MK, Mohamad H, Zin MZ (2020) Catechin profile and hypolipidemic activity of *Morinda citrifolia* leaf water extract. Heliyon 6:e04337
299. Orhan IE (2012) *Centella asiatica* (L.) Urban: from traditional medicine to modern medicine with neuroprotective potential. Evid-Based Compl Alt Med 2012:946259
300. Shukurova MK, Myint D, Yi SS, Saw OM, Watanabe KN (2021) Morphological description and ethnobotanical review of the orphan crop Myin-Hkwa (*Centella asiatica* L.) from Myanmar. Front Sustain Food Syst 5:680862
301. Polonsky J (1951) Chemical constitution of asiaticoside and especially of asiatic acid. VI. Inclusion of asiatic acid in the α-amyrine series. C R Hebd Seances Acad Sci 233:671

302. Sahu NP, Roy SK, Mahato SB (1989) Spectroscopic determination of structures of triterpenoid trisaccharides from *Centella asiatica*. Phytochemistry 28:2852
303. Günther B, Wagner H (1996) Quantitative determination of triterpenes in extracts and phytopreparations of *Centella asiatica* (L.) Urban Phytomed 3:59
304. Matsuda H, Morikawa T, Ueda H, Yoshikawa M (2001) Medicinal foodstuffs. XXVII. Saponin constituents of Gotu Kola (2): structures of new ursane- and oleanane-type triterpene oligoglycosides, centellasaponins B, C, and D, from *Centella asiatica* cultivated in Sri Lanka. Chem Pharm Bull (Tokyo) 49:1368
305. Jiang ZY, Zhang XM, Zhou J, Chen JJ (2005) New triterpenoid glycosides from *Centella asiatica*. Helv Chim Acta 88:297
306. Yu Q-L, Duan H-Q, Takaishi Y, Gao W-Y (2006) A novel triterpene from *Centella asiatica*. Molecules 11:661
307. Rumalla CS, Ali Z, Weerasooriya AD, Smillie TJ, Khan IA (2010) Two new triterpene glycosides from *Centella asiatica*. Planta Med 76:1018
308. Weng XX, Zhang J, Gao W, Cheng L, Shao Y, Kong DY (2012) Two new pentacyclic triterpenoids from *Centella asiatica*. Helv Chim Acta 95:255
309. Shao Y, Yang ODW, Gao W, Cheng L, Weng XX, Kong DY (2014) Three new pentacyclic triterpenoids from *Centella asiatica*. Helv Chim Acta 97:992
310. Hamid K, Ng I, Tallapragada VJ, Váradi L, Hibbs DE, Hanrahan J, Groundwater PW (2016) An investigation of the differential effects of ursane triterpenoids from *Centella asiatica*, and their semisynthetic analogues, on GABA A receptors. Chem Biol Drug Des 386
311. Wu ZW, Li WB, Zhou J, Liu X, Wang L, Chen B, Wang MK, Ji L, Hu WC, Li F (2020) Oleanane-and ursane-type triterpene saponins from *Centella asiatica* exhibit neuroprotective effects. J Agric Food Chem 68:6977
312. Ren B, Luo W, Xie M, Zhang M (2021) Two new triterpenoid saponins from *Centella asiatica*. Phytochem Lett 44:102
313. Dang Y, Liu T, Liu Y, Li J, Jing Y, Yang M, Zhang H, Jiang M, Wu H, Yang W, Li N, Zhang P (2024) Anti-photoaging activity of triterpenoids isolated from *Centella asiatica*. Phytochemistry 228:114246
314. Kabir AU, Samad MB, D'Costa NM, Akhter F, Ahmed A, Hannan JMA (2014) Anti-hyperglycemic activity of *Centella asiatica* is partly mediated by carbohydrase inhibition and glucose-fiber binding. BMC Compl Alt Med 14:31
315. Maulidiani AF, Khatib A, Perumal V, Suppaiah V, Ismail A, Hamid M, Shaari K, Lajis NH (2016) Metabolic alteration in obese diabetes rats upon treatment with *Centella asiatica* extract. J Ethnopharmacol 180:60
316. Oyenihi AB, Chegou NN, Oguntibeju OO, Masola B (2017) *Centella asiatica* enhances hepatic antioxidant status and regulates hepatic inflammatory cytokines in type 2 diabetic rats. Pharm Biol 55:1671
317. Oyenihi AB, Langa SOP, Mukaratirwa S, Masola B (2019) Effects of *Centella asiatica* on skeletal muscle structure and key enzymes of glucose and glycogen metabolism in type 2 diabetic rats. Biomed Pharmacother 112:108715
318. Rahman S, Jamal MA, Parvin A, Al-Mamun M, Islam R (2011) Antidiabetic activity of *Centella asiatica* (L.) Urbana in alloxan induced type 1 diabetic model rats. J Biosci (Rajshari) 19:23
319. Ramachandran V, Saravanan R (2013) Efficacy of asiatic acid, a pentacyclic triterpene on attenuating the key enzymes activities of carbohydrate metabolism in streptozotocin-induced diabetic rats. Phytomedicine 20:230
320. Tan SC, Rajendran R, Bhattamisra SK, Krishnappa P, Davamani F, Chitra E, Ambu S, Furman B, Candasamy M (2023) Effect of madecassoside in reducing oxidative stress and blood glucose in streptozotocin-nicotinamide-induced diabetes in rats. J Pharm Pharmacol 75:1034
321. Swapna K, Sathibabu Uddandrao VV, Parim B, Ravindarnaik R, Suresh P, Ponnusamy P, Balakrishnan S, Vadivukkarasi S, Harishankar N, Reddy KP, Nivedha PR, Saravanan G (2019) Effects of asiatic acid, an active constituent in *Centella asiatica* (L.): restorative perspectives of streptozotocin-nicotinamide induced changes on lipid profile and lipid metabolic enzymes in diabetic rats. Comp Clin Pathol 28:1321

322. Yu QL, Duan HQ, Gao WY, Takaishi Y (2007) A new triterpene and a saponin from *Centella asiatica*. Chin Chem Lett 18:62
323. Masola B, Oguntibeju OO, Oyenihi AB (2018) *Centella asiatica* ameliorates diabetes-induced stress in rat tissues via influences on antioxidants and inflammatory cytokines. Biomed Pharmacother 101:447
324. Oyenihi AB, Opperman M, Alabi TD, Mpahleni B, Masola B (2020) *Centella asiatica* alleviates diabetes-induced changes in fatty acid profile and oxidative damage in rat testis. Andrologia 52:13751
325. Xu C, Xia L, Xu D, Liu Y, Jin P, Zhai M, Mao Y, Wang Y, Wen A, Yang J, Yang L (2024) Cardioprotective effects of asiaticoside against diabetic cardiomyopathy: activation of the AMPK/Nrf2 pathway. J Cell Mol Med 28:18055
326. Sasikala S, Lakshminarasaiah S, Naidu MD (2015) Effect of *Centella asiatica* (Linn.) leaves on selective mitochondrial and cytosolic enzymes in streptozotocin induced diabetic rats. Int J Pharm Pharm Sci 7:322
327. Palupi FD, Wasita B, Nuhriawangsa AMP (2019) The potency of *Centella asiatica* in protecting organs of rats with type 2 diabetes mellitus. Bali Med J 8:316
328. Hayati F, Chabib L, Sekarraras FD, Faizah WS (2021) Antihyperglycemic activity of *Centella asiatica* (L.) Urb. leaf ethanol extract SNEDDS in zebrafish (*Danio rerio*). Open Chem 19:184
329. Muchtaromah B, Firdaus AMK, Ansori ANM, Duhita MR, Minarno EB, Hayati A, Ahmad M, Analisa I (2024) Effect of pegagan (*Centella asiatica*) nanoparticle coated with chitosan on the cytokine profile of chronic diabetic mice. Narra J 4:697
330. Ameer OZ, Salman IM, Asmawi MZ, Ibraheem ZO, Yam MF (2012) *Orthosiphon stamineus*: traditional uses, phytochemistry, pharmacology, and toxicology. J Med Food 15:678
331. Almatar M, Rahmat Z (2014) Identifying the developmental stages and optimizing the sample preparation for anatomical study of *Orthosiphon stamineus*. J Appl Pharm Sci 4:66
332. Awang-Kanak F, Bakar MFA, Mohamed M (2018) Ethnobotanical survey on plants used as traditional salad food (ulam) in Kampung Taun Gusi, Kota Belud Sabah, Malaysia. In: Proceedings of the AIP conference 2002:020024
333. Adewale Ahmed I (2022) Ethnomedicinal uses of some common Malaysian medicinal plants. In: El-Shemy H (Ed) Natural drugs from plants. IntechOpen, London
334. Awale S, Tezuka Y, Banskota AH, Kadota S (2003) Siphonols A–E: novel nitric oxide inhibitors from *Orthosiphon stamineus* of Indonesia. Bioorg Med Chem Lett 13:31
335. Awale S, Tezuka Y, Shimoji S, Taira K, Kadota S (2002) Secoorthosiphols A-C: three highly oxygenated *seco*-isopimarane-type diterpenes from *Orthosiphon stamineus*. Tetrahedron Lett 43:1473
336. Awale S, Tezuka Y, Banskota AH, Adnyana IK, Kadota S (2003) Highly-oxygenated isopimarane-type diterpenes from *Orthosiphon stamineus* of Indonesia and their nitric oxide inhibitory activity. Chem Pharm Bull 51:268
337. Awale S, Tezuka Y, Banskota AH, Kouda K, Tun KM, Kadota S (2002) Four highly oxygenated isopimarane-type diterpenes of *Orthosiphon stamineus*. Planta Med 68:286
338. Awale S, Tezuka Y, Banskota AH, Kouda K, Tun KM, Kadota S (2001) Five novel highly oxygenated diterpenes of *Orthosiphon stamineus* from Myanmar. J Nat Prod 64:592
339. Tezuka Y, Stampoulis P, Banskota AH, Awale S, Tran KQ, Saiki I, Kadota S (2000) Constituents of the Vietnamese medicinal plant *Orthosiphon stamineus*. Chem Pharm Bull 48:1711
340. Takeda Y, Matsumoto T, Terao H, Shingu T, Futatsuishi Y, Nohara T, Kajimoto T (1993) Orthosiphol D and E, minor diterpenes from *Orthosiphon stamineus*. Phytochemistry 33:411
341. Hossain MA, Ismail Z (2013) Isolation and characterization of triterpenes from the leaves of *Orthosiphon stamineus*. Arab J Chem 6:295
342. Sriplang K, Adisakwattana S, Rungsipipat A, Yibchok-Anun S (2007) Effects of *Orthosiphon stamineus* aqueous extract on plasma glucose concentration and lipid profile in normal and streptozotocin-induced diabetic rats. J Ethnopharmacol 109:510
343. Nalamolu KR, Bethala K, Sisinthy S, Rajeswari K (2014) Antidiabetic activity of *Orthosiphon stamineus* Benth. roots in streptozotocin-induced type 2 diabetic rats. Asian J Pharm Clin Res 7:149

344. Alsarhan A, Sultana N, Chua LS, Khashroom A, Hailat M, Abu Dayyih W, Tamimi L, Zakareia Z, Aburumman A, Awad R (2022) Antidiabetic effect of aqueous extract of *Orthosiphon stamineus* Benth. on alloxan-induced diabetic rats. Sapporo Igaku Zasshi 56:10
345. Mohamed EAH, Mohamed AJ, Asmawi MZ, Sadikun A, Ebrika OS, Yam MF (2011) Anti-hyperglycemic effect of *Orthosiphon stamineus* Benth. leaves extract and its bioassay-guided fractions. Molecules 16:3787
346. Nguyen PH, Tuan HN, Hoang DT, Vu QT, Pham MQ, Tran MH, To DC (2019) Glucose uptake stimulatory and PTP1B inhibitory activities of pimarane diterpenes from *Orthosiphon stamineus* Benth. Biomolecules 9:859
347. Bassalat N, Kadan S, Melamed S, Yaron T, Tietel Z, Karam D, Kmail A, Masalha M, Zaid H (2023) In vivo and in vitro antidiabetic efficacy of aqueous and methanolic extracts of *Orthosiphon stamineus* Benth. Pharmaceutics 15:945
348. Mohamed EAH, Siddiqui MJA, Ang LF, Sadikun A, Chan SH, Tan SC, Asmawi MZ, Yam MF (2012) Potent α-glucosidase and α-amylase inhibitory activities of standardized 50% ethanolic extracts and sinensetin from *Orthosiphon stamineus* Benth. as anti-diabetic mechanism. BMC Compl Alt Med 12:176
349. Yuniarto A, Purwani H, Juanda D, Setiawan F, Kurnia I (2015) Kumis kucing (*Orthosiphon stamineus* [Benth.]) leaves ethanol extract as anti-obesity agent in high-fat diet-induced obese mice. Asian J Pharm Clin Res 8:234
350. Seyedan A, Alshawsh MA, Alshagga MA, Mohamed Z (2017) Antiobesity and lipid lowering effects of *Orthosiphon stamineus* in high-fat diet-induced obese mice. Planta Med 83:684
351. Yam MF, Tan CS, Shibao R (2018) Vasorelaxant effect of sinensetin via the NO/sGC/cGMP pathway and potassium and calcium channels. Hypertens Res 41:787
352. Cicero AFG, De Sando V, Izzo R, Vasta A, Trimarco A, Borghi C (2012) Effect of a combined nutraceutical containing *Orthosiphon stamineus* effect on blood pressure and metabolic syndrome components in hypertensive dyslipidaemic patients: a randomized clinical trial. Compl Ther Clin Pract 18:190
353. Supari F (2002) Lowering blood pressure effect of *Apium graviolens* (Seledri) and *Orthosiphon stamineus* Benth. (kumis kucing) in mild and moderate hypertension. Med J Indonesia 11:195
354. Mohamed EAH, Yam MF, Ang LF, Mohamed AJ, Asmawi MZ (2013) Antidiabetic properties and mechanism of action of *Orthosiphon stamineus* Benth. bioactive sub-fraction in streptozotocin-induced diabetic rats. J Acupunct Meridian Stud 6:31
355. Fogacci F, Degli Esposti D, Di Micoli A, Fiorini G, Veronesi M, Borghi C, Cicero AF (2023) Effect of dietary supplementation with Diuripres® on blood pressure, vascular health, and metabolic parameters in individuals with high-normal blood pressure or stage I hypertension: the CONDOR randomized clinical study. Phytother Res 37:4851
356. Samling B, Assim Z, Tong WY, Leong CR, Rashid AS, Kamal NMNNS, Muhamad M, Tan WN (2021) *Cynometra cauliflora* L.: an indigenous tropical fruit tree in Malaysia bearing essential oils and their biological activities. Arab J Chem 14:103302
357. National Parks (2024) *Cynometra cauliflora* https://www.nparks.gov.sg/florafaunaweb/flora/2/8/2833
358. Wahyuni DK, Pradana RAP, Rahmawati CT, Ansori ANM, Kharisma VD, Parikesit AA, Suhargo L (2023) The importance of medicinal plant and herb concoctions used by the Osing tribe in East Java, Indonesia for treating COVID-19 symptoms. J Namibian Stud 38:1212
359. Nofrianti N, Utaminingrum W, Hartanti D (2021) Traditional use of medicinal plants in Baturraden, Central Java. J Jamu Indonesia 6:42
360. Hasan CC, Ferdinal F, Sari T (2021) Namnam (*Cynometra cauliflora*) leaves methanol extract: phytochemical screening, antioxidant capacity and toxicity test. Asian Fed Pharm Sci (AFPS) conference submission
361. Wahab NZA, Azizul A, Badya N, Ibrahim N (2021) Antiviral activity of an extract from leaves of the tropical plant *Cynometra cauliflora*. Pharmacogn J 13:752
362. Sumarlin LO, Suprayogi A, Rahminiwati M, Satyaningtijas A, Nugraha AT, Sukandar D, Pangestika H, Pratiwi L (2020) Identification of compounds flavonoids namnam leaf extract (*Cynometra cauliflora*) as inhibiting α-glucosidase. J Phys Conf Ser 1594:012005

363. Fathoni A, Sumarlin LO, Hanief F, Sukandar D (2021) Isolation and cytotoxic activity of the β-carotene combination of trigona honey and namnam leaves extract (*Cynometra cauliflora*). Elkawnie 7:52
364. Nurbidayah N, Amalia N (2023) Antibacterial activity of namnam fruit juice from South Kalimantan against *Staphylococcus aureus* and *Escherichia coli*. J Voc Health Stud 7:48
365. Anliza S, Hamtinia RN (2022) Efektivitas ekstrak etanol daun namnam (*Cynometra cauliflora* L) sebagai antibakteri pada formulasi sediaan gel hand sanitizer. J Ilmu Kefarmasian 3:2
366. Anliza S, Hamtini H, Rachmawati N (2024) Pengaruh variasi HPMC dan carbopol pada formulasi sedian gel ekstrak daun namnam (*Cynometra cauliflora* L.) terhadap methicillin-resistant *Staphylococcus aureus* (MRSA). Lumbung Farmasi: J Ilmu Kefarmasian 5:70
367. Khalil RA, Mohamad SSA, Rahman ANRH, Ikhsan NAK, Yunus NM, Ajibola OO, Amin MCIBM (2022) Characterisation of endophytic bacteria from nam-nam plants (*Cynometra cauliflora*) for antibacterial activity and production of plant growth promoting factors. Malaysian UENI Appl Biol 51:119
368. Wahab NZA, Badya N, Ibrahim N, Kamarudin MKA (2019) Phytochemistry and antibacterial activity of *Cynometra cauliflora*. Indian J Public Health Res Dev 10:765
369. Sukchai K (2022) *Cynometra cauliflora* flower extract on antibacterial and antioxidant activity. Princess Naradhiwas Univ J 14:158
370. Putri CAK (2019) Aktivitas antibakteri ekstrak daun namnam (*Cynometra cauliflora* L.) dalam basis gel karbopol terhadap *Staphylococcus aureus* dan *Pseudomonas aeruginosa*. Dissertation, University of Atma Jaya, Yogyakarta. Indonesia
371. Sumarlin LO, Hadera M, Chalid SY, Sukandar D (2018) Aktivitas antioksidan kombinasi madu monoflora dengan ekstrak daun namnam (*Cynometra cauliflora* L.). Alchemy: J Chem Sci 6:10
372. Mallawaarachchi MALN, Madhujith T (2022) Antioxidant properties of selected underutilized fruit species of Sri Lanka after simulated oral and gastro-intestinal digestion. J Nutr Food Sci 1:34
373. Somasiri RPIV, Herath HM, Ratnayake R, Senanayake RASP (2021) Phytochemical screening and antioxidant potential of three underutilized fruit species in Sri Lanka. In: Proceedings of the international conference on applied and pure sciences, University of Kelaniya, Sri Lanka
374. Abeysuriya HI, Bulugahapitiya VP, Loku Pulukkuttige J (2020) Total vitamin C, ascorbic acid, dehydroascorbic acid, antioxidant properties, and iron content of underutilized and commonly consumed fruits in Sri Lanka. Int J Food Sci 2020:1
375. Somsap OA, Piboonpol G, Payanglee K, Daengrot C (2019) Antibacterial activity of Nam-Nam fruit extract. Princess Naradhiwas Univ J 11:156
376. Suprayogi A, Rahminiwati M, Satyaningtijas A, Hajar H, Wulandari M (2023) Study of in vivo antidiabetic activity of namnam leaves (*Cynometra cauliflora*) extract in Sprague Dawley rat. al-Kimiya 10:20
377. Komarudin D, Putri R, Putri R (2021) Uji efektivitas ekstrak dekok daun namnam (*Cynometra cauliflora* L.) terhadap penurunan kadar gula darah pada tikus putih jantan (*Rattus norvegicus*) yang diinduksi aloksan. ISTA Online Technol J 2:12
378. Anliza S, Rachmawati N (2021) Cytotoxic activity of ethanol extract in namnam leaves (*Cynometra cauliflora* L.) to HeLa cell. Walisongo J Chem 4:107
379. Santiparadon M, Payanglee K, Somsap OA, Pengmusaw, H, Seumae K, Pintarasri S (2024) The effect of *Cynometra cauliflora* Linn. fruit extract-induced congenital anomalies on chick embryos. Trends Sci 21:7454
380. Seyedan A, Mohamed Z, Alshagga MA, Koosha S, Alshawsh MA (2019) *Cynometra cauliflora* Linn. attenuates metabolic abnormalities in high-fat diet-induced obese mice. J Ethnopharmacol 236:173
381. Mukherjee S, Karati D (2023) Exploring the phytochemistry, pharmacognostic properties, and pharmacological activities of medically important plant *Momordica charantia*. Pharmacol Res Mod Chin Med 6:100226
382. Torre VVE, Guarniz WS, Correa SC, Razco CL, Siche R (2020) Antimicrobial activity and chemical composition of *Momordica charantia*: a review. Pharmacogn J 12:213

383. Gayathry KS, John JA (2022) A comprehensive review on bitter gourd (*Momordica charantia* L.) as a gold mine of functional bioactive components for therapeutic foods. Food Prod Proc Nutr 4:10
384. Zakaria M, Ali Mohd M (2015) Traditional Malay medicinal plants. Institi Terjemahan Buku Malaysia, Kuala Lumpur
385. Wiart C (2002) Medicinal plants of Southeast Asia, 2nd edn. Pearson Malaysia Sdn Bhd, Kuala Lumpur
386. Rawat V, Gupta M, Shivnath N, Gupta P, Srivastava SK, Arshad M (2018) Phytochemical analysis of bitter melon juice; antiproliferative and apoptosis inducing activity on human osteosarcoma cells. Int J Sci Eng Res 10:1
387. Sun L, Zhang X, Dong L, Zhang C, Guo P, Wu C (2021) The triterpenoids of the bitter gourd (*Momordica charantia*) and their pharmacological activities: a review. J Food Compn Anal 96:103726
388. Ota A, Ulrih NP (2017) An overview of herbal products and secondary metabolites used for management of type 2 diabetes. Front Pharmacol 8:00436
389. Çiçek SS (2022) *Momordica charantia* L.—diabetes-related bioactivities, quality control, and safety considerations. Front Pharmacol 13:904643
390. Mahamat O, Flora H, Tume C, Kamanyi A (2020) Immunomodulatory activity of *Momordica charantia* L. (Cucurbitaceae) leaf diethyl ether and methanol extracts on *Salmonella typhi*-infected mice and LPS-induced phagocytic activities of macrophages and neutrophils. Evid-Based Compl Alt Med 2020:5248346
391. Chen F, Huang G, Huang H (2021) Preparation, analysis, antioxidant activities in vivo of phosphorylated polysaccharide from *Momordica charantia.* Carbohydr Polym 252:117179
392. Oyelere SF, Ajayi OH, Ayoade TE, Pereira GBS, Owoyemi BCD, Ilesanmi AO, Akinyemi OA (2022) A detailed review on the phytochemical profiles and anti-diabetic mechanisms of *Momordica charantia.* Heliyon 8:e09253
393. Bora AFM, Kouame KJEP, Li X, Liu L, Pan Y (2023) New insights into the bioactive polysaccharides, proteins, and triterpenoids isolated from bitter melon (*Momordica charantia*) and their relevance for nutraceutical and food application: a review. Int J Biol Macromol 231:123173
394. Ramoza SA, Aminin ALN, Cahyono B (2024) Biotransformation of bitter gourd (*Momordica charantia*) by *Lactobacillus plantarum* and its bioactivities. Int Food Res J 31:253
395. Singh CP, Maheshwari KK (2023) An overview of *Momordica charantia*: a neutraceutical approach for depression. Int J Pharm Sci Med 8:67
396. Muronga M, Quispe C, Tshikhudo PP, Msagati TAM, Mudau FN, Martorell M, Salehi B, Razis AFA, Sunusi U, Kamal RM, Rad SJ (2021) Three selected edible crops of the genus *Momordica* as potential sources of phytochemicals: biochemical, nutritional, and medicinal values. Front Pharmacol 12:625546
397. Palanisamy S, David R, Madav E, Kannan VD, Rajendran V, Sampath S, Ahmed MZ, Alqahtani AS, Kazmi S, Asaithambi P (2024) Green fabrication of silver nanoparticles using leaf extract of tropical vine *Momordica charantia*: spectral characterization and in vitro cytotoxicity evaluation on human breast cancer cells. Mater Technol 39:2304428
398. Jatale AP, Jaiswal GB, Kabra PG, Mahajan CV, Deshmukh SP (2024) *Momordica dioica*: a medicinal plant, its benefits and pharmacological activity. GSC Biol Pharm Sci 26:014
399. Riaz G, Chopra R (2018) A review on phytochemistry and therapeutic uses of *Hibiscus sabdariffa* L. Biomed Pharmacother 102:575
400. Yaacob M, Ahmad EIE, Hussain Y (2006) Manual penanaman rosel. MARDI, Serdang, Selangor, Malaysia, p 28
401. Ahmad EIE, Yaacob M, Hussain Y (2005) Rosel (*Hibiscus sabdariffa*). In: Musa Y, MGM and MP (ed) Penanaman tumbuhan ubatan & beraroma. MARDI, Serdang, p 77
402. Dasuki UA (2001) Hibiscus L. In: van Valkenburg JLCH, Bunyapraphatsara N (eds) Plant resources of South-East Asia: medicinal and poisonous plants 2, 2nd edn. Backhuys Publishers, Leiden, Netherlands, p 297

403. Abdourazak MA, Boukar CMM, Kéllou KKB (2024) Roselle (*Hibiscus sabdariffa* L.): overview of its biology, ecology, socio-economic importance and cultivation constraints in West Africa. World J Adv Pharm Life Sci 10:47
404. Loganathan A, Francis NS (2024) Physiological and biochemical outcomes of herbal medicine use in the treatment of hypertension. In: Izah SC, Ogwu MC, Akram M (eds) Herbal medicine phytochemistry. Springer Nature Switzerland AG, Cham, Switzerland, p 1
405. Mageed AAM, Abdellatif MM, Ali YH, Elsayed DA, Intisar KS, Tahon MAA (2024) Anti-inflammatory and antioxidant immunomodulatory effects of some natural common plants. Seybold Rep J 19:91
406. Ari M, Alu S, Longtep B, Alum I, Usman A, Kwassau S, Onu J (2015) Nutrient composition and phytochemical screening of crushed, toasted and fermented roselle (*Hibiscus sabdariffa* L) seeds. Annu Res Rev Biol 7:39
407. Burunkar VA, Nemade LS, Bhagwat RR, Kadam MS, Kamat VA, Tikam SS (2024) Herbal harmony: a synopsis on diverse herbs and their anticancer attributes. Biosci Biotechnol Res Asia 21:519
408. El Bilali H (2024) State and contours of research on roselle (*Hibiscus sabdariffa* L.) in Africa. Open Agric 9:0336
409. McKay DL, Chen C-YO, Saltzman E, Blumberg JB (2010) *Hibiscus sabdariffa* L. tea (tisane) lowers blood pressure in prehypertensive and mildly hypertensive adults. J Nutr 140:298
410. Chew LY, Teng SK, Neo YP, Sim YY, Chew SC (2024) The potential of roselle (*Hibiscus sabdariffa*) plant in industrial applications: a promising source of functional compounds. J Oleo Sci 73:275
411. Gaya Bako I, Bako IG, Mabrouk MA, Abubakar A (2009) Antioxidant effect of ethanolic seed extract of *Hibiscus sabdariffa* Linn. (Malvaceae) alleviates the toxicity induced by chronic administration of sodium nitrate on some haematological parameters in Wistar rats. Adv J Food Sci Technol 1:39
412. Azzahra F (2024) Kombinasi senyawa antikanker dengan tanaman genus *Hibiscus*: narrative review. Pharm Genius 3:93
413. Nguyen DTC, Nguyen NTT, Nguyen TTT, Van TT (2024) Recent advances in the biosynthesis of ZnO nanoparticles using floral waste extract for water treatment, agriculture and biomedical engineering. Nanoscale Adv 6:4047
414. Sahu AR, Nayak NR, Ekka NJ (2024) A mini-review on phytochemical screening, biological activity, and therapeutic capability of *Hibiscus*: an ornamental plant species. In: Chachad D, Mishra S, Mahishi P, Sahu AR (eds) Research and review in plant science, vol IV. Bhumi Publishing, Kothapur, India, p 72
415. Yob NJ, Jofrry SM, Affandi MMRMM, Teh LK, Salleh MZ, Zakaria ZA (2011) *Zingiber zerumbet* (L.) Smith: a review of its ethnomedicinal, chemical, and pharmacological uses. Evid Based Compl Alt Med 2011:543216
416. Tan JW, Israf DA, Tham CL (2018) Major bioactive compounds in essential oils extracted from the rhizomes of *Zingiber zerumbet* (L.) Smith: a mini-review on the anti-allergic and immunomodulatory properties. Front Pharmacol 9:652
417. Baby S, Dan M, Thaha ARM, Johnson AJ, Kurup R, Balakrishnapillai P, Lim CK (2009) High content of zerumbone in volatile oils of *Zingiber zerumbet* from southern India and Malaysia. Flavour Fragr J 24:301
418. Koga AY, Beltrame FL, Pereira AV (2016) Several aspects of *Zingiber zerumbet*: a review. Rev Bras Farmacogn 26:385
419. Chan JSW, Lim XY, Japri N, Ahmad IF, Tan TYC (2023) *Zingiber zerumbet*: a scoping review of its medicinal properties. Planta Med 90:204
420. Lechat-Vahirua I, François P, Menut C, Lamaty G, Bessiere JM (1993) Aromatic plants of French Polynesia. I. Constituents of the essential oils of rhizomes of three Zingiberaceae: *Zingiber zerumbet* Smith, *Hedychium coronarium* Koenig and *Etlingera cevuga* Smith. J Essent Oil Res 5:55
421. Yu F, Okamto S, Nakasone K, Adachi K, Matsuda S, Harada H, Misawa N, Utsumi R (2008) Molecular cloning and functional characterization of α-humulene synthase, a possible key

enzyme of zerumbone biosynthesis in shampoo ginger (*Zingiber zerumbet* Smith). Planta 227:1291
422. Khalid MH, Akhtar MN, Mohamad AS, Perimal EK, Akira A, Israf DA, Lajis N, Sulaiman MR (2011) Antinociceptive effect of the essential oil of *Zingiber zerumbet* in mice: possible mechanisms. J Ethnopharmacol 137:345
423. Ruslay S, Abas F, Shaari K, Zainal Z, Maulidiani SH, Israf DA, Lajis NH (2007) Characterization of the components present in the active fractions of health gingers (*Curcuma xanthorrhiza* and *Zingiber zerumbet*) by HPLC-DAD-ESIMS. Food Chem 104:1183
424. Akbar S (2020) Handbook of 200 medicinal plants: a comprehensive review of their traditional medical uses and scientific justifications. Springer International Publishing, Cham, Switzerland
425. Dev S (1960) Studies in sesquiterpenes—XVI: zerumbone, a monocyclic sesquiterpene ketone. Tetrahedron 8:171
426. Nigam CI, Levi L (1963) Column and gas chromatographic analysis of oil of wild ginger identification and estimation of some new constituents. Can J Chem 41:1726
427. Damodaran NP, Dev S (1968) Studies in sesquiterpenes—XXXIII: structure of humulene epoxide-I and humulene epoxide-II. Tetrahedron 24:4123
428. Dai DN, Thang TD, Chau LTM, Ogunwande IA (2013) Chemical constituents of the root essential oils of *Zingiber rubens* Roxb. and *Zingiber zerum*bet (L.) Smith. Am J Plant Sci 4:7
429. Songsiang U, Pitchuanchom S, Boonyarat C, Hahnvajanawong C, Yenjai C (2010) Cytotoxicity against cholangiocarcinoma cell lines of zerumbone derivatives. Eur J Med Chem 45:3794
430. Kitayama T (2011) Attractive reactivity of a natural product, zerumbone. Biosci Biotechnol Biochem 75:199
431. Kumar SCS, Srinivas P, Negi PS, Bettadaiah BK (2013) Antibacterial and antimutagenic activities of novel zerumbone analogues. Food Chem 141:1097
432. Bhuiyan NI, Chowdhury JU, Begum J (2009) Chemical investigation of the leaf and rhizome essential oils of *Zingiber zerumbet* (L.) Smith from Bangladesh. Bangladesh J Pharmacol 4:9
433. Sulaiman MR, Tengku Mohamad TAS, Shaik Mossadeq WM, Moin S, Yusof M, Mokhtar AF, Zakaria ZA, Israf DA, Lajis N (2010) Antinociceptive activity of the essential oil of *Zingiber zerumbet*. Planta Med 76:107
434. Chien TY, Huang SKH, Lee CJ, Tsai PW, Wang CC (2016) Antinociceptive and anti-inflammatory effects of zerumbone against mono-iodoacetate-induced arthritis. Int J Mol Sci 17:20249
435. Jang DS, Han AR, Park G, Jhon GJ, Seo EK (2004) Flavonoids and aromatic compounds from the rhizomes of *Zingiber zerumbet*. Arch Pharm Res 27:386
436. Akhtar NMY, Jantan I, Arshad L, Haque MA (2019) Standardized ethanol extract, essential oil and zerumbone of *Zingiber zerumbet* rhizome suppress phagocytic activity of human neutrophils. BMC Compl Alt Med 19:331
437. Nakamura Y, Yoshida C, Murakami A, Ohigashi H, Osawa T, Uchida K (2004) Zerumbone, a tropical ginger sesquiterpene, activates phase II drug metabolizing enzymes. FEBS Lett 572:245
438. Li L, Kong L, Song H (2017) The therapeutic effect of zerumbone on chronic gastritis via antioxidant mechanisms. Exp Ther Med 14:2505
439. Girisa S, Shabnam B, Monisha J, Fan L, Halim CE, Arfuso F, Ahn KS, Sethi G, Kunnumakkara AB (2019) Potential of zerumbone as an anti-cancer agent. Molecules 24:743
440. Albaayit SFA, Maharjan R, Abdullah R, Noor MHM (2022) Evaluation of anti-methicillin-resistant *Staphylococcus aureus* property of zerumbone. J Appl Biomed 20:15
441. Wang M, Niu J, Ou L, Deng B, Wang Y, Li S (2019) Zerumbone protects against carbon tetrachloride (CCl_4)-induced acute liver injury in mice via inhibiting oxidative stress and the inflammatory response: Involving the TLR4/NF-κB/COX-2 pathway. Molecules 24:101964
442. Rogerio AP, Andrade EL, Leite DFP, Figueiredo CP, Calixto JB (2009) Preventive and therapeutic anti-inflammatory properties of the sesquiterpene α-humulene in experimental airways allergic inflammation. Br J Pharmacol 158:1074

443. Tzeng TF, Lu HJ, Liou SS, Chang CJ, Liu IM (2014) Lipid-lowering effects of zerumbone, a natural cyclic sesquiterpene of *Zingiber zerumbet* Smith, in high-fat diet-induced hyperlipidemic hamsters. Food Chem Toxicol 69:132
444. Pereira CAA, Pereira ALG, Noveletto JVM, Jordão CC, Pavarina AC (2023) Zerumbone disturbs the extracellular matrix of fluconazole-resistant *Candida albicans* biofilms. J Fungi 9:50576
445. Kim M, Miyamoto S, Yasui Y, Oyama T, Murakami A, Tanaka T (2009) Zerumbone, a tropical ginger sesquiterpene, inhibits colon and lung carcinogenesis in mice. Int J Cancer 124:264
446. Taha MME, Abdul AB, Abdullah R, Abdelwahab SI, Mohan S (2010) Potential chemoprevention of diethylnitrosamine-initiated and 2-acetylaminofluorene-promoted hepatocarcinogenesis by zerumbone from the rhizomes of the subtropical ginger (*Zingiber zerumbet*). Chem Biol Interact 186:295
447. Choi SH, Lee YJ, Seo WD, Lee HJ, Nam JW, Lee YJ, Kim J, Seo EK, Lee YS (2011) Altered cross-linking of HSP27 by zerumbone as a novel strategy for overcoming HSP27-mediated radioresistance. Int J Radiat Oncol Biol Phys 79:1196
448. Kähkönen MP, Hopia AI, Vuorela HJ, Rauha JP, Pihlaha K, Kujala TS, Heinonen M (1999) Antioxidant activity of plant extracts containing phenolic compounds. J Agric Food Chem 47:3954
449. Shieh YH, Huang HM, Wang CC, Lee CC, Fan CK, Lee YL (2015) Zerumbone enhances the Th1 response and ameliorates ovalbumin-induced Th2 responses and airway inflammation in mice. Int Immunopharmacol 24:383
450. Nam SY, Chung CK, Seo JH, Rah SY, Kim HM, Jeong HJ (2014) The therapeutic efficacy of α-pinene in an experimental mouse model of allergic rhinitis. Int Immunopharmacol 23:273
451. Mitoshi M, Kuriyama I, Nakayama H, Miyazato H, Sugimoto K, Kobayashi Y, Jippo T, Kuramochi K, Yoshida H, Mizushina Y (2014) Suppression of allergic and inflammatory responses by essential oils derived from herbal plants and citrus fruits. Int J Mol Med 33:1643
452. Gopalsamy B, Farouk AAO, Mohamad TAST, Sulaiman MR, Perimal EK (2017) Antiallodynic and antihyperalgesic activities of zerumbone via the suppression of IL-1β, IL-6, and TNF-α in a mouse model of neuropathic pain. J Pain Res 10:2605
453. Tzeng TF, Hong TY, Tzeng YC, Liou SS, Liu IM (2015) Consumption of polyphenol-rich *Zingiber zerumbet* rhizome extracts protects against the breakdown of the blood-retinal barrier and retinal inflammation induced by diabetes. Nutrients 7:7821
454. Shori AB (2015) Screening of antidiabetic and antioxidant activities of medicinal plants. J Integr Med 13:297
455. Rukachaisirikul T, Siriwattanakit P, Sukcharoenphol K, Wongvein C, Ruttanaweang P, Wongwattanavuch P, Suksamrarn A (2004) Chemical constituents and bioactivity of *Piper sarmentosum*. J Ethnopharmacol 93:173
456. Sun X, Chen W, Dai W, Xin H, Rahmand K, Wang Y, Zhang J, Zhang S, Xu L, Han T (2020) *Piper sarmentosum* Roxb.: a review on its botany, traditional uses, phytochemistry, and pharmacological activities. J Ethnopharmacol 263:112897
457. Zakaria ZA, Patahuddin H, Mohamad AS, Israf DA, Sulaiman MR (2010) In vivo antinociceptive and anti-inflammatory activities of the aqueous extract of the leaves of *Piper sarmentosum*. J Ethnopharmacol 128:42
458. Mathew S, Nair GPM (2004) *Piper sarmentosum* Roxb.—an addition to the flora of Andaman Islands. Curr Sci 87:141
459. Ugusman A, Zakaria Z, Hui CK, Nordin NAMM (2010) *Piper sarmentosum* increases nitric oxide production in oxidative stress: a study on human umbilical vein endothelial cells. Clinics 65:709
460. Atiax E, Ahmad F, Sirat HM, Arbain D (2011) Antibacterial activity and cytotoxicity screening of Sumatran Kaduk (*Piper sarmentosum* Roxb.). Iran J Pharmacol Ther 10:1
461. Hussain K, Hashmi FK, Latıf A, Ismail Z, Sadikun A (2012) A review of the literature and latest advances in research of *Piper sarmentosum*. Pharm Biol 50:1045
462. Arun Waman A, Bohra P, Chakraborty G (2019) Vegetative propagation of *Piper sarmentosum* Roxb.—a medicinally important species. Curr Agric Res J 7:46

463. Syed RF, Rahman SA (2016) *Piper sarmentosum* Roxb.: a mini review of ethnobotany, phytochemistry and pharmacology. J Anal Pharm Res 2:31
464. Seyyedan A, Yahya F, Kamarolzaman MFF, Suhaili Z, Desa MNM, Khairi HM, Somchit MN, Fatimah CA, Teh LK, Salleh MZ, Zakaria ZA (2013) Review on the ethnomedicinal, phytochemical and pharmacological properties of *Piper sarmentosum*: scientific justification of its traditional use. TANG 3:e19
465. Taweechaisupapong S, Singhara S, Lertsatitthanakorn P, Khunkitti W (2010) Antimicrobial effects of *Besenbergia pandurata* and *Piper sarmentosum* leaf extracts on planktonic cells and biofilm of oral pathogens. Pak J Pharm Sci 23:224
466. Rititid W, Ruangsang P, Reanomongkol W, Malinee W (2007) Studies of the anti-inflammatory and antipyretic activities of the methanolic extract of *Piper sarmentosum* Roxb. leaves in rats. Songklanakarin J Sci Technol 29:6
467. Chanwitheesuk A, Teerawutgulrag A, Rakariyatham N (2005) Screening of antioxidant activity and antioxidant compounds of some edible plants of Thailand. Food Chem 92:491
468. Sireeratawong S, Vannasiri SM, Sritiwong DVMS, Itharat A, Jaijoy KM (2010) Anti-inflammatory, anti-nociceptive and antipyretic effects of the ethanol extract from root of *Piper sarmentosum* Roxb. J Med Assoc Thai 93:1
469. Ong HC, Norzalina J (1999) Malay herbal medicine in Gemencheh, Negri Sembilan, Malaysia. Fitoterapia 70:10
470. Hussain K, Ismail Z, Sadikun A, Ibrahim P (2009) Proximate and qualitative analysis of different parts of *Piper sarmentosum*, and quantification of total amides in various extracts. Pharmacogn Res 1:113
471. Archibold AAD, Santana AI, Gupta MP (2018) Ethnomedical uses and pharmacological activities of most prevalent species of genus *Piper* in Panama: a review. J Ethnopharmacol 217:63
472. Qin W, Huang S, Li C, Chen S, Peng Z (2010) Biological activity of the essential oil from the leaves of *Piper sarmentosum* Roxb. (Piperaceae) and its chemical constituents on *Brontispa longissima* (Gestro) (Coleoptera: Hispidae). Pestic Biochem Physiol 96:132
473. Chieng TC, Assim ZB, Fasihuddin BA (2008) Toxicity and antitermite activities of the essential oils from *Piper sarmentosum*. Malays J Anal Sci 12:234
474. Hematpoor A, Liew SY, Azirun MS, Awang K (2017) Insecticidal activity and the mechanism of action of three phenylpropanoids isolated from the roots of *Piper sarmentosum* Roxb. Sci Rep 7:12576
475. Hieu LD, Thang TD, Hoi TM, Ogunwande IA (2014) Chemical composition of essential oils from four Vietnamese species of *Piper* (Piperaceae). J Oleo Sci 63:211
476. Ee GCL, Lim CM, Lim CK, Rahmani M, Shaari K, Bong CFJ (2009) Alkaloids from *Piper sarmentosum* and *Piper nigrum*. Nat Prod Res 23:1416
477. Tuntiwachwuttikul P, Phansa P, Pootaeng-On Y, Charles W (2006) Chemical constituents of the roots of *Piper sarmentosum*. Chem Pharm Bull 54:149
478. Bokesch HR, Gardella RS, Rabe DC, Bottaro DP, Linehan WM, McMahon JB, McKee TC (2011) A new hypoxia inducible factor-2 inhibitory pyrrolinone alkaloid from roots and stems of *Piper sarmentosum*. Chem Pharm Bull 59:1178
479. Pan L, Matthew S, Lantvit DD, Zhang X, Ninh TN, Chai H, Carcache de Blanco EJ, Soejarto DD, Swanson SM, Kinghorn AD (2011) Bioassay-guided isolation of constituents of *Piper sarmentosum* using a mitochondrial transmembrane potential assay. J Nat Prod 74:2193
480. Miean KH, Mohamed S (2001) Flavonoid (myricetin, quercetin, kaempferol, luteolin, and apigenin) content of edible tropical plants. J Agric Food Chem 49:3106
481. Yang SX, Sun QY, Yang FM, Hu GW, Luo JF, Wang YH, Long CL (2013) Sarmentosumols A to F, new mono-and dimeric alkenylphenols from *Piper sarmentosum*. Planta Med 79:693
482. Estai AM, Suhaimi F, Shuid AN, Das S, Abdullah S, Soelaiman IN (2012) Biomechanical evaluation of fracture healing following administration of *Piper sarmentosum* in ovariectomised rats. Afr J Pharm Pharmacol 6:148
483. Nirwana SI, Suhana MRE, Fadziyah MAS, Farihah HS, Fairus A, Alfakri MNM (2012) *Piper sarmentosum* improves bone structure and biomechanical strength of rats given excess glucocorticoid. J Pharm Res Int 2:168

484. Horcajada MN, Habauzit V, Trzeciakiewicz A, Morand C, Gil-Izquierdo A, Mardon J, Lebecque P, Davicco MJ, Chee WSS, Coxam V, Offord E (2008) Hesperidin inhibits ovariectomized-induced osteopenia and shows differential effects on bone mass and strength in young and adult intact rats. J Appl Physiol 104:648
485. Shi YN, Liu FF, Jacob MR, Li XC, Zhu HT, Wang D, Cheng RR, Yang CR, Xu M, Zhang YJ (2017) Antifungal amide alkaloids from the aerial parts of *Piper flaviflorum* and *Piper sarmentosum*. Planta Med 83:143
486. Ab Raman SFS, Sijam K, Omar D (2016) Chemical composition of *Piper sarmentosum* extracts and antibacterial activity against the plant pathogenic bacteria *Pseudomonas fuscovaginae* and *Xanthomonas oryzae* pv. *oryzae*. Plant Dis Protect 121:237
487. Mohd Ramli ES, Suhaimi F, Ahmad F, Shuid AN, Ima-Nirwana MN (2013) *Piper sarmentosum*: a new hope for the treatment of osteoporosis. Curr Drug Targets 14:1675
488. Yeo ETY, Wong KWL, See ML, Wong KY, Gan SY, Chan EWL (2018) *Piper sarmentosum* Roxb. confers neuroprotection on beta-amyloid (Aβ)-induced microglia-mediated neuroinflammation and attenuates tau hyperphosphorylation in SH-SY5Y cells. J Ethnopharmacol 217:187
489. Makchuchit S, Rattarom R, Itharat A (2017) The anti-allergic and anti-inflammatory effects of Benjakul extract (a Thai traditional medicine), its constituent plants and its some pure constituents using in vitro experiments. Biomed Pharmacother 89:1018
490. Li Q, Qu FL, Gao Y, Jiang YP, Rahman K, Lee KH, Han T, Qin LP (2017) *Piper sarmentosum* Roxb. produces antidepressant-like effects in rodents, associated with activation of the CREB-BDNF-ERK signaling pathway and reversal of HPA axis hyperactivity. J Ethnopharmacol 199:9
491. Ariffin ZSH, Wan Omar WHH, Safian MF, Ariffin ZZ, Senafi S, Megat Abdul Wahab R (2009) Intrinsic anticarcinogenic effects of *Piper sarmentosum* ethanolic extract on a human hepatoma cell line. Cancer Cell Int 9:6
492. Yong Y, Matthew S, Wittwer J, Pan L, Shen Q, Kinghorn AD, Swanson SM, Carcache de Blanco EJ (2013) Dichamanetin inhibits cancer cell growth by affecting ROS-related signaling components through mitochondrial-mediated apoptosis. Anticancer Res 33:5349
493. Peungvicha P, Thirawarapan SS, Temsiririrkkul R, Watanabe H, Kumar Prasain J, Kadota S (1998) Hypoglycemic effect of the water extract of *Piper sarmentosum* in rats. J Ethnopharmacol 60:27
494. Hematpoor A, Paydar M, Liew T, Sivaspthy Y, Mohebli N, Looi CY, Wong WF, Azirin MS, Awang K (2018) Phenylpropanoids isolated from *Piper sarmentosum* Roxb. induce apoptosis in breast cancer cells through reactive oxygen species and mitochondrial dependent pathways. Chem-Biol Interact 279:210
495. Thent ZC, Lin TS, Das S, Zakaria Z (2012) Histological changes in the heart and the proximal aorta in experimental diabetic rats fed with *Piper sarmentosum*. Afr J Trad Compl Alt Med 9:396
496. Lee K, Lee JH, Kim SI, Cho MH, Lee J (2014) Anti-biofilm, anti-hemolysis, and anti-virulence activities of black pepper, cananga, myrrh oils, and nerolidol against *Staphylococcus aureus*. Appl Microbiol Biotechnol 98:9447
497. Chaimanee V, Thongtue U, Sornmai N, Songsri S, Pettis JS (2017) Antimicrobial activity of plant extracts against the honeybee pathogens, *Paenibacillus larvae* and *Ascosphaera apis* and their topical toxicity to *Apis mellifera* adults. J Appl Microbiol 123:1160
498. Hafizah AH, Zaiton Z, Zulkhairi A, Mohd Ilham A, Nor Anita MMN, Zaleha AM (2010) *Piper sarmentosum* as an antioxidant on oxidative stress in human umbilical vein endothelial cells induced by hydrogen peroxide. J Zhejiang Univ Sci B 11:357
499. Hematpoor A, Paydar M, Liew SY, Sivasothy Y, Mohebali N, Looi CY, Wong WF, Azirun MS, Awang K (2018) Phenylpropanoids isolated from *Piper sarmentosum* Roxb. induce apoptosis in breast cancer cells through reactive oxygen species and mitochondrial-dependent pathways. Chem Biol Interact 279:210
500. Amran AA, Zakaria Z, Othman F, Das S, Al-Mekhlafi HM, Raj S, Nordin NAM (2012) Effect of methanolic extract of *Piper sarmentosum* leaves on neointimal foam cell infiltration in rabbits fed with high cholesterol diet. EXCLI J 11:274

501. Araújo-Vilges KM, de Oliveira SV, de Couto SCP, Fokoue HH, Romero GAS, Kato MJ, Romeiro LAS, Leite JRSA, Kuckelhaus SAS (2017) Effect of piplartine and cinnamides on *Leishmania amazonensis*, *Plasmodium falciparum* and on peritoneal cells of Swiss mice. Pharm Biol 55:1601
502. Othman AN, Zainudin NFS, Zaidin UH, Shamsi S (2021) A comparative study on the larvicidal effects of *Piper sarmentosum* (Kaduk) leaves extracts against *Aedes aegypti*. Perktanika J Sci Technol 29:2811
503. Anwar F, Ashraf M, Bhanger MI (2005) Interprovenance variation in the composition of *Moringa oleifera* oilseeds from Pakistan. J Am Oil Chem Soc 82:45
504. Popoola JO, Obembe OO (2013) Local knowledge, use pattern and geographical distribution of *Moringa oleifera* Lam. (Moringaceae) in Nigeria. J Ethnopharmacol 150:682
505. Gopalakrishnan L, Doriya K, Kumar DS (2016) *Moringa oleifera*: a review on nutritive importance and its medicinal application. Food Sci Hum Wellness 5:49
506. Santos A, Luz L, Pontual E, Napoleão T, Paiva P, Coelho L (2015) *Moringa oleifera*: resource management and multiuse life tree. Adv Res 4:388
507. Pandey A, Pandey RD, Tripathi P, Gupta PP, Haider J, Bhatt S, Singh AV (2012) *Moringa oleifera* Lam. (Sahijan) a plant with a plethora of diverse therapeutic benefits: an updated retrospection. Med Arom Plants 1:1
508. Prajapati C, Ankola M, Upadhyay TK, Sharangi AB, Alabdallah NM, Al-Saeed FA, Muzammil K, Saeed M (2022) *Moringa oleifera*: miracle plant with a plethora of medicinal, therapeutic, and economic importance. Horticulturae 8:492
509. Pareek A, Pant M, Gupta MM, Kashania P, Ratan Y, Jain V, Pareek A, Chuturgoon AA (2023) *Moringa oleifera*: an updated comprehensive review of its pharmacological activities, ethnomedicinal, phytopharmaceutical formulation, clinical, phytochemical, and toxicological aspects. Int J Mol Sci 24:2098
510. Stohs SJ, Hartman MJ (2015) Review of the safety and efficacy of *Moringa oleifera*. Phytother Res 29:796
511. Aekthammarat D, Pannangpetch P, Tangsucharit P (2019) *Moringa oleifera* leaf extract lowers high blood pressure by alleviating vascular dysfunction and decreasing oxidative stress in L-NAME hypertensive rats. Phytomedicine 54:9
512. Bhat SK, Joy EA (2014) Antianxiety effect of ethanolic extract of leaves of *Moringa oleifera* in Swiss albino mice. Arch Med Health Sci 2:5
513. Balogun TA, Buliaminu KD, Chukwudozie OS, Tiamiyu ZA, Idowu TJ (2021) Anticancer potential of *Moringa oleifera* on BRCA-1 gene: systems biology. Bioinform Biol Insights 15:11779322211010704
514. Tahkur RS, Soren G, Pathapati RM, Buchineni M (2016) Diuretic activity of *Moringa oleifera* leaves extract in Swiss albino rats. Pharma Innov J 5:8
515. Deveraj VC, Asad M, Prasad S (2007) Effect of leaves and fruits of *Moringia oleifera* on gastric and duodenal ulcers. Pharm Biol 45:332
516. Karadi RV, Gadge NB, Alagawadi KR, Savadi RV (2006) Effect of *Moringa oleifera* Lam. root-wood on ethylene glycol induced urolithiasis in rats. J Ethnopharmacol 105:306
517. Ghasi S, Nwobodo E, Ofili JO (2000) Hypocholesterolemic effects of crude extract of leaf of *Moringa oleifera* Lam. in high-fat diet fed Wistar rats. J Ethnopharmacol 69:21
518. Mahajan SG, Mali RG, Mehta AA (2007) Protective effect of ethanolic extract of seeds of *Moringa oleifera* Lam. against inflammation associated with development of arthritis in rats. J Immunotoxicol 4:39
519. Bhattacharya A, Tiwari P, Sahu PK, Kumar S (2018) A review of the phytochemical and pharmacological characteristics of *Moringa oleifera*. J Pharm Bioallied Sci 10:181
520. Tayo GM, Pone GW, Komtangi MC, Yondo G, Ngangout AM, Mbida M (2014) Anthelminthic activity of *Moringa oleifera* leaf extracts evaluated in vitro on four developmental stages of *Haemonchus contortus* from goats. Am J Plant Sci 5:1702
521. Meireles D, Gomes J, Lopes L, Hinzmann M, Machado J (2020) A review of properties, nutritional and pharmaceutical applications of *Moringa oleifera*: integrative approach on conventional and traditional Asian medicine. Adv Tradl Med 20:495

522. Liu WL, Wu BF, Shang JH, Zhao YL, Huang AX (2022) *Moringa oleifera* seed ethanol extract and its active component kaempferol potentiate pentobarbital-induced sleeping behaviours in mice via a GABAergic mechanism. Pharm Biol 60:810
523. Gothai S, Arulselvan P, Tan WS, Fakurazi S (2016) Wound healing properties of ethyl acetate fraction of *Moringa oleifera* in normal human dermal fibroblasts. J Intercult Ethnopharmacol 5:1
524. Paikra BK, Dhongade HKJ, Gidwani B (2017) Phytochemistry and pharmacology of *Moringa oleifera* Lam. J Pharmacopuncture 20:194
525. Woldeyohannes MG, Eshete GT, Abiye AA, Hailu AE, Huluka SA, Tadesse WT (2022) Antidiarrheal and antisecretory effect of 80% hydromethanolic leaf extract of *Moringa stenopetala* Baker f. in mice. Biochem Res Int 2022:5768805
526. Zhu Y, Yin Q, Yang Y (2020) Comprehensive investigation of *Moringa oleifera* from different regions by simultaneous determination of 11 polyphenols using UPLC-ESI-MS/MS. Molecules 25:25030676
527. Bajpai M, Pande A, Tewari SK, Prakash D (2005) Phenolic contents and antioxidant activity of some food and medicinal plants. Int J Food Sci Nutr 56:287
528. Gopinath LR, Jeevitha S, Gokiladevi T, Archaya S (2017) Isolation and identification of therapeutic compounds from *Moringa oleifera* and its antimicrobial activity. IOSR J Pharm Biol Sci 12:1
529. Alia F, Putri M, Anggtaeni N, Syamsunamo MRAA (2022) The potency of *Moringa oleifera* as protective agent in cardiac damage and dysfunction. Front Pharmacol 12:724439
530. Borgonovo G, De Petrocellis L, Moriello AS, Bertoli S, Leone A, Battezzati A, Mazzini S, Bassoli A (2020) Moringin, a stable isothiocyanate from *Moringa oleifera*, activates the somatosensory and pain receptor TRPA1 channel in vitro. Molecules 25:040976
531. Sahakitpichan P, Mahidol C, Disadee W, Ruchirawat S, Kanchanapoom T (2011) Unusual glycosides of pyrrole alkaloid and 4′-hydroxyphenylethanamide from leaves of *Moringa oleifera*. Phytochemistry 72:791
532. Saini RK, Shetty NP, Giridhar P (2014) Carotenoid content in vegetative and reproductive parts of commercially grown *Moringa oleifera* Lam. cultivars from India by LC-APCI-MS. Eur Food Res Technol 238:971
533. Liao PC, Lai MH, Hsu KP, Kuo YH, Chen J, Tsai MC, Li CX, Yin XJ, Jeyashoke N, Chao LKP (2018) Identification of β-sitosterol as in vitro anti-inflammatory constituent in *Moringa oleifera*. J Agric Food Chem 66:10748
534. Matic I, Guidi A, Kenzo M, Mattei M, Galgani A (2018) Investigation of medicinal dietary plants: review of traditionally supplements: on *Moringa oleifera*. J Public Health Afr 9:191
535. Tan WS, Arulselvan P, Karthivashan G, Fakurazi S (2015) *Moringa oleifera* flower extract suppresses the activation of inflammatory mediators in lipopolysaccharide-stimulated RAW 264.7 macrophages via NF-κB pathway. Mediators Inflammat 2015:720171
536. Abdel-Daim MM, Khalil SR, Awad A, Zeid EHA, El-Aziz RA, El-Serehy HA (2020) Ethanolic extract of *Moringa oleifera* leaves influences NF-κB signaling pathway to restore kidney tissue from cobalt-mediated oxidative injury and inflammation in rats. Nutrients 12:41031
537. Vongsak B, Mangmool S, Gritsanapan W (2015) Antioxidant activity and induction of mRNA expressions of antioxidant enzymes in HEK-293 cells of *Moringa oleifera* leaf extract. Planta Med 81:1084
538. El-Hadary AE, Ramadan MF (2019) Antioxidant traits and protective impact of *Moringa oleifera* leaf extract against diclofenac sodium-induced liver toxicity in rats. J Food Biochem 43:12704
539. Das AN, Nirala SK, Shukla S, Mathur R (2015) Co-administration of adjuvants along with *Moringa oleifera* attenuates beryllium-induced oxidative stress and histopathological alterations in rats. Pharm Biol 53:1465
540. Mohd Fisall UF, Ismail NZ, Adebayo IA, Arsad H (2021) Dichloromethane fraction of *Moringa oleifera* leaf methanolic extract selectively inhibits breast cancer cells (MCF7) by induction of apoptosis via upregulation of Bax, p53 and caspase 8 expressions. Mol Biol Rep 48:4465

541. Soliman MM, Al-Osaimi SH, Hassanmohamed E, Aldhahrani A, Alkhedaide A, Althobaiti F, Mohamed WA (2020) Protective impacts of *Moringa oleifera* leaf extract against methotrexate-induced oxidative stress and apoptosis on mouse spleen. Evid-Based Compl Alt Med 2020:6738474
542. Kou X, Li B, Olayanju JB, Drake JM, Chen N (2018) Nutraceutical or pharmacological potential of *Moringa oleifera* Lam. Nutrients 10:343
543. Jaiswal D, Kumar Rai P, Kumar A, Mehta S, Watal G (2009) Effect of *Moringa oleifera* Lam. leaves aqueous extract therapy on hyperglycemic rats. J Ethnopharmacol 123:392
544. Bakre AG, Aderibigbe AO, Ademowo OG (2013) Studies on neuropharmacological profile of ethanol extract of *Moringa oleifera* leaves in mice. J Ethnopharmacol 149:783
545. Chuang PH, Lee CW, Chou JY, Murugan M, Shieh BJ, Chen HM (2007) Anti-fungal activity of crude extracts and essential oil of *Moringa oleifera* Lam. Bioresour Technol 98:232
546. Khan H, Jaiswal V, Kulshreshtha S, Khan A (2019) Potential angiotensin converting enzyme inhibitors from *Moringa oleifera*. Recent Pat Biotechnol 13:239
547. Martínez-González CL, Martínez L, Martínez-Ortiz EJ, González-Trujano ME, Déciga-Campos M, Ventura-Martínez R, Díaz-Reval I (2017) *Moringa oleifera*, a species with potential analgesic and anti-inflammatory activities. Biomed Pharmacother 87:482
548. Ali Redha A, Perna S, Riva A, Petrangolini G, Peroni G, Nichetti M, Iannello G, Naso M, Faliva MA, Rondanelli M (2021) Novel insights on anti-obesity potential of the miracle tree, *Moringa oleifera*: a systematic review. J Funct Foods 84:104600
549. Adeyi AO, Ajisebiola SB, Adeyi EO, Alimba CG, Okorie UG (2020) Antivenom activity of *Moringa oleifera* leaves against pathophysiological alterations, somatic mutation and biological activities of *Naja nigricollis* venom. Sci Afr 8:e00356
550. Mehwish HM, Riaz Rajoka MS, Xiong Y, Zheng K, Xiao H, Anjin T, Liu Z, Zhu Q, He Z (2022) *Moringa oleifera*—a functional food and its potential immunomodulatory effects. Food Rev Int 38:1533
551. Chin CY, Jalil J, Ng PY, Ng SF (2018) Development and formulation of *Moringa oleifera* standardised leaf extract film dressing for wound healing application. J Ethnopharmacol 212:188
552. Toh SC, Lihan S, Bunya SB, Leong SS (2023) In vitro antimicrobial efficacy of *Cassia alata* (Linn.) leaves, stem, and root extracts against cellulitis causative agent *Staphylococcus aureus*. BMC Compl Med Ther 23:85
553. Oladeji OS, Adelowo FE, Oluyori AP, Bankole DT (2020) Ethnobotanical description and biological activities of *Senna alata*. Evid-Based Compl Alt Med 2020:2580259
554. Otto RBD, Ameso S, Onegi B (2014) Assessment of antibacterial activity of crude leaf and root extracts of *Cassia alata* against *Neisseria* gonorrhea. Afr Health Sci 14:840
555. Yon JAL, Lee SK, Keng JW, Chow SC, Liew KB, Teo SS, Mossadeq SWM, Marriott PJ, Akowuah GA, Ming LC, Goh BH, Chew YL (2023) *Cassia alata* (Linnaeus) Roxburgh for skin: natural remedies for atopic dermatitis in Asia and their pharmacological activities. Cosmetics 10:5
556. Bharathi DR, Mani RK, Ajay BV, Ajay BV, Pooja RC, Shankar J, Kumar MK, Mahesh C (2022) *Cassia alata*: phytopharmacological, traditional, and medicinal considerations. World J Curr Med Pharm Res 4:147
557. El-Mahmood M, Doughari JH (2008) Phytochemical screening and antibacterial evaluation of the leaf and root extracts of *Cassia alata* Linn. Afr J Pharm Pharmacol 2:124
558. Adedoyin TA, Nkechi JO, Imoru OJ, Kola AE, Titilayo B (2015) Evaluation of phytochemical components of aqueous root extracts of *Cassia alata* and its pharmacological effects on isolated rat. Niger J Pharm Sci 4:1
559. Fatmawati S, Yuliana PAS, Bakar AMF (2020) Chemical constituents, usage and pharmacological activity of *Cassia alata*. Heliyon 6:e04396
560. Chatsiriwej N, Wungsintaweekul J, Panichayupakaranant P (2006) Anthraquinone production in *Senna alata* root cultures. Pharm Biol 44:416
561. Keng JW, Lee SK, Sang SH, Liew KB, Teo SS, Mossadeq WMSM, Chow SC, Akowuah GA, Lee SK, Mai CW, Chew YL (2024) *Cassia alata* and its phytochemicals: a promising natural strategy in wound recovery. Sci 6:1

562. Abille N, Mae J, (2015) A review on the medicinal plant *Cassia alata* (L.) Roxb. (Fabaceae). Biological Sciences Department, College of Science & Computer Studies De La Salle University—Dasmarinas, Cavite, Philippines, p 1
563. Meenupriya J, Vinisha AS, Priya P (2014) *Cassia alata* and *Cassia auriculata*—review of their bioactive potential. World J Pharm Sci 2:1760
564. Edegbo E, Okolo MLO, Adegoke AS, Omatola CA, Idache BM, Abraham JO, Akor ME, Zakari DA, Alaba AZ, Omale S, Yahaya A, Musa Q, Muhammed D (2023) Phytochemical screening and antifungal activity of *Cassia alata* (Linn.) crude leaf extracts. Afr J Microbiol Res 17:176
565. Legaspi CLB, Lazarte MCC (2020) The phytochemical content and the in vitro antifungal properties of *Senna alata* (Linn.) Roxb.: a review. Acta Med Philipp 54:86
566. Colin MN, Claudiana NSE, Kaffah AU, Hasanah AN, Megantara S (2024) Review on *Cassia alata* bioactive compounds: in silico, in vitro, and in vivo studies. Drug Des Devel Ther 18:4427
567. Villaseñor S (2009) Cassiaindoline, a new analgesic and anti-inflammatory alkaloid from *Cassia alata*. Z Naturforsch C J Biosci 64:335
568. Ademola IO, Eloff JN (2011) Ovidical and larvicidal acivity of *Cassia alata* leaf acetone extract and fractions on *Haemonchus contortus*: in vitro studies. Pharm Biol 49:539
569. Mou K, Dash P (2016) A comprehensive review on *Gynura procumbens* leaves. Int J Pharmacogn 3:167
570. Tan HL, Chan KG, Pusparajah P, Lee LH, Goh BH (2016) *Gynura procumbens*: an overview of the biological activities. Front Pharmacol 7:191
571. Yusoff MM, Misran A, Ahmed OA, Majid WHDW, Wahab PEM, Ahmad NF (2019) Tropical agricultural science *Gynura procumbens*: agronomic practices and future prospects in Malaysia. Pertanika J Trop Agric Sci 42:421
572. Ashraf K (2019) An updated phytochemical and pharmacological review on *Gynura procumbens*. Asian J Pharm Clin Res 9
573. Bari MS, Khandokar L, Haque E, Romano B, Capasso R, Seidel V, Haque MA, Rashid MA (2021) Ethnomedicinal uses, phytochemistry, and biological activities of plants of the genus *Gynura*. J Ethnopharmacol 271:113834
574. Kim HH, Ha SE, Vetrivel P, Bhosale PB, Kim SM, Kim GS (2021) Potential antioxidant and anti-inflammatory function of *Gynura procumbens* polyphenols ligand. Int J Mol Sci 22:168716
575. Tan JN, Saffian SM, Buang F, Jubri Z, Jantan I, Husain K, Fauzi NM (2020) Antioxidant and anti-inflammatory effects of genus *Gynura*: a systematic review. Front Pharmacol 11:1
576. Jobaer MA, Ashrafi S, Ahsan M, Hasan, CM, Rashid MA, Islam SN, Masud MM (2023) Phytochemical and biological investigation of an indigenous plant of Bangladesh, *Gynura procumbens* (Lour.) Merr.: drug discovery from nature. Molecules 28:104186
577. Kim MJ, Lee HJ, Wiryowidagdo S, Kim HK (2006) Antihypertensive effects of *Gynura procumbens* extract in spontaneously hypertensive rats. J Med Food 9:587
578. Hoe SZ, Kamaruddin MY, Lam SK (2007) Inhibition of angiotensin-converting enzyme activity by a partially purified fraction of *Gynura procumbens* in spontaneously hypertensive rats. Med Princ Pract 16:203
579. Hoe SZ, Lee CN, Mok SL, Kamaruddin, MY, Lam AZ (2011) *Gynura procumbens* Merr. decreases blood pressure in rats by vasodilatation via inhibition of calcium channels. Clinics 66:143
580. Abrika OSS, Yam MF, Asmawi MZ, Sadikun A, Dieng H, Hussain EA (2013) Effects of extracts and fractions of *Gynura procumbens* on rat atrial contraction. J Acupunct Meridian Stud 6:199
581. Poh TF, Ng HK, Hoe SZ, Lam SK (2013) *Gynura procumbens* causes vasodilation by inhibiting angiotensin II and enhancing bradykinin actions. J Cardiovasc Pharmacol 61:378
582. Zhang XF, Tan BK (2000) Effects of an ethanolic extract of *Gynura procumbens* on serum glucose, cholesterol and triglyceride levels in normal and streptozotocin-induced diabetic rats. Singapore Med J 41:9

583. Algariri K, Meng KY, Atangwho IJ, Asmawi MZ, Sadikun A, Murugaiyah V, Ismail N (2013) Hypoglycemic and anti-hyperglycemic study of *Gynura procumbens* leaf extracts. Asian Pac J Trop Biomed 3:358
584. Bohari SP, Khozirah S, Nordin L (2006) Glucose uptake: stimulatory activity of *Gynura procumbens* in 3T3-F442A adipocytes. In: Malaysian medicinal plant: chemistry and biological activity, UNIMAS and Malaysian Natural Products Society, Kota Samarahan, Malaysia
585. Sari KRP, Nugroho SAE (2015) Effect of herbal combination of *Andrographis paniculata* (Burm. f.) Ness and *Gynura procumbens* (Lour.) methanolic extracts in alloxan-induced hyperglycemic rats. Int Food Res J 22:1332
586. Sani HA, Darus NA, Noor MM, Ismail II (2008) *Gynura procumbens* leaves aqueous extract decreased blood glucose level and increased sperm quality in diabetic-induced rats. Sains Malays 37:435
587. Hakim P, Sani A, Noor MM (2008) Effects of *Gynura procumbens* extract and glibenclamide on sperm quality and specific activity of testicular lactate dehydrogenase in streptozotocin-induced diabetic rats. Malays J Biochem Mol Biol 16:10
588. Agustina D, Haryana S, Supartinah A (2006) Anticarcinogenesis effect of *Gynura procumbens* (Lour) Merr on tongue carcinogenesis in 4NQO-induced rat. Dental J 39:126
589. Wang H, Zhou JW, Fu DH, Zhou Y (2013) *Gynura procumbens* ethanolic extract suppresses osteosarcoma cell proliferation and metastasis in vitro. Oncol Lett 6:113
590. Jenie RI, Meiyanto E (2007) Co-chemotherapy of sambung nyawa (*Gynura procumbens* (Lour.) Merr.) leaves ethanolic extract and doxorubicin on breast cancer cell. Majalah Farm Indonesia 18:81
591. Jenie RI, Meiyanto E, Murwanti R (2006) Antiangiogenic effect of sambung nyawa leaves (*Gynura procumbens* (Lour.) Merr.) ethanolic extract on chick embryo chorioallantoic membrane (CAM). Indonesian J Pharm:50
592. Hamid, Saufi MB, Nik Musaadah, M. (2004) AGRIS - International system for agricultural science and technology: study on antidiabetic properties of *Gynura procumbens* Merr. In: Therapeutique; toxicite; Azucar en sangre. Universiti Malaysia Sabah, Kota Kinabalu, Sabah (Malaysia), p 65
593. Jarikasem S, Charuwichitratana S, Siritantikorn S, Chantratita W, Iskander M, Frahm AW, Jiratchariyakul W (2013) Antiherpetic effects of *Gynura procumbens*. Evid-Based Compl Alt Med 2013:1
594. Maw SS, Mon MM, Oo ZK (2011) Study on antioxidant and antitumor activities of some herbal extracts. Int J Med Health Biomed Bioeng Pharm Eng 2013:394865
595. Kumar S, Pandey AK (2013) Chemistry and biological activities of flavonoids: an overview. Sci World J 162750:1
596. Mahmood AA, Mariod AA, Al-Bayaty F, Abdel-Wahab SI (2010) Anti-ulcerogenic activity of *Gynura procumbens* leaf extract against experimentally induced gastric lesions in rats. J Med Plants Res 4:685
597. Lee HJ, Lee BC, Chung JH, Wiryowidagdo W (2007) Inhibitory effects of an aqueous extract of *Gynura procumbens* on human mesangial cell proliferation. Kor J Physiol Pharmacol 11:145
598. Iskander MN, Song Y, Coupar IM, Jiratchariyakul W (2002) Antiinflammatory screening of the medicinal plant *Gynura procumbens*. Plant Foods Human Nutr 57:233
599. Dwijayanti DR, Rifa'i M (2014) Immunomodulator testing on ethanol extract of *Gynura procumbens* leaves to *Mus musculus* adaptive immune system: in vitro study. Life Sci 4:10
600. Dwijayanti DR, Rifa'i M (2015) *Gynura procumbens* ethanolic extract promotes lymphocyte activation and regulatory T cell generation in vitro. J Trop Life Sci 5:14
601. Susanti D, Sirat HM, Ahmad F, Ali RM, Aimi N, Kitajima M (2007) Antioxidant and cytotoxic flavonoids from the flowers of *Melastoma malabathricum* L. Food Chem 103:710
602. Hanafiah RM, Aisyah S, Ghafar SAA, Ibrahim Z (2018) Anti-bacterial activities of *Melastoma malabathricum* stem bark fractions against *Streptococcus mutans*. Malays J Sci Health Technol 1:52

603. Joffry SM, Yob NJ, Rofiee MS, Redzuan M (2012) *Melastoma malabathricum* (L.) Smith ethnomedicinal uses, chemical constituents, and pharmacological properties: a review. Evid-Based Compl Alt Med 2012:258434
604. Samad NA, Kamal NNSNM, Yahaya N, Aziz MYB, Zain NNM, Yusoff NAM, Lim V (2018) Ethnobotanical, phytochemical, and pharmacological aspects of *Melastoma* sp. Malays J Med Health Sci 14:152
605. Hosni S, Gani SSA, Orsat V, Hassan M, Abdullah S (2023) Ultrasound-assisted extraction of antioxidants from *Melastoma malabathricum* Linn.: modeling and optimization using Box–Behnken design. Molecules 28:487
606. Taha H, Diris MN, Basri AM, Metali F, Ahmad N, Taha H (2016) Phytochemicals and antimicrobial activities of *Melastoma malabathricum* and *Melastoma beccarianum* leaf crude extracts. Res J Phytochem 11:35
607. Mazura MP, Susanti D, Rasadah MA (2007) Anti-inflammatory action of components from *Melastoma malabathricum*. Pharm Biol 45:372
608. Lee NYS, Khoo WKS, Adnan MA, Mahalingam TP, Fernandez AR, Jeevaratnam K (2016) The pharmacological potential of *Phyllanthus niruri*. J Pharm Pharmacol 68:953
609. Meilani R, Asra R, Rivai H (2020) Reviews on ethnopharmacology, phytochemistry, and pharmacology of Meniran (*Phyllanthus niruri* L.). World J Pharm Pharm Sci 9:144
610. Dev A, Sharma A (2023) Pharmacological properties of the magical plant *Phyllanthus niruri* Linn: a review. PEXACY Int J Pharm Sci 2:1
611. Dahanayake J, Kamal Perera P, Galappaththy P, Maheshika Dahanayake J, Arawwawala M (2020) A mini review on therapeutic potential of *Phyllanthus niruri* L. Trends Phytochem Res 4:101
612. Rosidah (2023) A mini-review: potential of *Phyllanthus niruri* L. as an immunostimulator in fish aquaculture. Int J Fish Aquat Stud 11:26
613. Mao X, Wu LF, Guo HL, Chen WJ, Cui YP, Qi Q, Li S, Liang WY, Yang GH, Shao YY, Zhu D, She GM, You Y, Zhang LZ (2016) The genus *Phyllanthus*: an ethnopharmacological, phytochemical, and pharmacological review. Evid-Based Compl Alt Med 2016:7584952
614. Narendra K, Swathi J, Sowjanya KM, Satya K (2012) *Phyllanthus niruri*: a review on its ethnobotanical, phytochemical and pharmacological profile. J Pharm Res 5:4681
615. Nara T, Gleye J, Cerval E, Stanistan E (1977) Flavonoids of *Phyllanthus niruri, Phyllanthus urinaria, Phyllanthus orbiculatus*. Plantes Méd Phytothér 11:82
616. Chauhan JS, Sultan M, Srivastava SK (1977) Two new glycoflavones from the roots of *Phyllanthus niruri*. Planta Med 32:217
617. Row LR, Srinivasulu C (1964) New lignans from *Phyllanthus niruri* Linn. Tetrahedron Lett 24:1557
618. Rajeshkumar NV, Joy KL, Kuttan G, Ramsewak RS, Nair MG, Kuttan R (2002) Antitumour and anticarcinogenic activity of *Phyllanthus amarus* extract. J Ethnopharmacol 81:17
619. Wei WX, Gong XG, Ishrud O, Pan YJ (2002) New lignan isolated from *Phyllanthus niruri* Linn. Structure elucidation by NMR spectroscopy. Bull Kor Chem Soc 23:896
620. Anjaneyulu ASR, Rao KJ, Row LR, Subrahmanyam C (1973) Crystalline constituents of Euphorbiaceae-XII: isolation and structural elucidation of three new lignans from the leaves of *Phyllanthus niruri* Linn. Tetrahedron 29:1291
621. Huang YL, Chen CC, Ou JC (1992) Isolintetralin: a new lignan from *Phyllanthus niruri*. Planta Med 58:473
622. Satyanarayana P, Venkateswarlu S (1991) Isolation, structure and synthesis of new diarylbutane lignans from *Phyllanthus niruri*: synthesis of 5′-desmethoxy-niranthin and an antitumour extractive. Tetrahedron 47:8931
623. Singh B, Agrawal PK, Thakur RS (1989) A new lignan and a new neolignan from *Phyllanthus niruri*. J Nat Prod 52:48
624. Ishimaru K, Yoshimatsu K, Yamakawa T, Kamada H, Shimomura K (1992) Phenolic constituents ın tıssue cultures of *Phyllanthus niruri*. Phytochemistry 31:2015
625. Yong Li, Li X, Wang JK, Kamada H, Shimomura K (2017) Anti-hepatitis B viral activity of *Phyllanthus niruri* L. (Phyllanthaceae) in HepG2/C3A and SK-HEP-1 cells. Trop J Pharm Res 16:1873

626. Ooi KL, Loh SI, Sattar MA, Muhammad TST, Sulaiman SF (2015) Cytotoxic, caspase-3 induction and in vivo hepatoprotective effects of phyllanthin, a major constituent of *Phyllanthus niruri*. J Funct Foods 14:236
627. Sathisha A, Udupa L, Rathnakar UP, Pal PG, Acharya SD, Shastry R (2009) Anti-inflammatory and analgesic activity of *Phyllanthus niruri* in rodent models. Indian Drugs 46:50
628. Porto CRC, Soares LAL, Souza TP, Petrovick PR, Lyra IL, Araújo Júnior RF, Langassner SMZ, Ferreira AAA, Guerra GCB (2013) Anti-inflammatory and antinociceptive activities of *Phyllanthus niruri* spray-dried standardized extract. Rev Bras Farmacogn 23:138
629. Harish R, Shivanandappa T (2006) Antioxidant activity and hepatoprotective potential of *Phyllanthus niruri*. Food Chem 95:180
630. Giribabu N, Rao PV, Kumar KP, Muniandy S, Swapna Rekha S, Salleh N (2014) Aqueous extract of *Phyllanthus niruri* leaves displays in vitro antioxidant activity and prevents the elevation of oxidative stress in the kidney of streptozotocin-induced diabetic male rats. Evid Based Compl Alt Med 2014:834815
631. Jose J, Sudhakaran S, Kumar S, Jayaraman S, Variyar J (2014) Study of in vitro immunomodulatory effect of flavonoid isolated from *Phyllanthus niruri* on human blood lymphocytes and evaluation of its antioxidant potential. Int J Pharmacogn Phytochem Res 6:284
632. Ashraf K, Halim H, Lim SM, Ramasamy K, Sultan S (2020) In vitro antioxidant, antimicrobial and antiproliferative studies of four different extracts of *Orthosiphon stamineus*, *Gynura procumbens* and *Ficus deltoidea*. Saudi J Biol Sci 27:417
633. Corner E (1969) The complex of *Ficus deltoidea*; a recent invasion of the Sunda shelf. Philos Trans R Soc Lond B, Biol Sci 256:281
634. Ken Fern (2022) Tropical Plants Database: *Ficus deltoidea*. https://tropical.theferns.info/viewtropical.php?id=Ficus+deltoidei
635. Bunawan H, Amin NM, Bunawan SN, Baharum SN, Noor MN (2014) *Ficus deltoidea* Jack: a review on its phytochemical and pharmacological importance. Evid-Based Compl Alt Med 2014:902734
636. Sin MH, Mamat AS, Aslam MS, Ahmad MS (2017) Total phenolic content and anti-oxidant potential of *Ficus deltoidea* using green and non-green solvents. J Pharm Negat Res 8:15
637. Ariffin NHM, Hasham R, Hamzah MAAM, Park CS (2024) Skin hydration modulatory activities of *Ficus deltoidea* extract. Fitoterapia 172:105755
638. Burkill IH, Haniff M (1930) Malay village medicine. Garden's bulletin, vol 6, part 2. Botanic Gardens, Singapore
639. Sulaiman MR, Hussain MK, Zakaria ZA, Somchit MN, Moin S, Mohamad AS, Israf DA (2008) Evaluation of the antinociceptive activity of *Ficus deltoidea* aqueous extract. Fitoterapia 79:557
640. Mat-Salleh K, Latif A (2002) Dikotiledon: subkelas hamamelidae. *Ficus deltoidea* Jack. In: Tumbuhan Ubatan Malaysia. Pusat Pengurusan Penyelidikan, Universiti Kebangsaan Malaysia, p 184
641. Hakiman M, Maziah M (2009) Non enzymatic and enzymatic antioxidant activities in aqueous extract of different *Ficus deltoidea* accessions. J Med Plant Res 3:120
642. Dzolin S (2014) Biological activities of four varieties of *Ficus deltoidea* (Jack) and correlation to their flavonoid content. Thesis, Universiti Teknologi MARA, Selangor, Malaysia
643. Hanafi MMM, Alqahtani OSO, Gui L, Yaakob H, Prieto JM (2022) Phytochemical fingerprint and biological activities of three Malaysian *Ficus deltoidea* cultivars. J Nat Prod Disc 2022 1:1
644. Ramamurthy S, Kumarappan C, Dharmalingam SR, Sangeh JK (2014) Phytochemical, pharmacological and toxicological properties of *Ficus deltoidea*: a review of a recent research. Annu Res Rev Biol 4:2357
645. Jamal NAM (2017) Chemical constituents from *Ficus deltoidea* var. *kunstleri* Corner and its antidiabetic property. Thesis, Universiti Pendidikan Sultan Idris, Penang, Malaysia
646. Lip JM, Hisham DN, Zaidi JA, Musa Y, Ahmad AW, Normah A, Sharizan A (2009) Isolation and identification of moretenol from *Ficus deltoidea* leaves (Pemencilan dan pengenalpastian moretenol dari daun mas cotek (*Ficus deltoidea*)). J Trop Agric Food Sci 37:195

647. Suryati NH, Dachriyanus LMNH (2011) Structure elucidation of antibacterial compound from *Ficus deltoidea* Jack leaves. Indonesian J Chem 11:67
648. Choo CY, Sulong NY, Man F, Wong TW (2012) Vitexin and isovitexin from the leaves of *Ficus deltoidea* with in-vivo α-glucosidase inhibition. J Ethnopharmacol 142:776
649. Farsi E, Shafaei A, Hor SY, Khadeer Ahamed MB, Yam MF, Attitalla IH, Asmawi MZ, Ismail Z (2011) Correlation between enzyme inhibitory effects and antioxidant activities of standardized fractions of methanolic extract obtained from *Ficus deltoidea* leaves. Afr J Biotechnol 10:15184
650. Adam Z, Khamis S, Ismail A, Hamid M (2012) *Ficus deltoidea*: a potential alternative medicine for diabetes mellitus. Evid Based Compl Alt Med 2012:632763
651. Lee KW, Kim YJ, Lee HJ, Lee CY (2003) Cocoa has more phenolic phytochemicals and a higher antioxidant capacity than teas and red wine. J Agric Food Chem 51:7292
652. Abdsamah O, Zaidi NT, Sule AB (2012) Antimicrobial activity of *Ficus deltoidea* Jack (Mas Cotek). Pak J Pharm Sci 25:675
653. Azizan N, Mohd Said S, Abidin ZZ, Jantan I (2017) Composition and antibacterial activity of the essential oils of *Orthosiphon stamineus* Benth. and *Ficus deltoidea* Jack against pathogenic oral bacteria. Molecules 22:2135
654. Jamal P, Karim IA, Abdullah E, Raus RA, Hashim YZ (2011) Phytochemical screening for antibacterial activity of potential Malaysian medicinal plants. Afr J Biotechnol 10:18795
655. Zahra MASF, Mahmood AA, Hapipah MA, Suzita MN, Salmah I (2009) Anti-ulcerogenic activity of aqueous extract of *Ficus deltoidea* against ethanol-induced gastric mucosal injury in rats. Res J Med Sci 3:42
656. National Parks (2022) Flora & Fauna web: *Clinacanthus nutans* (Burm. f.) Lindau. https://www.nparks.gov.sg/florafaunaweb/flora/5/6/5655
657. Haida Z, Hakiman M (2019) A review of therapeutic potentials of *Clinacanthus nutans* as source for alternative medicines. Sains Malays 48:2683
658. Aslam MS, Ahmad MS, Mamat AS (2014) A review on phytochemical constituents and pharmacological activities of *Clinacanthus nutans*. Int J Pharm Pharm Sci 7:30
659. Yusuf MSNA, Mood CCNA, Ahmad NH, Sandai D, Lee CK, Lim V (2020) Optimization of biogenic synthesis of silver nanoparticles from flavonoid-rich *Clinacanthus nutans* leaf and stem aqueous extracts. R Soc Open Sci 7:200065
660. Quattrocchi U (2012) CRC World dictionary of medicinal and poisonous plants: common names, scientific names, eponyms, synonyms, and etymology, 1st edn. CRC Press, Boca Raton, FL, USA
661. Kuo X, Herr DR, Ong WY (2021) Anti-inflammatory and cytoprotective effect of *Clinacanthus nutans* leaf but not stem extracts on 7-ketocholesterol induced brain endothelial cell injury. Neuromol Med 23:176
662. Andasari SD, Mustofa CH, Zahra S (2020) Standarisasi spesifik dan non spesifik ekstrak etil asetat daun dandang gendis (*Clinacanthus nutans*). MOTORIK J Kesehatan Sekolah Tinggi Ilmu Kesehatan Muhammadiyah Klaten 15:70
663. Kosai P, Sirisidthi K, Jiraungkoorskul W (2016) Evaluation of total phenolic compound and cytotoxic activity of *Clinacanthus nutans*. Indian J Pharm Sci 78:283
664. Alam A, Ferdosh S, Ghafoor K, Hakim A, Juraimi AS, Khatib A, Sarker ZI (2016) *Clinacanthus nutans*: a review of the medicinal uses, pharmacology and phytochemistry. Asian Pac J Trop Med 9:402
665. Fong SY, Piva T, Urban S, Huynh T (2014) Genetic homogeneity of vegetatively propagated *Clinacanthus nutans* (Acanthaceae). J Med Plants Res 8:903
666. Zulkipli IN, Rajabalaya R, Idris A, Sulaiman NA, David SR (2017) *Clinacanthus nutans*: a review on ethnomedicinal uses, chemical constituents and pharmacological properties. Pharm Biol 55:1093
667. Chia TY, Gan CY, Murugaiyah V, Hashmi SF, Fatima T, Ibrahim L, Abdulla MH, Alswailmi FK, Johns EJ, Ahmad A (2022) A narrative review on the phytochemistry, pharmacology and therapeutic potentials of *Clinacanthus nutans* (Burm. f.) Lindau leaves as an alternative source of future medicine. Molecules 27:1

668. Sookmai W, Ekalaksananan T, Pientong C, Sakdarat S, Kongyingyoes B (2011) The anti-papillomavirus infectivity of *Clinacanthus nutans* compounds. Srinagarind Med J 26:240
669. Pieroni A, Vandebroek I (2007) Traveling cultures and plants: the ethnobiology and ethnopharmacy of human migrations, 1st edn. Berghahn Books, New York, Oxford
670. Watson RR, Preedy VR (2008) Botanical medicine in clinical practice. CABI, São Paulo
671. Ailiah R (2020) Rawatan alternatif moden bagi penyakit kanser: Sabah Snake Grass. https://b17nitrilosides.blogspot.com/2011/12/sabah-snake-grass-extract.html
672. Chelyn JL, Omar MH, Mohd Yousof NSA, Ranggasamy R, Wasiman MI, Ismail Z (2014) Analysis of flavone C-glycosides in the leaves of *Clinacanthus nutans* (Burm. f.) Lindau by HPTLC and HPLC-UV/DAD. Sci World J 2014:724267
673. Huang D, Guo W, Gao J, et al (2015) *Clinacanthus nutans* (Burm. f.) Lindau ethanol extract inhibits hepatoma in mice through upregulation of the immune response. Molecules 20:17405
674. Teshima K-I, Kaneko T, Ohtani K, Kasai R, Lhieochaiphant S, Picheansoonthon C, Yamasaki K (1998) Sulfur-containing glucosides from *Clinacanthus nutans*. Phytochemistry 48:831
675. Tu SF, Liu RH, Cheng YB, Hsu YM, Du YC, El-Shazly M, Wu YC, Chang FR (2014) Chemical constituents and bioactivities of *Clinacanthus nutans* aerial parts. Molecules 19:20382
676. Sakdarat S, Shuyprom A, Ayudhya TDN, Waterman PG, Karagiainis G (2008) Chemical composition investigation of the *Clinacanthus nutans* Lindau leaves. Semantic Scholar Corpus ID:21316952
677. Sakdarat S, Shuyprom A, Pientong C, Ekalaksananan T, Thongchai S (2009) Bioactive constituents from the leaves of *Clinacanthus nutans* Lindau. Bioorg Med Chem 17:1857
678. Dampawan P (1976) Studies of the chemical constituents of *Clinacanthus nutans* (Acanthaceae) and *Zingiber cassumunar* Roxb. Thesis, Mahidol University, Bangkok, Thailand
679. Kangsamaksin T, Chaithongyot S, Wootthichairangsan C, Hanchaina R, Tangshewinsirikul C, Svasti J (2017) Lupeol and stigmasterol suppress tumor angiogenesis and inhibit cholangiocarcinoma growth in mice via downregulation of tumor necrosis factor-α. PLoS One 12:e0189628
680. Kamarudin MNA, Sarker MMR, Kadir HA, Ming LC (2017) Ethnopharmacological uses, phytochemistry, biological activities, and therapeutic applications of *Clinacanthus nutans* (Burm. f.) Lindau: a comprehensive review. J Ethnopharmacol 206:245
681. Nazanin S, Crawford GH (2006) Botanical briefs: Ylang-Ylang Oil—extracts from the tree *Cananga odorata*. Cutis 77:149
682. Nurhayani FO, Wulandari AS, Suharsi KT (2019) Morphology and anatomy of the fruit and seed of *Cananga odorata* (Lam.) Hook. f. et Thompson. Biodiversitas 20:1112
683. Makrjras OC (2009) *Cananga odorata*. Agofor Database 4:1
684. Ong HC (2004) Tumbuhan Liar: Khasiat & Kegunaan Lain. Utusan Publi Distr Sdn. Bhd. http://myagric.upm.edu.my/id/eprint/11083
685. Manner HI, Elevitch CR (2006) *Cananga odorata* (ylang-ylang) species profiles for Pacific Island agroforestry. www.traditionaltree.org
686. Turner IM, Veldkamp JF (2009) A history of *Cananga* (Annonaceae). Gard Bull Singapore 61:189
687. Adnan M, Zainuddin AF, Hamzah MA, Moorthy M, Isa MZM (2018) Koleksi Pokok Taman Botani Kepong. Institut Penyelidikan Perhutanan Malaysia (FRIM), Malaysia 20:1
688. Yaacob M, Kadir AA, Hasan Z (2009) Tumbuhan Ubatan Popular. Institut Penyelidikan dan Kemajuan Pertanian Malaysia (MARDI) Serdang, Selingor, Malaysia, p 182
689. Parrotta JA (2014) *Cananga odorata*. Enzyklopädie der Holzgewächse: Handbuch und Atlas der Dendrologie. Wiley-VCH Verlag GmbH, Weinheim, Germany
690. Qin XW, Hao CY, He SZ, Wu G, Tan LH, Xu F, Hu RS (2014) Volatile organic compound emissions from different stages of *Cananga odorata* flower development. Molecules 19:8965
691. Tan LTH, Lee LH, Yin WF, Chan CK, Kadir AH, Chan KG, Goh BH (2015) Traditional uses, phytochemistry, and bioactivities of *Cananga odorata* (Ylang-Ylang). Evid-Based Compl Alt Med 2015:896314
692. Gaydou EM, Randriamiharisoa R, Bianchini JP (1986) Composition of the essential oil of Ylang-Ylang (*Cananga odorata* Hook. fil. et Thomson *forma genuina*) from Madagascar. J Agric Food Chem 34:481

693. Stashenko E, Martinez JR, MacKu C, Shibamoto T (1993) HRGC and GC–MS analysis of essential oil from Colombian ylang-ylang (*Cananga odorata* Hook fil. et Thomson, *forma genuina*). J High Res Chromatogr 16:441
694. Brokl M, Fauconnier ML, Benini C, Lognay G, Jardin P, Focant JF (2013) Improvement of Ylang-Ylang essential oil characterization by GC×GC-TOFMS. Molecules 18:1783
695. Giang PM, Son PT (2016) GC and GC-MS analysis of the fresh flower essential oil of *Cananga odorata* (Lam.) Hook. f. et Th. var. *fruticosa* (Craib) J Sincl. Am J Ess Oils Nat Prod 4:9
696. Jin J, Kim MJ, Dhandapani S, Tjhang JG, Yin JL, Wong L, Sarojam R, Chua NH, Jang IC (2015) The floral transcriptome of ylang ylang (*Cananga odorata* var. *fruticosa*) uncovers biosynthetic pathways for volatile organic compounds and a multifunctional and novel sesquiterpene synthase. J Exp Bot 66:3959
697. Rao JUM, Giri GS, Hanumaiah T, Rao KVJ (1986) Sampangine, a new alkaloid from *Cananga odorata*. J Nat Prod 49:346
698. Woo SH, Reynolds MC, Sun NJ, Cassady JM, Snapka RM (1997) Inhibition of topoisomerase II by liriodenine. Biochem Pharmacol 54:467
699. Hsieh T, Chang F, Wu Y (1999) The constituents of *Cananga odorata*. J Chin Chem Soc 46:607
700. Hsieh TJ, Chang FR, Chia YC, Chen CY, Chiu HF, Wu YC (2001) Cytotoxic constituents of the fruits of *Cananga odorata*. J Nat Prod 64:616
701. Ragasa CY, Tamboong B, Rideout J (2003) Secondary metabolites from *Jasminum sambac* and *Cananga odorata*. ACGC Chem Res Commun 16:40
702. Caloprisco E, Fourneron JD, Faure R, Demarne FE (2002) Unusual lactones from *Cananga odorata* (Annonaceae). J Agric Food Chem 50:78
703. Matsunami K, Nagashima J, Sugimoto S, Otsuka H, Takeda Y, Lhieochaiphant D, Lhieochaiphant S (2010) Megastigmane glucosides and an unusual monoterpene from the leaves of *Cananga odorata* var. *odorata*, and absolute structures of megastigmane glucosides isolated from *C. odorata* var. *odorata* and *Breynia officinalis*. J Nat Med 64:460
704. Matsumoto T, Nakamura S, Nakashima S, Fujimoto K, Yoshikawa M, Otha T, Ogawa K, Matsuda H (2014) Lignan dicarboxylates and terpenoids from the flower buds of *Cananga odorata* and their inhibitory effects on melanogenesis. J Nat Prod 77:990
705. Matsumoto T, Nakamura S, Fujimoto K, Ohta T, Ogawa K, Yoshikawa M, Matsuda H (2014) Structure of constituents isolated from the flower buds of *Cananga odorata* and their inhibitory effects on aldose reductase. J Nat Prod 68:709
706. Nagashima J, Matsunami K, Otsuka H, Lhieochaiphant D, Lhieochaiphant S (2010) The unusual canangafruticosides A–E: five monoterpene glucosides, two monoterpenes and a monoterpene glucoside diester of the aryldihydronaphthalene lignan dicarboxylic acid from leaves of *Cananga odorata* var. *fruticosa*. Phytochemistry 71:1564
707. Rahman MM, Lopa SS, Sadik G, Rashid H, Islam R, Khondkar P, Alam AHMK, Rashid MA (2005) Antibacterial and cytotoxic compounds from the bark of *Cananga odorata*. Fitoterapia 76:758
708. Kusuma IW, Murdiyanto AET, Syafrizal KY (2014) Antimicrobial and antioxidant properties of medicinal plants used by the Bentian tribe from Indonesia. Food Sci Human Wellness 3:191
709. Niaci S (2021) Phytochemical properties of methanol and ethyl acetate extracts of *Cananga odorata* flowers and their pharmacological activities. J Med Pharm Allied Sci 10:3874
710. Indrakumar I, Selvi V, Gomathi R, Karpagam S (2012) Evaluation of antimicrobial activity of *Cananga odorata* (Lam.) Hook. f. & Thomson leaf extract: an in vitro study. Mintage J Pharm Med Sci 1:21
711. Murbach Teles Andrade BF, Nunes Barbosa L, da Silva PI, Fernandes Júnior A (2014) Antimicrobial activity of essential oils. J Essent Oil Res 26:34
712. Lee JH, Lee JS (2010) Chemical composition and antifungal activity of plant essential oils against *Malassezia furfur*. Microbiol Biotechnol Lett 38:315
713. Kon K, Rai M (1970) Antibacterial activity of *Thymus vulgaris* essential oil alone and in combination with other essential oils. Nusantara Biosci 4:202

714. Bilcu M, Grumezescu A, Oprea A, Popescu R, Mogoşanu G, Hristu R, Stanciu G, Mihailescu D, Lazar V, Bezirtzoglou E, Chifiriuc M (2014) Efficiency of vanilla, patchouli and Ylang Ylang essential oils stabilized by iron Oxide@C14 nanostructures against bacterial adherence and biofilms formed by *Staphylococcus aureus* and *Klebsiella pneumoniae* clinical strains. Molecules 19:17943
715. Sacchetti G, Maietti S, Muzzoli M, Scaglianti M, Manfredini S, Radice M, Bruni R (2005) Comparative evaluation of 11 essential oils of different origin as functional antioxidants, antiradicals and antimicrobials in foods. Food Chem 91:621
716. Wang HF, Wang YK, Yih KH (2008) DPPH free-radical scavenging ability, total phenolic content, and chemical composition analysis of forty-five kinds of essential oils. J Cosmet Sci 59:509
717. Choi EM, Hwang JK (2005) Screening of Indonesian medicinal plants for inhibitor activity on nitric oxide production of RAW264.7 cells and antioxidant activity. Fitoterapia 76:194
718. Phasomkusolsil S, Soonwera M (2011) Efficacy of herbal essential oils as insecticide against *Aedes aegypti* (Linn.), *Culex quinquefasciatus* (Say) and *Anopheles dirus* (Peyton and Harrison). Southeast Asian J Trop Med Public Health 42:1083
719. Siriporn P, Mayura S (2012) The effects of herbal essential oils on the oviposition-deterrent and ovicidal activities of *Aedes aegypti* (Linn.), *Anopheles dirus* (Peyton and Harrison) and *Culex quinquefasciatus* (Say). Trop Biomed 29:138
720. Phasomkusolsil S, Soonwera M (2011) Comparative mosquito repellency of essential oils against *Aedes aegypti* (Linn.), *Anopheles dirus* (Peyton and Harrison) and *Culex quinquefasciatus* (Say). Asian Pac J Trop Biomed 1:S113
721. Fradin MS, Day JF (2002) Comparative efficacy of insect repellents against mosquito bites. New Engl J Med 347:13
722. Mayura S, Siriporn P (2014) Mosquito repellent from Thai essential oils against dengue fever mosquito (*Aedes aegypti* (L.)) and filarial mosquito vector (*Culex quinquefasciatus* (Say)). Afr J Microbiol Res 8:1819
723. Soonwera M, Phasomkusolsil S (2015) Efficacy of Thai herbal essential oils as green repellent against mosquito vectors. Acta Trop 142:127
724. Verbel OJ, Ballestas TI, Gallardo CK, Stashenko EE (2013) Essential oils applied to the food act as repellents toward *Tribolium castaneum*. J Stored Prod Res 55:145
725. Gallardo CK, Verbel OJ, Stashenko EE (2011) Repellent activity of essential oils and some of their individual constituents against *Tribolium castaneum* Herbst. J Agric Food Chem 59:1690
726. Vera SS, Zambrano DF, Sanchez MSC, Sanabria RF, Stashenko EE, Duque Luna JE (2014) Essential oils with insecticidal activity against larvae of *Aedes aegypti* (Diptera: Culicidae). Parasitol Res 113:2647
727. Cheng J, Yang K, Zhao NN, Wang XG, Wang SY, Liu ZL (2012) Composition and insecticidal activity of the essential oil of *Cananga odorata* leaves against *Sitophilus zeamais* Motschulsky (Coleoptera: Curculionidae). J Med Plants Res 6:3568
728. Efdi M, Ohguchi K, Akao Y, Nozawa Y, Koketsu M, Ishihara H (2007) *N-trans*-Feruloyltyramine as a melanin biosynthesis inhibitor. Biol Pharm Bull 30:1972
729. Lin CC, Yang CH, Wu PS, Kwan CC, Chen YS (2011) Antimicrobial, anti-tyrosinase and antioxidant activities of aqueous aromatic extracts from forty-eight selected herbs. J Med Plants Res 5:6203
730. Wei A, Shibamoto T (2010) Antioxidant/lipoxygenase inhibitory activities and chemical compositions of selected essential oils. J Agric Food Chem 58:7218
731. Peana AT, D'Aquila PS, Panin F, Serra G, Pippia P, Moretti MDL (2002) Anti-inflammatory activity of linalool and linalyl acetate constituents of essential oils. Phytomedicine 9:721
732. Yasmeen A, Maniyar CH, Devi, J (2015) Evaluation of anti-inflammatory activity of ethanolic extract of *Cananga odorata* Lam. in experimental animals. Int J Basic Clin Pharmacol 4:354
733. Pujiarti R, Ohtani Y, Widowati TB, Wahyudi W, Kasmudjo K, Herath NK, Wang CN (2017) Effect of *Melaleuca leucadendron*, *Cananga odorata* and *Pogostemon cablin* oil odors on human physiological responses. Wood Res J 4:100

734. Hongratanaworakit T, Buchbauer G (2004) Evaluation of the harmonizing effect of Ylang-Ylang oil on humans after inhalation. Planta Med 70:632
735. Jung DJ, Cha JY, Kim SE, Ko IG, Jee YS (2013) Effects of Ylang-Ylang aroma on blood pressure and heart rate in healthy men. J Exerc Rehabil 9:250
736. Moss M, Hewitt S, Moss L, Wesnes K (2008) Modulation of cognitive performance and mood by aromas of peppermint and Ylang-ylang. Int J Neurosc 118:59
737. Ishiguchi A, Saitou A, Suenaga K, Ohta K, Matsuura M (2008) 14. Effects of odors on electroencephalogram and subjective alterations. Clin Neurophysiol 119:e78
738. Watanabe S, Hara K, Ohta K, Iino H, Miyajima M, Matsuda A, Hara M, Maehara T, Matsuura M, Matsushima E (2013) Aroma helps to preserve information processing resources of the brain in healthy subjects but not in temporal lobe epilepsy. Seizure 22:59
739. Anitha P, Indira M (2007) Impact of feeding ethanolic extract of root bark of *Cananga odorata* (Lam.) on reproductive functions in male rats. Ind J Exp Biol 44:976
740. Pankajakshy A, Madambath I (2009) Spermatotoxic effects of *Cananga odorata* (Lam.): a comparison with gossypol. Fertil Steril 91:2243
741. Ridhwan MJM, Musa'adah NM, Ahmad R, Rashid A, Ismail NH (2011) In vitro antihyperglycaemic effects of some Malaysian plants. J Trop For Sci 23:467
742. Indrasetiawan P, Utsubo AC, Hanafi M, Hartati S, Sri Wahyuni T, Kameoka M, Yano Y, Hotta H, Hayashi Y (2019) Antiviral activity of *Cananga odorata* against hepatitis B virus. Kobe J Med Sci 65:71
743. Sriboonlert J, Munkong W, Rintawut S, Paladkhua S, Suwongsa R, Kirisattayakul W (2023) *Cananga odorata* aromatherapy reduces anxiety in unexperienced patients hospitalized for interventional neuroradiology procedures: a randomized control trial. J Evid-Based Integr Med 28:2515690X221150527
744. Sedky NK, Kader ANM, Issa MY, Abdelhady MMM, Shamma SN, Bakowsky U, Fahmy SA (2023) Co-delivery of Ylang Ylang oil of *Cananga odorata* and oxaliplatin using intelligent pH-sensitive lipid-based nanovesicles for the effective treatment of triple-negative breast cancer. Int J Mol Sci 24:8392
745. Yahaya H, Ahmad PN (2005) Limau purut (*Citrus hystrix*). In: Penanaman tumbuhan ubatan & beraroma. MARDI, Serdang, Selangor, Malaysia, p 108
746. Lim TK (2012) *Citrus hystrix*. In: Edible medicinal and non-medicinal plants. Springer Netherlands, Dordrecht, p 634
747. Abirami A, Nagarani G, Siddhuraju P (2014) In vitro antioxidant, anti-diabetic, cholinesterase and tyrosinase inhibitory potential of fresh juice from *Citrus hystrix* and *C. maxima* fruits. Food Sci Human Wellness 3:16
748. Othman MS, Hassan M, Nahar L, Basar N, Jamil S, Sarker S (2016) Essential oils from the Malaysian Citrus (Rutaceae) medicinal plants. Medicines 3:13
749. Youkwan J, Sutthivaiyakit S, Sutthivaiyakit P (2010) Citrusosides A–D and furanocoumarins with cholinesterase inhibitory activity from the fruit peels of *Citrus hystrix*. J Nat Prod 73:1879
750. Anuchapreeda S, Anzawa R, Viriyaadhammaa N, Neimkhum W, Chaiyana W, Okonogi S, Usuki T (2020) Isolation and biological activity of agrostophillinol from kaffir lime (*Citrus hystrix*) leaves. Bioorg Med Chem Lett 30:127256
751. Murakami A, Gao G, Kim OK, Omura M, Yano M, Ito C, Furukawa H, Jiwajinda S, Koshimizu K, Ohigashi H (1999) Identification of coumarins from the fruit of *Citrus hystrix* DC. as inhibitors of nitric oxide generation in mouse macrophage RAW 264.7 cells. J Agric Food Chem 47:333
752. Sun S, Phrutivorapongkul A, Dibwe DF, Balachandran C, Awale S (2018) Chemical constituents of Thai *Citrus hystrix* and their antiausterity activity against the PANC-1 human pancreatic cancer cell line. J Nat Prod 81:1877l
753. Umran NSS, Mohamed S, Lau SF, Mohd Ishak NI (2020) *Citrus hystrix* leaf extract attenuated diabetic-cataract in STZ-rats. J Food Biochem 44:e13258
754. Butryee C, Sungpuag P, Chitchumroonchokchai C (2009) Effect of processing on the flavonoid content and antioxidant capacity of *Citrus hystrix* leaf. Int J Food Sci Nutr 60:162

755. Sadasivam M, Kumarasamy C, Thangaraj A, Govindan M, Kasirajan G, Vijayan V, Devadasan V, Chia-Her L, Madhusudhanan GR, Ramaraj T, Palathurai Subramaniam M (2018) Phytochemical constituents from dietary plant *Citrus hystrix*. Nat Prod Res 32:1721
756. Murakami A, Nakamura Y, Koshimizu K, Ohigashi H (1995) Glyceroglycolipids from *Citrus hystrix*, a traditional herb in Thailand, potently inhibit the tumor-promoting activity of 12-*O*-tetradecanoylphorbol 13-acetate in mouse skin. J Agric Food Chem 43:2779
757. Panthong K, Srisud Y, Rukachaisirikul V, Towatana HN, Voravuthikunchai SP, Tewtrakul S (2013) Benzene, coumarin and quinolinone derivatives from roots of *Citrus hystrix*. Phytochemistry 88:79
758. Waikedre J, Dugay A, Barrachina I, Herrenknecht C, Cabalion P, Fournet A (2010) Chemical composition and antimicrobial activity of the essential oils from New Caledonian *Citrus macroptera* and *Citrus hystrix*. Chem Biodivers 7:871
759. Wongsariya K, Phanthong P, Bunyapraphatsara N, Srisukh V, Chomnawang MT (2014) Synergistic interaction and mode of action of *Citrus hystrix* essential oil against bacteria causing periodontal diseases. Pharm Biol 52:273
760. Borusiewicz M, Trojanowska D, Paluchowska P, Janeczko ZW, Petitjean M, Budak A (2017) Cytostatic, cytotoxic, and antibacterial activities of essential oil isolated from *Citrus hystrix*. ScienceAsia 43:96
761. Pumival P, Tadtong S, Athikomkulchai S, Chittasupho C (2020) Antifungal activity and the chemical and physical stability of microemulsions containing *Citrus hystrix* DC. leaf oil. Nat Prod Commun 2020:15
762. Ulhaq ZS, Hendyatama TH, Hameed F, Santosaningsih D (2021) Antibacterial activity of *Citrus hystrix* toward *Salmonella* spp. infection. Enferm Infecc Microbiol Clin (English ed) 39:283
763. Singhapol C, Tinrat S (2020) Virulence genes analysis of *Vibrio parahaemolyticus* and anti-vibrio activity of the *Citrus* extracts. Curr Microbiol 77:1390
764. Chit-aree L, Unpaprom Y, Ramaraj R, Thirabunyanon M (2023) Valorization and biorefinery of kaffir lime peels waste for antifungal activity and sustainable control of mango fruit anthracnose. Biomass Convers Biorefin 13:10735
765. Sreepian PM, Rattanasinganchan P, Sreepian A (2023) Antibacterial efficacy of *Citrus hystrix* (makrut lime) essential oil against clinical multidrug-resistant methicillin-resistant and methicillin-susceptible *Staphylococcus aureus* isolates. Saudi Pharm J 31:1094
766. Nararak J, Sathantriphop S, Kongmee M, Bangs MJ, Chareonviriyaphap T (2017) Excito-repellency of *Citrus hystrix* DC. leaf and peel essential oils against *Aedes aegypti* and *Anopheles minimus* (Diptera: Culicidae), vectors of human pathogens. J Med Entomol 54:178
767. Nishan M, Subramanian P, Saahatish R (2017) Toxicity of *Citrus aurantifolia* and *Citrus hystrix* against *Aedes albopictus* larvae. Int J Biosci 10:48
768. Butryee C, Kupradinun P (2008) Antioxidant capacity of *Citrus hystrix* leaf using in vitro methods and their anticlastogenic potential using the erythrocyte micronucleus assay in the mouse. Toxicol Lett 180:S79
769. Ali M, Akhter R, Narjish SN, Shahriar M, Bhuiyan MA (2015) Studies of preliminary phytochemical screening, membrane stabilizing activity, thrombolytic activity and in-vitro antioxidant activity of leaf extract of *Citrus hystrix*. Int J Pharm Sci Res 6:2367
770. Ratanachamnong P, Chunchaowarit Y, Namchaiw P, Niwaspragrit C, Rattanacheeworn P, Jaisin Y (2023) HPLC analysis and in vitro antioxidant mediated through cell migration effect of *C. hystrix* water extract on human keratinocytes and fibroblasts. Heliyon 9:e13068
771. Tunjung WAS, Cinatl J, Michaelis M, Smales CM (2015) Anti-cancer effect of Kaffir Lime (*Citrus hystrix* DC.) leaf extract in cervical cancer and neuroblastoma cell lines. Procedia Chem 14:465
772. Ho Y, Suphrom N, Daowtak K, Potup P, Thongsri Y, Usuwanthim K (2020) Anticancer effect of *Citrus hystrix* DC. leaf extract and its bioactive constituents citronellol and citronellal on the triple negative breast cancer MDA-MB-231 cell line. Pharmaceuticals 13:476
773. Anuchapreeda S, Chueahongthong F, Viriyaadhammaa N, Panyajai P, Anzawa R, Tima S, Ampasavate C, Saiai A, Rungrojsakul M, Usuki T, Okonogi S (2020) Antileukemic

cell proliferation of active compounds from Kaffir Lime (*Citrus hystrix*) leaves. Molecules 25:1300

774. Sammi SR, Mishra DP, Trivedi S, Smita SS, Nagar A, Tandon S, Pandey R (2016) *Citrus hystrix*-derived 3,7-dimethyloct-6-enal and 3,7-dimethyloct-6-enyl acetate ameliorate acetylcholine deficits. RSC Adv 6:68870
775. Pattarachotanant N, Tencomnao T (2020) *Citrus hystrix* extracts protect human neuronal cells against high glucose-induced senescence. Pharmaceuticals 13:283
776. Koh D, Ong CN (1999) Phytophotodermatitis due to the application of *Citrus hystrix* as a folk remedy. Br J Dermatol 140:737
777. Hongratanaworakit T, Buchbauer G (2007) Chemical composition and stimulating effect of *Citrus hystrix* oil on humans. Flavour Fragr J 22:443
778. Azmil ARI, Ahmad AW, Omar T (2005) Kesum (*Polygonum minus* Huds). In: Penanaman tumbuhan ubatan & beraroma. MARDI, Serdang, Salangor, Malaysia, p 102
779. Bunawan H, Talip N, Noor NM (2011) Foliar anatomy and micromorphology of *Polygonum minus* Huds. and their taxonomic implications. Austral J Crop Sci 5:123
780. Ong HC, Nordiana M (1999) Malay ethno-medico botany in Machang, Kelantan, Malaysia. Fitoterapia 70:502
781. Anonymous (2002) Compendium of medicinal plants used in Malaysia. HMRC-IMR, Kuala Lumpur
782. Hassan WW (2007) Healing herbs of Malaysia. Federal Land Development Authority (FELDA), Kuala Lumpur
783. Rashid AN, Hussan F, Hamid A, Ridzuan ANR, Halim SASA, Jalil ANA, Najib NHM, Teoh SL, Budin SB (2020) *Polygonum minus* essential oil modulates cisplatin-induced hepatotoxicity through inflammatory and apoptotic pathways. EXCLI J 19:1246
784. Ali AM, Mackeen MM, Ei-Sharkawy SH, Hamid JA, Ismail NH, Ahmad FH, Lajis NH (1996) Antiviral and cytotoxic activities of some plants used in Malaysian indigenous medicine. Pertanika J Soc Sci Humanit 19:129
785. Christapher P, Xin T, Kiun C, Leng L, Fu N, Yuan G, Parasuraman S, Vikneswaran M (2015) Evaluation of analgesic, anti-inflammatory, antipyretic and antiulcer effect of aqueous and methanol extracts of leaves of *Polygonum minus* Huds. (Polygonaceae) in rodents. Arch Med Health Sci 3:12
786. Abdullah MZ, Mohd Ali J, Abolmaesoomi M, Abdul-Rahman PS, Hashim OH (2017) Anti-proliferative, in vitro antioxidant, and cellular antioxidant activities of the leaf extracts from *Polygonum minus* Huds.: effects of solvent polarity. Int J Food Prop 20:846
787. Urones JG, Marcos IS, Pérez BG, Barcala PB (1990) Flavonoids from *Polygonum minus*. Phytochemistry 29:3687
788. Ahmad R, Sahidin I, Taher M, Low C, Noor NM, Sillapachaiyaporn C, Chuchawankul S, Sarachana T, Tencomnao T, Iskandar F, Rajab NF, Baharum SN (2018) Polygonumin A, a newly isolated compound from the stem of *Polygonum minus* Huds. with potential medicinal activities. Sci Rep 8:4202
789. Ahmad R, Rosandy AR, Sahidin I, Ghani ANS, Noor NM, Baharum SN (2022) Bioassay analysis and molecular docking study revealed the potential medicinal activities of active compounds polygonumins B, C and D from *Polygonum minus* (*Persicaria minor*). Plants 12:59
790. Vimala S, Rohana S, Rashih AA, Juliza M (2011) Antioxidant evaluation in Malaysian medicinal plant: *Persicaria minor* (Huds.) leaf. Sci J Med Clin Trials 1:9
791. Wasman SQ, Mahmood AA, Salehhuddin H, Zahra AA, Salmah I (2010) Cytoprotective activities of *Polygonum minus* aqueous leaf extract on ethanol-induced gastric ulcer in rats. J Med Plants Res 4:2658
792. Ibrahim WWI, Sidik K, Kuppusamy UR (2010) A high antioxidant level in edible plants is associated with genotoxic properties. Food Chem 122:1139
793. Qader SW, Abdulla MA, Chua LS, Najim N, Zain MM, Hamdan S (2011) Antioxidant, total phenolic content and cytotoxicity evaluation of selected Malaysian plants. Molecules 16:3433

794. Korulkin DY, Muzychkina MRA (2015) Essential oils of *Polygonum* L. plants growing in Kazakhstan and their antibacterial and antifungal activity. World Acad Sci Eng Technol 9:757
795. Hodges RE, Minich DM (2015) Modulation of metabolic detoxification pathways using foods and food-derived components: a scientific review with clinical application. J Nutr Metab 2015:760689
796. Thanapakiam G, Lim SY, Grace KYX, Loo SW, Amutha VV, Pavitra L, Neoh HF, Ahmad HR, Christapher PV (2016) PP25—effect of aqueous and methanol extracts *Polygonum minus* leaves on drug-induced hepatotoxicity in rats. In: Proceedings of the 3rd Regional conference on biosensors, biodiagnostics, biochips and biotechnology. AIMST University, Bedong, Kedah, Malaysia
797. Christapher PV, Parasuraman S, Asmawi Mohd Z, Murugaiyah V (2017) Acute and subchronic toxicity studies of methanol extract of *Polygonum minus* leaves in Sprague Dawley rats. Reg Toxicol Pharmacol 86:33
798. Hassan KZ, Noor HM and Kader J (2015) Antibacterial efficacy of three different extracts of *Polygonum minus* (Huds.). In: International conference on waste management, ecology and biological sciences, Kuala Lumpur, p 41
799. Johnny L, Yusuf UK, Nulit R (2011) Antifungal activity of selected plant leaves crude extracts against a pepper anthracnose fungus, *Colletotrichum capsici* (Sydow) Butler and Bisby (Ascomycota: Phyllachorales). Afr J Biotechnol 10:4157
800. Dewi S, Asseggaf SN, Natalia D, Mahyarudin M (2019) Efek ekstrak etanol daun Kesum (*Polygonum minus* Huds.) sebagai antifungi terhadap *Trichophyton rubrum*. J Kesehatan Andalas 198:203
801. Qader SW, Abdulla MA, Chua LS, Sirat HM, Hamdan S (2012) Pharmacological mechanisms underlying gastroprotective activities of the fractions obtained from *Polygonum minus* in Sprague Dawley rats. Int J Mol Sci 13:1481
802. Nurul H, Ruzita H, Aronal AP (2010) The antioxidant effects of *Cosmos caudatus* and *Polygonum minus* in refrigerated duck meatballs. Int Food Res J 17:893
803. Zakaria N, Baba S, Ku Bahaudin KNA, Hamdana S (2015) Total phenolic content and antioxidant activity of pure and formulated extracts of kesum (*Polygonum minus*), bawang putih (*Allium sativum*), pegaga (*Centella asiatica*), and ulam raja (*Cosmos caudatus*). Malays J Fund Appl Sci 179:183
804. Khalid NM, Babji AS (2018) Antioxidative and antihypertensive activities of selected Malaysian ulam (salad), vegetables and herbs. J Food Res 27:37
805. Hussin M, Abdul Hamid A, Abas F, Ramli NS, Jaafar AH, Roowi S, Majid AN, Dek PMS (2019) NMR-based metabolomics profiling for radical scavenging and anti-aging properties of selected herbs. Molecules 24:3208
806. George A, Chinnappan S, Chintamaneni M, Kotak CV, Choudhary Y, Kueper T, Radhakrishnan AK (2014) Anti-inflammatory effects of *Polygonum minus* (Huds.) extract (Lineminus^{TM}) in in-vitro enzyme assays and carrageenan induced paw edema. BMC Compl Alt Med 14:355

Nurhuda Manshoor a Malaysian academic, obtained her Ph.D. degree in Natural Product Chemistry from Universiti Teknologi MARA Malaysia (UiTM), supervised by Professor Jean-Frédéric Faizal Weber. She has been a Lecturer at the Faculty of Pharmacy, UiTM, since 2009 and a Research Fellow at the Atta-ur-Rahman Institute for Natural Product Discovery (AuRIns), Malaysia, since 2011. She has undertaken postdoctoral fellowships at the University of Geneva, Switzerland, and Tokushima Bunri University, Japan. She recently completed a sabbatical at the Royal Botanic Gardens, Kew, UK, collaborating on phytochemical research. Dr. Manshoor specializes in pharmaceutical analysis and pharmacognosy, and teaches undergraduate courses on analytical techniques, spectroscopy, organic chemical reactions, and pharmacognosy, emphasizing the study of natural medicines and phytochemicals. Her research interests focus on phytochemistry, particularly oligostilbenes and phenolic compounds in plants of the family Dipterocarpaceae, and the development of modern chromatographic methodologies for isolating plant secondary metabolites. She is also involved in dereplication strategies for rapid compound identification using LC-MS and database comparisons and metabolomic fingerprinting to identify plant constituents using chemometric analysis and mass spectrometric data. Dr. Manshoor is a member of the Malaysian Natural Product Society, the Groupe Polyphénols, and the Malaysian Metabolomics Society. Her contributions reflect a commitment to advancing pharmaceutical education and research while promoting the study and application of natural products.

Fatimah Salim is an Associate Professor specializing in natural product chemistry. She holds B.Sc., M.Sc., and Ph.D. degrees in Natural Product Chemistry from Universiti Teknologi MARA (UiTM), Malaysia, where she also completed a postdoctoral fellowship. Currently, she serves as the Director of the Atta-ur-Rahman Institute for Natural Product Discovery (AuRIns), a Centre of Excellence at UiTM. Her research focuses on identifying and characterizing natural products from plants and microorganisms using advanced chromatographic and spectroscopic techniques, including LC-MS molecular networking. Dr. Salim explores Malaysian ethnobotanical plants for applications in pharmaceuticals, functional foods, and agriculture. She holds leadership roles in the Malaysian Institute of Chemistry (IKM), the Malaysian Natural Product Society (MNPS), and the Academy of Sciences Malaysia (ASM). She is also a member of the Society for Medicinal Plant and Natural Product Research (GA). Dr. Salim has received several awards, including the 2021 Academic Young Figure Award from UiTM Dengkil and the 2023 MNPS Young Researcher Excellence Award. In 2022, she was named one of the "20 Most Influential Women" by AURA Mingguan Wanita for her public engagement in chemistry. Dr. Salim has published 55 journal articles and four book chapters in the field of natural product chemistry.

Che Puteh Osman is a Malaysian academic and obtained her Ph.D. degree in Natural Product Chemistry from Universiti Teknologi MARA. In 2013, she became a Senior Lecturer at the Universiti Teknologi MARA Shah Alam. She joined the Atta-ur-Rahman Institute for Natural Product Discovery as an Associate Fellow in 2016 and became a Research Fellow in 2018, and is currently the Deputy Director. She teaches natural product chemistry and advanced spectroscopy and has supervised more than 10 postgraduate students. Her research interests include the identification of bioactive compounds from natural products, herbal product analysis, and computational chemistry. She works on the purification and structural elucidation of complex natural products using chromatography and organic spectroscopy, especially NMR, IR, and MS. She has published in over 30 indexed journals, and has presented more than 20 conference proceedings on natural products. Dr. Osman has obtained several awards during her academic career including International Travel Grant Awards and her University's Academic Award. She is a member of the Malaysian Institute of Chemistry and Assistant Secretary of the Malaysian Natural Products Society. She occasionally writes for outlets including social media and newspapers and also appears in the mass media to spread awareness on natural products and traditional medicine.

Norizan Ahmat is a Professor at the Universiti Teknologi MARA. She obtained her B.S. degree in Chemistry from Arkansas State University, USA and then continued her M.Sc. and Ph.D. degrees at the Universiti Kebangsaan Malaysia in the field of Chemistry and Phytochemistry, respectively. She became a Fellow the of Atta-ur-Rahman Institute for Natural Product Discovery (AuRIns) in 2013. She is now an Associate Fellow of AuRIns. She teaches organic chemistry, spectroscopy, and other chemistry-related subjects. She has graduated more than 30 postgraduate students (M.Sc. and Ph.D.) as the supervisor or co-supervisor. She is a life member of the Malaysian Natural Product Society (MNPS) and a member of the Society for Medicinal Plants and Natural Product Research (GA) and the Phytochemical Society of Asia (PSA). Her research is in the field of phytochemistry (isolation, elucidation, bioassay and molecular networking) on the plant families Dipterocarpaceae, Euphorbiaceae, and others. Prof. Ahmat has published several publications related to natural products, including their synthesis as well as bioassay testing in "Phytochemistry", the "European Journal of Medicinal Chemistry", the "Journal of Ethnopharmacology", "Fitoterapia", "Bioorganic Chemistry", "Bioorganic and Medicinal Chemistry", and "Biosystematics and Ecology".

Aisyah Salihah Kamarozaman graduated from Universiti Kebangsaan Malaysia with a Bachelor's (Honors) degree in Science (Chemistry) in 2006 and a M.Sc. (Chemistry) degree in 2010. In 2011, she was appointed as a Lecturer at Universiti Teknologi MARA. She then pursued her studies in Natural Products and Synthesis and received her Ph.D. (Science) degree from Universiti Teknologi MARA in 2019. She participated in a Mobility Program at Tokushima Bunri University, Japan, conducting research utilizing chromatographic and spectroscopic instrumentation. She currently serves as a Senior Lecturer and holds a position as an Associate Fellow of the Atta-ur-Rahman Institute for Natural Product Discovery (AuRIns). She is affiliated with the Malaysian Natural Product Society and the Society for Medicinal Plant and Natural Product Research. Her research area is mainly on the isolation, structural elucidation, and synthesis of phytochemicals, especially stilbenoids from the Dipterocarpaceae and Euphorbiaceae plant families, and bioassay studies. Dr. Kamarozaman has received several awards, namely, the Icon for Research, Innovation and Publication, the Proactive Researcher (Science and Technology), and a Travel Grant and Best Presenter Awards. She has been a supervisor or co-supervisor for six Ph.D. and M.Sc. students. She has approximately 50 publications, encompassing articles in indexed journals, meeting proceedings, book chapters, and newsletters.

Nurul 'Izzah Mohd Sarmin is a Senior Lecturer at the Universiti Teknologi MARA, Malaysia. She obtained her Ph.D. degree in Biomedical Science from the National University of Malaysia, where she successfully isolated a novel endophytic actinomycete, *Streptomyces kebangsaanensis*, from a Malaysian ethnomedicinal plant. She is a Research Fellow at the Atta-ur-Rahman Institute for Natural Product Discovery (AuRIns). Her research interests in the anti-biofilm activity of endophytic actinomycetes and medicinal plants against multi-species oral biofilms, focus on the role of quorum-sensing systems in regulating microbial resistance mechanisms, to reduce cases of antimicrobial resistance. She underwent training as a Visiting Research Fellow at King's College London, elucidating the antibiofilm activity of plant extracts against multi-species cariogenic consortia biofilms. She is currently working on the isolation of antibiofilm compounds from dipterocarpaceous plants and peptides from endophytic bacteria. She was a member of the Malaysia One Health University Network (MyOHUN), for which the mission is to focus on a holistic approach including antimicrobial resistance. She has published her research findings in well-regarded journals, supervised undergraduate and postgraduate students, won several awards in invention and innovation competitions, and has contributed to the body of knowledge in the field of natural product studies.

Sharifah Aminah Syed Mohamad obtained her first degree in Microbiology from the University of Malaya and took her M.Sc. degree from Glasgow Caledonian University in Biomedical Sciences. She obtained her Ph.D. degree in Molecular Biology and Biotechnology from the University of Sheffield. She has been a Senior Lecturer at the Faculty of Applied Sciences, UiTM, since 1997 and a Senior Research Fellow at the Atta-ur-Rahman Institute for Natural Product Discovery (AuRIns) since 2022. Her academic background in molecular biology and microbial biotechnology allows her to teach postgraduate courses on molecular microbiology and undergraduate courses on microbial biotechnology, emphasizing gene manipulation principles and strategies in microbial host systems and the applications of microbial biotechnology in various sectors. Her research interests focus on antibiotic resistance, antimicrobial studies, and microbial pathogenicity, particularly on the increasing antibiotic resistance in emerging bacterial foodborne pathogens, the adhesion/invasion mechanisms of foodborne pathogens to intestinal cell lines in ensuring their intracellular survival, the antimicrobial activity of plant and microbial bioactive compounds against foodborne pathogens, and the inhibition potential of these compounds towards adhesion/invasion of foodborne pathogens. Dr Sharifah Aminah is also interested in the neuroprotective activity of plant and microbial bioactive compounds on oxidative stress-induced toxicity in SH-SY5Y human neuroblastoma cells. She has a number of professional affiliations, including lifetime membership in the Asian Federation of Biotechnology—Malaysian Chapter. She has been the supervisor or co-supervisor to 17 Ph.D. students and postdoctoral fellows and has co-authored around 50 publications in national and international scientific journals.

The manufacturer's authorised representative in the EU is Springer Nature Customer Service Centre GmbH, Europaplatz 3, 69115 Heidelberg, Germany. If you have any concerns regarding our products, please contact ProductSafety@springernature.com

Printed and bound by CPI Group (UK) Ltd, Croydon, CR0 4YY
30/09/2026
02225507-0006